Applications of Lie Algebras to Hyperbolic and
Stochastic Differential Equations

# Mathematics and Its Applications

Managing Editor:

M. HAZEWINKEL
*Centre for Mathematics and Computer Science, Amsterdam, The Netherlands*

Volume 466

# Applications of Lie Algebras to Hyperbolic and Stochastic Differential Equations

*by*

Constantin Vârsan

*Institute of Mathematics,*
*Romanian Academy,*
*Bucharest, Romania*

SPRINGER-SCIENCE+BUSINESS MEDIA, B.V.

A C.I.P. Catalogue record for this book is available from the Library of Congress.

ISBN 978-94-010-5970-1      ISBN 978-94-011-4679-1 (eBook)
DOI 10.1007/978-94-011-4679-1

*Printed on acid-free paper*

To Wilhelmina

# Contents

Preface                                                              ix

Introduction                                                          1

1  Gradient Systems in a Lie Algebra                                  5
   1.1  Preliminaries . . . . . . . . . . . . . . . . . . . . . . .    5
   1.2  Gradient systems in $F_n$ and Der $(R^n)$ . . . . . . . . .   11
   1.3  Gradient Systems Determined by a Lie Algebra . . . . . . .    17

2  Representation of a Gradient System                               25
   2.1  Finite–Dimensional Lie Algebra . . . . . . . . . . . . . .   25
   2.2  The Maximal Rank Lie Algebra . . . . . . . . . . . . . . .   33
   2.3  Integral Manifolds . . . . . . . . . . . . . . . . . . . .   36
   2.4  Some applications . . . . . . . . . . . . . . . . . . . . .   40

3  F. G. O. Lie Algebras                                            49
   3.1  Lie algebras finitely generated over orbits . . . . . . . .   49
   3.2  Nonsingularity of the gradient system . . . . . . . . . . .   55
   3.3  Some Applications . . . . . . . . . . . . . . . . . . . . .   69

4  Applications                                                     77
   4.1  Systems of Semiliniar Equations . . . . . . . . . . . . . .   77
   4.2  Stochastic Differential Equations . . . . . . . . . . . . .   83
   4.3  Systems of Hyperbolic equations . . . . . . . . . . . . . .   91
   4.4  Finite–Dimensional Nonlinear Filters . . . . . . . . . . .   97
   4.5  Affine Control Systems . . . . . . . . . . . . . . . . . .  108
   4.6  Integral Representation of Solutions . . . . . . . . . . .  109
   4.7  Decomposition of affine control systems . . . . . . . . . .  113

5  Stabilization and Related Problems                              117
   5.1  Equivalent Controllable Systems . . . . . . . . . . . . . .  117
   5.2  Approximations, Small Controls . . . . . . . . . . . . . .  128

5.3   Nonlinear Control Systems . . . . . . . . . . . . . . . . . . . . . 144
5.4   Stabilization of Affine Control Systems . . . . . . . . . . . . 147
5.5   Controlled Invariant Lie Algebras . . . . . . . . . . . . . . . 158
5.6   Stochastic differential equations . . . . . . . . . . . . . . . . 175
     5.6.1   Singularly perturbed equations viewed as controlled
            equations  . . . . . . . . . . . . . . . . . . . . . . . . . 176
     5.6.2   Bounded solutions for singularly perturbed system
            $(\alpha), (\beta)$  . . . . . . . . . . . . . . . . . . . . . . . . . 179
     5.6.3   Characterization of the dynamical limits . . . . . . . . 182

**Appendix**                                                                                          **197**

**Bibliography**                                                                                      **235**

**Subject Index**                                                                                     **237**

# Preface

The main part of the book is based on a one semester graduate course for students in mathematics.

I have attempted to develop the theory of hyperbolic systems of differential equations in a systematic way, making as much use as possible of gradient systems and their algebraic representation. However, despite the strong similarities between the development of ideas here and that found in a Lie algebras course this is not a book on Lie algebras. The order of presentation has been determined mainly by taking into account that algebraic representation and homomorphism correspondence with a full rank Lie algebra are the basic tools which require a detailed presentation. I am aware that the inclusion of the material on algebraic and homomorphism correspondence with a full rank Lie algebra is not standard in courses on the application of Lie algebras to hyperbolic equations. I think it should be. Moreover, the Lie algebraic structure plays an important role in integral representation for solutions of nonlinear control systems and stochastic differential equations yelding results that look quite different in their original setting. Finite-dimensional nonlinear filters for stochastic differential equations and, say, decomposability of a nonlinear control system receive a common understanding in this framework. A special attention is paid to geometric methods applied to affine control systems which, in a way, can be taken as nonholonomic contraints appearing in mechanics and therefore directly connected with hyperbolic differential equations analyzed in the first part of the book. The prerequisites are modest; the book can be read by students who are familiar with basic computational tools provided they have the patience for some detailed explanation. I have deliberately restricted the generality in such a way as to make it possible.

To illustrate, I have used the exercices to introduce some additional definitions and results that appear to be applicable to the problems of interest here.

I have not tried to make the reference list complete in any sense. The items that are listed have been of direct use to me in one way or another in learning the subject or preparing the manuscript. I apologize in advance to those authors whose relevant work is not included.

I am indebted to many people who have made suggestions in the early development of the manuscript.

I want to particularly acknowledge Aristide Halanay whose influence has been proved fruitfull.

For typing and retyping the earlier forms of the manuscript I want to thank especially Anghelache Mariana.

Bucharest, December, 1995                                          Constantin Vârsan

# Introduction

These lecture notes develop the theory of hyperbolic systems of differential equations by a differential geometric analysis of the associated gradient system. The main tools are Lie algebras, algebraic representation of the gradient systems, and their associated integral manifolds.

We begin with recalling the relationship between solutions of gradient systems and their representation as a composition of deterministic flows.

For a given smooth Lipschitz continous function (or vector field on $R^n$) $Y(y) \in R^n$, $y \in R^n$, let $G(t)(x)$ be the solution of an ordinary differential equation starting from $x \in R^n$ at time 0.

(a)
$$\frac{dG}{dt}(t) = Y(G(t)), \quad G(0) = x.$$

It is easily seen that the flow $G(t)(x)$ has the following properties

(b) $Y(G(t)) \neq 0$ foreach t if $Y(G(0)) \neq 0$

(c) For each $t$ the map $G(t) : R^n \rightarrow R^n$ is a diffeomorphism.

(d) $(t, x) \rightarrow G(t)(x)$ gives a smooth map from $R \times R^n \rightarrow R^n$.

Similarly, for a finite set $\{g_1, \cdot, g_m\}$ of vector fields on $R^n$ let $G_i(t)(x)$ be the flow generated by $g_i$.

We may, and do, consider the corresponding composition of flows

$$G(p) = G_1(t_1) \circ \cdot \circ G_m(t_m)(x)$$

for $p = (t_1, \cdot, t_m) \in R^m$.

Replacing $t \in R$ with $p \in R^m$ does not change the properties (c) and (d) but (a) and (b) must be strongly adjusted when the given vector fields are not commuting, i.e.,

$$[g_i, g_j](y) = \frac{\partial g_i}{\partial y} g_j(y) - \frac{\partial g_j}{\partial y}(y) g_i(y) \neq 0$$

for some $y$. The property (a) means that we can recover the original vector field $Y$ along the flow $G(t)(x)$ starting with the flow itself, and for $G(p)$ it becomes

$(\alpha)$            whether is possible to find $q_1(p), \cdot, q_m(p) \in R^m$ such that

$$\frac{\partial G}{\partial p}(p)q_i(p) = g_i(G(p)), \ p \in R^m, \ i = 1, \cdot, m,$$

The property (b) has an obvious analog for $G(p)$

$(\beta)$            dim span $\{g_1(G(p)), \cdots, g_m(G(p))\}$
              $= $ dim span $\{g_1(G(0)), \cdots, g_m(G(0))\} \ (\forall) \ p \in R^m$

The answer for the problems $(\alpha)$ and $(\beta)$ is yes if $\{g_1, \cdot, g_m\}$ determine a Lie algebra $L(g_1, \cdots, g_m)$ for which $\{g_1, \cdots, g_m\}$ is a system of generators, but the verification is by no means simple. One reason could be that the calculus of hyperbolic systems of differential equations is not as easy as the calculus of the ordinary differential equations.

Approaches to the above problems are summarized by the following two types:

a) We regard the composition of flows $G(p)$, $p \in R^m$ as the solution in a gradient system

$$\frac{\partial y}{\partial t_1} = g_1(y), \ \frac{\partial y}{\partial t_2} = X_2(t_1; y), \ldots, \ \frac{\partial y}{\partial t_m} = X_m(t_1, \cdot, t_{m-1}; y), y(0) = x$$

where the new vector fields with parameters $\{g_1, X_2(t_1), \cdots, X_m(t_1, \cdot, t_{m-1})\}$ are defined by exponential formal series and try to obtain the convergence of the formal series using the Fréchet space topology of

$$C^\infty(R^n; R^n) = \mathbf{F}_n$$

b) We associate the same gradient system of formal series and try to obtain the algebraic representation in $L(g_1, \cdot, g_m)$ using a weaker topology defined by a system of generators over orbits of $L(g_1, \cdot, g_m)$ starting from a fixed point.

In this book we will adopt the approach (b). An advantage of the latter approach might be that the algebraic representation of the gradient system along the solution can be obtained under a quite mild assumption on vector fields. In addition, it shows explicitly that the degeneracy of the Lie algebra at the initial point is preserved along the solution $G(p)$ and the problems $(\alpha)$ and $(\beta)$ receive a straight answer.

Actually, ($\alpha$) is an inverse problem for systems of hyperbolic equations but it may help to define explicitly integral manifolds for a distribution $\Lambda$ with singularities and as a consequence to find solutions for systems of hyperbolic differential equations.

This book consists of five chapters.

Chapter I contains the relationship between solutions for gradient systems and their realizations as solutions for hyperbolic differential equations. For the sake of simplicity we recall some properties of formal series in a Lie algebra, mainly defined by linear applications Vect($\mathbb{R}^n$) on $C^\infty(\mathbb{R}^n)$ and its subspace of derivations Der($\mathbb{R}^n$) which are the basic ingredients for a formal gradient system. Then some connections with function–solution of linear hyperbolic equations is analyzed. Here the classical, Frobenius theorem is established for both a gradient system of vector fields and the corresponding system of linear hyperbolic differential equations. Gradient systems determined by a Lie algebra conclude this part. In chapter II we prove that the solution for the problems similar to ($\alpha$) and ($\beta$) can be constructed if the associated gradient system is represented algebraically.

Finite generation over reals (f.g.r.), finite generation over orbits (f.g.o.) starting from a fixed $x_0 \in \mathbb{R}^n$ are the only two topologies which are used to achieve the goal and here the analysis is done focusing on (f.g.r.). A reason for that could be a more explicit form of the gradient system and of the nonsingular analytic representation as well.

It has to be remarked that the gradient system receives a global representation while solution may exists only locally.

Then the global nonsingular representation is taken as a basis for a homomorphism correspondence with a maximal rank Lie algebra determined by analytical vector fields; some implications of integral manifolds and a more concrete form of the homomorphism will be given in the last part where a stabilising feedback for affine control systems and solutions for linear hyperbolic equations are given.

Chapter III concerns differential geometry related to distributions with singularities determined by a Lie algebra $\Lambda$ which is finitely generated over orbits (f.g.o.). In the final part the integral manifolds of a f.g.o. Lie algebra are defined by means of a solution in a gradient system. The complexity of the problem is generated by the algebraic representation of the system which is no longer a global one but depends essentially on the local solution. It influences directly the homomorphism correspondence with a maximal rank Lie algebra determined by some locally-$C^\infty$ vector fields.

Perhaps it must be so since the f.g.o. property does not involve any finite dimension assumption for the Lie algebra. The applications included here are more or less consequences of the general results in Theorems 2 and 3.

Chapter IV. This part is fully devoted to applications encompassing quasilinear systems of hyperbolic differential equations and stochastic differential equations viewed as the limit of adapted (nonanticipative) control systems. The analysis is completed considering the integral representation of solutions and decomposability of affine control systems. As expected, integral manifolds and the homomorphism mapping of a Lie algebra with singularities constitute the main frame in which the above mentioned problems receive an answer.

Though affine control systems conclude the applications it is also possible to refer to them as a starting point for the gradient system analysis.

The analysis of stochastic differential equations is based on Langevin's approximation, which could be a good guide for a numerical treatment. In the last chapter of this book we analyze some relevant problems of control theory encompassing approximations, parametrization around a fixed flow and stabilization for affine control systems. As far as the stabilization problem is concerned it is strongly related to the theoretical tools developed in the previous chapters allowing one to perform the analysis on the associated controllable system, which is easily translated to the original system with singularities. In the final part entitled Appendix we collect some proofs and statements which may have implications in more general situations than those considered here. The preliminary results are selected to support Langevin's approximations and gradient systems on a smooth manifold.

# Chapter 1

# Gradient Systems in a Lie Algebra

## 1.1 Preliminaries

Let $\mathcal{A}$ be a real algebra and consider a formal series in the variable $t \in R$

$$a(t) = \sum_{k=0}^{\infty} \frac{t^k}{k!} a_k, \quad a_k \in \mathcal{A}, \quad a_0 \text{ is the initial element}$$

The set $\mathcal{A}(t)$ of all formal series of the variable $t$ in $\mathcal{A}$ is a real algebra if summation and multiplication by scalars $\alpha \in R$ are defined and the product of a two elements is given by the Cauchy rule

$$a(t_1) \cdot b(t_2) \stackrel{\triangle}{=} \sum_{k=0}^{\infty} \frac{t_1^i t_2^j}{k!} \left( \sum_{i+j=k} \frac{k!}{i!j!} a_i b_j \right) \quad \text{for } t_1 = t_2 = t$$

Usually, the original algebra $\mathcal{A}$ is embedded naturally into $\mathcal{A}(t)$ if $a_k = 0$, $k \geq 1$.

The linear application $\dfrac{d}{dt}$ on $\mathcal{A}(t)$ is defined

$$\frac{d}{dt}(a(t)) \stackrel{\triangle}{=} \sum_{k=1}^{\infty} \frac{t^{k-1}}{(k-1)!}(a_k) \in \mathcal{A}(t)$$

and $\text{Der}(\mathcal{A})$ consists of all linear mappings

$$L : \mathcal{A} \to \mathcal{A} \quad \text{such that}$$

$$L(a_1 \cdot a_2) = L(a_1) \cdot a_2 + a_1 \cdot L(a_2) \quad (\forall)\ a_1, a_2 \in \mathcal{A}.$$

By a direct computation we see that

$$\frac{d}{dt}(a(t) \cdot b(t)) = \frac{d}{dt}a(t)) \cdot b(t) + a(t) \cdot \left(\frac{d}{dt}b(t)\right) \quad \text{and}$$

$$\frac{d}{dt} \in \text{Der}\,(\mathcal{A}(t)).$$

Denote by $\mathcal{L}(\mathcal{A})$ the real algebra consisting of all linear application acting from $\mathcal{A}$ to $\mathcal{A}$, and for a $B \in \mathcal{L}(\mathcal{A})$ define

$\exp tB \in \mathcal{L}(\mathcal{A})(t)$ as
$$\exp tB = \sum_{k=0}^{\infty} \frac{t^k}{k!}B^k$$

The following properties are easily seen by a direct inspection

1) $\exp tB$ is commuting with $B$, $B(\exp tB) = (\exp tB)B$, and satisfies the equation $\frac{d}{dt}(\exp tB) = B(\exp tB)$;

2) The equation $\frac{d}{dt}(a(t)) = Ba(t)$ with respect to $a(t) \in \mathcal{A}(t)$ with $a(0) = a_0 \in \mathcal{A}$ has a unique solution $a(t) = (\exp tB)a_0$, where $B \in \mathcal{L}(\mathcal{A})$;

3) $B(t) = (\exp tB)B_0$ is the unique solution in $\mathcal{L}(\mathcal{A})(t)$ for $\frac{d}{dt}(B(t)) = BB(t)$, with $B(0) = B_0 \in \mathcal{L}(\mathcal{A})$, where $B \in \mathcal{L}(\mathcal{A})$;

4) For any two commutative $B_1, B_2 \in \mathcal{L}(\mathcal{A})$, i.e. $0 = [B_1, B_2] \stackrel{\triangle}{=} B_1 \circ B_2 - B_2 \circ B_1$, it holds $\exp t(B_1 + B_2) = (\exp tB_1)(\exp tB_2)$;

5) The linear map $\exp tB : \mathcal{A}(t) \to \mathcal{A}(t)$ is defined using the Cauchy rule

$$(\exp tB)(a(t)) = \sum_{k=0}^{\infty} \frac{t^k}{k!}\left(\sum_{i+j=k} \frac{k!}{i!j!}B^i a_j\right),$$

where $B \in \mathcal{L}(\mathcal{A})$ and $a(t) = \sum_{j=0}^{\infty}\frac{t^j}{j!}a_j \in \mathcal{A}(t)$; there holds $(\exp tB) \cdot (\exp t(-B)) = I$, where $I : \mathcal{A}(t) \to \mathcal{A}(t)$ is the identity map and for any $b(t) \in \mathcal{A}(t)$ there is an unique $a(t) \in \mathcal{A}(t)$ fulfilling $(\exp t\,B)a(t) = b(t)$.

6) Any $B \in \text{Der}(\mathcal{A})$ is extended naturally to $B \in \text{Der}(\mathcal{A}(t))$ and the following are true

(a)     $(\exp tB)(a_1 \cdot a_2) = (a_1 \cdot a_2) = ((\exp tB)a_1) \cdot ((\exp tB)a_2)$  for $t = 0$

(b)
$$\frac{d}{dt}((\exp tB)a_1) \cdot ((\exp tB)a_2) = B(\exp tB)a_1 \cdot ((\exp tB)a_2)$$
$$+((\exp tB)a_1) \cdot (B(\exp tB) \cdot a_2)$$
$$= B\{((\exp tB)a_1) \cdot ((\exp tB)a_2)\}$$

7) Using (6) we obtain $(\exp t\, B)(a_1 \cdot a_2) = ((\exp tB)a_1) \cdot ((\exp tB)a_2)$ for any $a_1, a_2 \in \mathcal{A}, B \in \text{Der}(\mathcal{A})$ and in particular

$$B^m(a_1 \cdot a_2) = \sum_{i+j=m} \frac{m!}{i!j!}(B^i a_1) \cdot (B^j a_2)$$

**Definition 1.**
*A real algebra $\mathcal{A}$ is a Lie algebra if the multiplication operation satisfies $ab = -ba$ and Jacobi's identity*

$$a(bc) + c(ab) + b(ca) = 0 \quad \text{for any } a, b, c \in \mathcal{A}$$

*Generally, Der $(\mathcal{A}) \subseteq \mathcal{L}(\mathcal{A})$ is a linear subspace and*
*a) if $b_1, b_2 \in$ Der $(\mathcal{A})$ then $[b_1, b_2] \in$ Der $(\mathcal{A})$, where $[b_1, b_2] \triangleq b_1 \cdot b_2 - b_2 \cdot b_1$*
*b) Der $(\mathcal{A}), \mathcal{L}(\mathcal{A})$ become Lie algebras if the product operation is redefined as $[b_1, b_2] \triangleq b_1 \cdot b_2 - b_2 \cdot b_1$, where $b_1 \cdot b_2$ is the usual composition of the two linear applications.*
8) In a real Lie algebra $\Lambda$ there is a natural mapping ad $: \Lambda \to \mathcal{L}(\Lambda)$ defined by ad $v(w) = [v, w]$ $(\forall)$ $v, w \in \Lambda$ and using Jacobi's identity we obtain

(a)
$$adv([w_1, w_2]) = [adv(w_1), w_2] + [w_1, adv(w_2)]$$

(b)
$$ad[v_1, v_2](w) = [adv_1, adv_2](w)$$

$$ad : \Lambda \to \text{Der} (\Lambda) \quad \text{is a homomorphism}$$

In what follows we use constantly the real algebra $C^\infty(R^n) = F$ consisting of all functions $f : R^n \to R$ which are differentiable to any order. Denote Vect $(R^n)$ the linear space consisting of all linear mappings from $F$ to $F$ and let $F_n \triangleq \prod_{k=1}^{n} F$. Each linear map $L : F \to F$ is naturally extended as $L : F_n \to F_n$ by

$$La = L\begin{pmatrix} a_1 \\ \vdots \\ a_n \end{pmatrix} = \begin{pmatrix} La_1 \\ \vdots \\ La_n \end{pmatrix} \in F_n \text{ if } a = \begin{pmatrix} a_1 \\ \vdots \\ a_n \end{pmatrix} \in F_n$$

Write Der $(R^n)$ for the Lie algebra Der $(F)$ and associate $a$ vector field $X \in F_n$ for each $\vec{X} \in \text{Der}(R^n)$ such that $X \triangleq \vec{X}I$, where $I : R^n \to R^n$ is the identity map. In what follows $F_n$ will be considered as a real Lie algebra taking the Lie bracket $[X, Y]$ as

$$[X, Y](x) = \frac{\partial X}{\partial x}(x)Y(x) - \frac{\partial Y}{\partial x}(x)X(x), \quad x \in R^n$$

It is easily seen that for $\vec{X}_i \in \text{Der}(R^n)$ we obtain $X_i \triangleq \vec{X}_i I$ in $F_n$ , $i = 1, 2$, but $[\vec{X}_1, \vec{X}_2]I = -[X_1, X_2]$.

The Lie algebras $\text{Der}(R^n)$ and $\mathcal{L}(\text{Der}(R^n))$ are not finite dimensional spaces and to obtain $\exp t\vec{X} : F \to F$, for $\vec{X} \in \text{Der}(R^n)$, or $\exp tL : \text{Der}(R^n) \to \text{Der}(R^n)$, for $L \in \mathcal{L}(\text{Der}(R^n))$, as a linear map we need to take care of a suitable topology.

Let $\vec{X} \in \text{Der}(R^n)$ be fixed and define $X \triangleq \vec{X}I \in F_n$. We may and do consider the following linear hyperbolic equation

$$(*) \quad \frac{\partial S}{\partial t}(t; x) = \sum_{i=1}^{n} X^i(x)\frac{\partial S}{\partial x^i}(t; x) \triangleq \vec{X}(S)(t; x) , \ t \in \mathcal{I}_a \triangleq (-a, a), \ a > 0$$

and we are looking for a solution $S \in C^\infty(\mathcal{I}_a \times R^n; R)$ fulfilling the Cauchy condition $S(0; \cdot) = \varphi(\cdot) \in F$. Similarly, for $L \in \mathcal{L}(\text{Der}(R^n))$, $L \triangleq ad\vec{X}$, where $\vec{X} \in \text{Der}(R^n)$, define $X \triangleq \vec{X}I \in F_n$ and consider the following Cauchy problem of a linear hyperbolic equation

$$(**) \qquad \frac{d}{dt}a(t) = [X, a(t)], \ a(0) = Y \in F_n, \ t \in (-a, a) \triangleq \mathcal{I}_a$$

A solution for $(**)$ means a function $a(t, x) \in R^n$, $(t, x) \in \mathcal{I} \times R^n$ such that $a \in C^\infty(\mathcal{I}_a \times R^n; R^n)$ and fulfilling $a(0; \cdot) = Y(\cdot)$ for a fixed $Y \in F_n$. Both hyperbolic equations $(*)$ and $(**)$ admit a unique solution if the following system of ordinary differential equations

$$(***) \qquad\qquad \frac{dy}{dt} = X(y), \ y(0) = x \in R^n$$

generates a flow $y = G(t; x)$, $t \in \mathcal{I}_a = (-a, a)$, $x \in R^n$ as the solution fulfilling $G(0; \cdot) = I$

The following is a rewriting of the flow properties.

**Proposition 1.**

*Let the vector field $X \in F_n$ generates a flow $G(t; \cdot)$, $t \in \mathcal{I}_a$, for $(***)$ and $H(\tau; x) \triangleq \left( \dfrac{\partial G}{\partial x}(\tau; x) \right)^{-1}$. Then the unique solution in $(*)$ is $S(t; x) = \varphi(G(t; x))$, $t \in \mathcal{I}_a$, $\varphi \in F$, and $a(t; x) = H(-t; x)Y(G(-t; x))$, $t \in \mathcal{I}_a$, $Y \in F_n$, is the corresponding unique solution in $(**)$.*

**Remark 1.**

*We conclude that $\exp t\vec{X} : F \to F$ and $\exp t ad \vec{X} :$ Der $(R^n) \to$ Der$(R^n)$ for $\vec{X} \in$ Der $(R^n)$, are defined as linear maps if the conditions in Proposition 1 are fulfilled and it takes place $\exp t \vec{X}(\varphi) = \varphi(G(t; \cdot))$, $\varphi \in F$, $t \in \mathcal{I}_a$, $\exp t ad \vec{X}(\vec{Y}) = H(-t; \cdot)Y(G(-t; \cdot))$, $t \in \mathcal{I}_a$, $Y = \vec{Y}I \in F_n$.*

*In addition, both $\exp t\vec{X}$ and $\exp t ad \vec{X}$ are continous maps provided usual topology of Frechet spaces on $F$ and respectively $F_n$ are used.*

**Remark 2.**

*As far as local properties are involved the conditions in Proposition 1 are changed as follows.*

*A open set $D \subseteq R^n$ replacing $R^n$ defines the local flow $G(t; x)$, $t \in I_a$, $x \in D$ of the vector field $X$ in $(***)$ and we obtain*

$$S(t; x) = \varphi(G(t; x)), \quad a(t; x) = H(-t; x) \ Y(G(-t; x)), \quad t \in I_a, \ x \in D$$

*accordingly.*

**Remark 3.**

*The properties $(1)$–$(7)$ listed for $B = ad\vec{X} \in \mathcal{L}(Der(R^n))$ hold true for the function version in Proposition 1 and $(7)$ can be written*

$$(\exp t \, ad \, \vec{X})[\vec{Y}_1, \vec{X}_2](x) = [(\exp t \, ad \, \vec{X})\vec{Y}_1, (\exp t \, ad \, \vec{X})\vec{Y}_2](x), \quad x \in R^n,$$

*or*

$$H(-t; x)[Y_1, Y_2](G(-t; x)) = [H(-t; \cdot)Y_1(G(-t; \cdot)), \ H(-t; \cdot)Y_2(G(-t; \cdot))](x)$$

*$x \in R^n$ for any $Y_1, Y_2 \in F_n$ where $H$ and $G$ are the $(n \times n)$ matrix and flow defined in Proposition 1.*

## Exercices

**Exercise 1.**
Prove that the solution for Cauchy problem of hyperbolic equations $(*)$ and $(**)$ is unique.
Hint:
We notice that if $V(t; x)$, $t \in \mathcal{I}a$, $x \in R^n$, is a solution for $(*)$ then $V(t; G(-t; z)) = h(z)$ for any $t \in \mathcal{I}a$, $z \in R^n$, and, in particular, for $t = 0$ we obtain $h(z) = \varphi(z)$, $z \in R^n$.
On the other hand, $G(-t; G(t; x)) = x$ and for $z = G(t; x)$ we obtain $V(t; x) = \varphi(G(t; x))$, $t \in \mathcal{I}a$, $x \in R^n$. Similarly, we notice that $H(-t; x)^{-1} \triangleq \dfrac{\partial G}{\partial x}(-t; x) = H(t; G(-t; x))$ and $[H(-t; y)]^{-1}b(t; y) = X(z)$, $(\forall)\ t \in \mathcal{I}a$, if $y = G(t; z)$ and $b$ is a solution of $(**)$; see

$$[H(-t; G(t; z))]^{-1} = H(t; z)$$

and

$$\frac{d}{dt} H(t, z)b(t; G(t; z)) = 0, \quad t \in \mathcal{I}a.$$

Therefore $b(t; x) = H(-t; x)X(G(-t; x))\ \ t \in \mathcal{I}a, x \in R^n$.

**Exercise 2.**
Let $X, Y_1, Y_2 \in F_n$ and consider the flow $G(\tau; x)\ \ \tau \in \mathcal{I}a,\ \ x \in R^n$, and the matrix $H(\tau; x) = \left[\dfrac{\partial G}{\partial x}(\tau; x)\right]^{-1}$ determined by the vector field $X$ as in Proposition 1.
Prove by a direct computation that

$$H(-t; x)[Y_1, Y_2](G(-t; x))$$
$$= [H(-t; \cdot))Y_1(G(-t; \cdot)), H(-t; \cdot)Y_2(G(-t; \cdot))](x)(\forall)\ t \in \mathcal{I}a, x \in R^n.$$

Hint:
The Lie bracket in the right hand side is computed using $H(-t; y) = \dfrac{\partial G}{\partial x}(t, G(-t; y))$ and the symmetry of the matrices $\dfrac{\partial^2 Gi}{\partial x^2}(t; x)$, $i = 1, \cdot, n$ where $G = (G_1, \cdot, G_n)$, will complete the verification.

**Exercise 3.**
Let $X, Y_1, Y_2 \in C^2(R^n; R^n)$ and consider $G(\tau, x), \tau \in I_a = (-a, a)$, $x \in D \subseteq R^n$ the local flow generated by the equation

$$\frac{dy}{dt} = X(y), \quad y(0) = x \in D,$$

$D$ an open set. Then denote $H(\tau; x) = \left(\dfrac{\partial G}{\partial x}(\tau; x)\right)^{-1}$ $\tau \in I_a, x \in D$, and show

$$H(-t; x)[Y_1, Y_2](G(-t; x)) = [H(-t; \cdot)Y_1(G(-t; \cdot)), H(-t; \cdot)Y_2(G(-t; \cdot))](x),$$

for any $t \in I_a$, $x \in D$ Hint. Use the same calculus as in exercise 2.

**Exercise 4.**
Let $X \in F_n$ or $X \in C^2(R^n, R^n)$ and associate $G(\tau, x)$, $\tau \in Ia = (-a, a)$, $x \in D \subseteq R^n$ its local flow. Define $H(\tau; x) = \left[\dfrac{\partial G}{\partial x}(\tau; x)\right]^{-1}$, $(\forall)\ \tau \in Ia$, $x \in D$ and verify $H(\tau; x)X(G(\tau; x)) = X(x), \tau \in Ia$, $x \in D$.

**Exercise 5.**
Let $X \in F_n$ and associate $G(\tau; x)$, $\tau \in Ia = (-a, a)$, $x \in D \subseteq R^n$ its local flow. Then $G(\tau; \cdot) \in C^\infty(D; R^n)$ and $G(\tau_1 + \tau_2; \cdot) = G(\tau_1) \circ G(\tau_2)$, if $\tau_1 + \tau_2 \in Ia$, where $G(\tau) \stackrel{\triangle}{=} G(\tau; \cdot)$

By definition $G(0) = I, I(x) = x$ for $x \in D$, and for $G(\tau; \cdot)$ we shall use the formal writting $G(\tau) = \exp \tau X$, $\tau \in Ia$. Notice the difference between $\exp \tau X$, for $X \in F_n$, and $\exp \tau \vec{X}$ for $\vec{X} \in \text{Der}\ (R^n)$.

In particular, $\exp \tau X : F_n \to C^\infty(D; R^n)$ is a linear application if

$$(\exp \tau X)Y \stackrel{\triangle}{=} Y(G(\tau)), Y \in F_n.$$

## 1.2 Gradient systems in $F_n$ and Der $(R^n)$

We begin by recalling the classical Frobenius theorem for gradient systems in $F_n$ and Der$(R^n)$. Let $X_i \in F_n$, $i = 1, \cdot, m$, be given and consider the following system

(1) $\qquad \dfrac{\partial y}{\partial t_i} = X_i(y), \quad i = 1, \cdots m, \quad y(0) = x \in V \subseteq R^n.$

A solution for (1) means a function $G(p; x) : D_m \times V \to R^n$ of class $C^2$ fulfilling (1) for any $p \stackrel{\triangle}{=} (t_1, \cdot, t_m) \in D_m \stackrel{\triangle}{=} \prod_l^m (-a_i, a_i)$ and $x \in V \subseteq R^n$ an open set. The system (1) is completely integrable if for any $x_0 \in R^n$ there exist a neighbourhood $V(x_0) \subseteq R^n$ and $G(p; x), (p, x) \in D_m \times V(x_0)$ a unique solution for (1) fulfilling $G(0; x) = x, x \in V(x_0)$.

**Theorem 1. (Frobenius)**

*Let $X_i \in F_n, i = 1, \cdot, m$ be given. Then the system (1) is completely integrable iff $[X_i, X_j] = 0$, $(\forall)$ $i, j \in \{1, \cdot, m\}$, where $[X_i, X_j] \stackrel{\Delta}{=} \dfrac{\partial X_i}{\partial j}(y)X_j(y) - \dfrac{\partial X_j}{\partial y}(y)X_i(y)$, is the Lie bracket. In addition, any local solution $G(p; x)$, $p \in D_m$, $x \in V(x_0)$, is given by $G(p; x) = G_1(t_1) \circ \cdots \circ G_m(t_m)(x)$, where $G_i(\tau)(x)$ is the local flow generated by $Xi$.*

Replacing $Xi \in F_n$ by $\vec{X}i \in \text{Der}(R^n), i = 1, \cdots, m$, the system (1) becomes a system of linear hyperbolic equations.

$$(2) \qquad \frac{\partial S}{\partial t_i}(p; x) = \left\langle \frac{\partial S}{\partial x}(p; x), X_i(x) \right\rangle \stackrel{\Delta}{=} (\vec{X}_i S)(p; x), \quad i = 1, \cdots, m,$$

$$S(0; x) = \varphi(x), \quad (\forall)\ x \in V \subseteq R^n \text{ for some } \varphi \in F$$

where

$$\vec{X}i \stackrel{\Delta}{=} \sum_{s=1}^{n} X_i^s \frac{\partial}{\partial x^s}.$$

A solution for (2) means a scalar function $S(p; x)$, $p \stackrel{\Delta}{=} (t_1, \cdots, t_m) \in D_m$, $x \in V \subseteq R^n$, of class $C^1$ fulfilling (2) for any $(p; x) \in D_m \times V$ and $S(0; x) = \varphi(x)$, $x \in V$, for some $\varphi \in F$.

We say that (2) is completely integrable if for any $\varphi \in F$, $x_0 \in R^n$ there exist an open set $V(x_0)) \subseteq R^n$, $D_m \stackrel{\Delta}{=} \prod_{i=1}^{m}(-a_i, a_i)$, and $S(p; x), p \in D_m$, $x \in V(x_0)$ solution for (2) with $S(0; x) = \varphi(x)$, $x \in V(x_0)$.

The Frobenius theorem for (2) is then stated as follows.

**Theorem 2.**

*Let $\vec{X}_i \in \text{Der}(R^n)$ and $X_i \stackrel{\Delta}{=} \vec{X}_i I \in F_n, i = 1, \cdots, m$. Then the system (2) is completely integrable iff $[X_i, X_j] = 0$ $(\forall), i, j \in \{1, \cdots, m\}$. In addition, the local solution with Cauchy condition $S(0, x) = \varphi(x)$, $x \in V$, is given by $S(p; x) = \varphi(G(p; x))$, where $G(p; x), p \in D_m, x \in V(x_0)$ is a local solution for (1).*

**Proof.**

The arguments are similar to those used for proving Theorem 1. Namely, for implication $\Rightarrow$ we use $\dfrac{\partial^2 S}{\partial t_i \partial t_j}(p; x) = \dfrac{\partial^2 S}{\partial t_j \partial t_i}(p; x)$ and from (2) we obtain

$$\frac{\partial^2 S}{\partial t_i \partial t_j}(p; \cdot) = \frac{\partial}{\partial t_i}(\vec{X}_j S(p; \cdot)) = \vec{X}_j \left( \frac{\partial S}{\partial t_i}(p, \cdot) \right) = \vec{X}_j(\vec{X}_i S)(p; \cdot)$$

and

$$\frac{\partial^2 S}{\partial t_j \partial t_i}(p; \cdot) = \frac{\partial}{\partial t_j}(\vec{X}_i S(p; \cdot)) = \vec{X}_i \left(\frac{\partial S}{\partial t_j}(p, \cdot)\right) = \vec{X}_i(\vec{X}_j S)(p; \cdot)$$

In particular, for $p = 0$ and $\varphi_k(x) = x^k, k = 1, \cdots, n$, where $x \triangleq (x^1, \cdots, x^n)$, we obtain $[X_i, X_j](x_0) = 0, i, j \in \{1, \cdots, m\}$, for an arbitrary $x_0 \in R^n$. For the implication " $\Leftarrow$ " we use the commutative property of the flows when the corresponding vector fields commute. Namely, supposing $[X_i, X_j] = 0 \ (\forall) \ i, j \in \{1, \cdots, m\}$

$$G_i(t_i) \circ G_j(t_j)(x) = G_j(t_j) \circ G_i(t_i)(x).$$

and the local solution $G(p; x)$ for (1) can be written $G(p; x) = G_{i_1}(t_{i_1}) \circ \cdots \circ G_{i_m}(t_{i_m})(x)$ for any permutation $\{i_1, \cdots, i_m\}$ of $\{1, \cdots, m\}$. Take $\varphi \in F$, $x_0 \in R^n$ arbitrarily and define

$$S(p; x) = \varphi(G(p; x)), \ p \in D_m \triangleq \prod_1^m (-a_i, a_i), x \in V(x_0) \subseteq R^n.$$

By definition $S(0; x) = \varphi(x)$, $x \in V(x_0)$, and the Cauchy condition is fulfilled for $S$. Moreover, the calculus of $\dfrac{\partial S}{\partial t_k}(p; x)$ for some $k \in \{1, \cdots, m\}$ is done choosing $S(p; x) = \Psi(G_k(t_k)(x))$, where the scalar function

$$\Psi(y) = \varphi(G_1(t_1) \circ \cdots \circ G_{k-1}(G_{k-1}) \circ G_{k+1}(t_{k+1}) \circ \cdots \circ G_m(t_m)(y)).$$

Using the semigroup property for $G_k(\tau)(\cdots)$, $\tau \in (-a_k, a_k)$, we obtain finally

$$\frac{\partial S}{\partial t_k}(p; x) = \vec{X}_k \Psi(G_k(t_k)(x)) = (\vec{X}_k S)(p; x),$$

and the proof is complete.

**Remark 1.**

*As has been stated, the complete integrability of (1) and (2) are determined by the commuting condition of the vector fields, i.e., $[X_i, X_j] = 0, i, j \in \{1, \cdots, m\}$. In addition, we may define a solution using the composition of flows $G(p; x)$ which does not depend on the ordering.*

In the remaining part of this section we shall recall the inverse problem of a gradient system. Namely, starting with some *noncommuting* vector fields $X_i \in F_n, i = 1, \cdots, m$, define a gradient system (depending on parameters) such that the composition of flows $G(p; x) = G(t_1) \circ, \cdots, \circ G_m(t_m)(x), p \in$

$D_m$, $x \in V \subseteq R^n$ is a local solution. As is known, this inverse problem obtain s a solution without any other additional hypotheses. The result will be stated in the vector field frame $F_n$ but the analogy with $\text{Der}(R^n)$ is easily seen comparing Theorem 1 and 2 from above.

**Definition 1.**

Let $X_j(p; y) \in R^n$ for $p \stackrel{\triangle}{=} (t_1, \cdot, t_m) \in D_m \stackrel{\triangle}{=} \prod_1^m (-a_i, a_i), y \in V \subseteq R^n$, be of class $C^1$ for $j = 1, \cdots, m$. We say that $X_j(p; y)$ defines a gradient system (or fulfills the Frobenius integrability condition) if

$$\frac{\partial X_j}{\partial t_i}(p; y) - \frac{\partial X_i}{\partial t_j}(p; x) = [X_i(p; \cdot), X_j(p; \cdot)](y), \ (\forall) \ i, j \in \{1, \cdot, m\}$$

where

$$[Z_1, Z_2](y) \stackrel{\triangle}{=} \frac{\partial Z_1}{\partial y}(y) Z_2(y) - \frac{\partial Z_2}{\partial y} Z_1(y) \ (Lie \ bracket)$$

**Theorem 3.**

Let $Y_j \in C^2(R^n; R^n), x_0 \in R^n$, $j = 1, \cdots m$, be fixed. Then there exist $D_m = \prod_1^m (-a_i, a_i)$, $V(x_0) \subseteq R^n$ and $X_j(p_j; y) \in R^n$, $p \in D_m, y \in V(x_0)$ of class $C^1, p_j \stackrel{\triangle}{=} (t_1, \cdot, t_{j-1}), X_1 = Y_1$, such that

$(c_1)$      $\frac{\partial y}{\partial t_1} = Y_1(y), \quad \frac{\partial y}{\partial t_2} = X_2(t_1; y), \cdots, \frac{\partial y}{\partial t_m} = X_m(t_1, \cdot, t_{m-1}; y)$

is a gradient system $\left( \frac{\partial X_j}{\partial t_i}(p_j; y) = [X_i(p_i; \cdot), \ X_j(p_i; \cdot)](y), \ i < j \right)$ and

$(c_2)$      $G(p; x) \stackrel{\triangle}{=} G_1(t_1) \circ \cdot \circ G_m(t_m)(x), \ p \in D_m, \ x \in V(x_0)$

is the solution for $(c_1)$ with Cauchy condition $y(0) = x \in V(x_0)$ where $G_i(t)(x)$ is the local flow generated by $Y_i$.

**Proof.**

For $m = 1$ the conclusions $(c_1)$ and $(c_2)$ are trivially fulfilled. Assume that for $(m - 1)$ vector fields $Y_j \in C^2(R^n; R^n)$ the conclusions $(c_1)$ and $(c_2)$ are satisfied, i.e., there exist

$$\widehat{X}_2(\widehat{y}) = Y_2(y), \quad \widehat{X}_3(t_2; y), \quad \widehat{X}_m(t_2, \cdots, t_{m-1}; y)$$

of class $C^1$ in

$$\widehat{p} = (t_2, \cdot, t_m) \in \prod_{i=2}^{m}(-a_i, a_i), \ y \in \widehat{V}(x_0) \subseteq R^n,$$

such that

(3) $\widehat{G}(\widehat{p}; x) \stackrel{\triangle}{=} G_2(t_2) \circ \cdots \circ G_m(t_m)(x), \ \widehat{p} \in \prod_{i=2}^{m}(-a_i, a_i), \ x \in \widehat{V}(x_0) \subseteq R^n.$

is the solution in a gradient system

(4) $\qquad \dfrac{\partial y}{\partial t_2} = Y_2(y), \ \dfrac{\partial y}{\partial t_3} = \widehat{X}_3(t_2; y), \cdots, \dfrac{\partial y}{\partial t_m} = \widehat{X}_m(t_2, \cdots, t_{m-1}; y)$

with Cauchy condition $y(0) = x \in \widehat{V}(x_0)$.

Rewriting that (4) is a gradient system we obtain

(5) $\qquad \dfrac{\partial \widehat{X}_j(\widehat{p}; y)}{\partial t_i} = \left[ \widehat{X}_i(\widehat{p}_i; \cdot), \widehat{X}_j(\widehat{p}_j; \cdot) \right](y), \ 2 \le i < j = 3, \cdots, m.$

for $\widehat{p}_i \stackrel{\triangle}{=} (t_2, \cdots, t_{i-1})$

Let vector fields $Y_j \in C^2(R^n; R^n), j = 1, \cdots m$, be given and denote
(6)

$$y = G(p; x) = G_1(t_1) \circ \widehat{G}(\widehat{p}; x), \ p = (t_1, \widehat{p}) \in \prod_{i=1}^{m}(-a_i, a_i), \ x \in V(x_0) \subseteq \widehat{V}(x_0)$$

where $G_1(t)(x)$ is the local flow generated by $Y_1$ and $\widehat{G}(\widehat{p}; x)$ is defined in (3).

By definition $\dfrac{\partial G}{\partial t_1}(p; x) = Y_1(G(p; x))$, and to prove that $G(p; x)$ in (6) fulfils a gradient system we write $G_1(-t_1; y) = \widehat{G}(\widehat{p}; x)$, (see (6)) and we obtain

(7) $\qquad \dfrac{\partial G_1}{\partial y}(-t_1; y) \dfrac{\partial y}{\partial t_j} = \dfrac{\partial \widehat{G}}{\partial t_j}(\widehat{p}; x), \ \text{for } j = 2, \cdots, m, \ \text{or}$

$$\dfrac{\partial y}{\partial t_j} = H_1(-t_1; y) \widehat{X}_j(\widehat{p}_j; G_1(-t_1; y)), \ j = 2, \cdots, m,$$

The matrix $H_1(\tau; y) \stackrel{\triangle}{=} \left[ \dfrac{\partial G_1}{\partial y}(\tau; y) \right]^{-1}$ satisfies

$$\dfrac{dH_1}{d\tau} = -H_1 \dfrac{\partial Y_1}{\partial x}(G_1(\tau; y)), H_1(0; y) = I_n$$

Denote
$$X_2(t_1; y) \triangleq H_1(-t_1; y)Y_2(G_1(-t_1; y))$$
$$X_j(t_1, \cdots, t_{j-1}; y) \triangleq H_1(-t_1; y)\widehat{X}_j(t_2, \cdots, t_{j-1}; G(-t_1; y)), j = 3, \cdots, m,$$

for
$$y \in V(x_0) \subseteq \widehat{V}(x_0), \quad p \triangleq (t_1, \cdots, t_m) \in \prod_1^m (-a_i, a_i)$$

and the system (7) becomes a gradient system

(8)        $\dfrac{\partial y}{\partial t_1} = Y_1(y), \dfrac{\partial y}{\partial t_2} = X_2(t_1; y), \cdots, \dfrac{\partial y}{\partial t_m} = X(t_1, \cdots, t_{m-1}; y).$

if we prove that

(9)        $\dfrac{\partial X_j}{\partial t_i}(p_j; y) = [X_i(p_i; \cdot), X_j(p_j; \cdot)](y)$  for any  $1 \le i < j = 2, \cdots, m.$

For $j = 2$ by a direct computation we obtain

(10)        $\dfrac{\partial X_2}{\partial t_1}(t_1; y) = H_1(-t_1; y)[Y_1, Y_2](G_1(-t_1; y))$

and using a basic property of the right hand side (see exercices 1.3 and exercices 1.4) we obtain

(11)        $\dfrac{\partial X_2}{\partial t_1}(t_1; y) = [Y_1(\cdot), X_2(t_1; \cdot)](y),$

which means (9) for $i = 1$ and $j = 2$.

Using the same arguments we obtain

(12)        $\dfrac{\partial X_j}{\partial t_1}(p_j; y) = [Y_1(\cdot), X_j(p_j; \cdot)](y),$ for any $j = 2, 3, \cdots, m$

which means that (9) is fulfilled for $i = 1$ and $j = 2, 3, \cdots, m$. It remains to show (9) for $j = 3, \cdots, m$, and $2 \le i < j$. Using (5) and

$$\dfrac{\partial X_j}{\partial t_i}(p_j; y) = H_1(-t_1; y)\dfrac{\partial \widehat{X}_j}{\partial t_i}(\widehat{p}_j; G_1(-t_1; y)),$$

we obtain

(13)        $\dfrac{\partial X_j}{\partial t_i}(p_j; y) = H_1(-t_1; y)[\widehat{X}_i(\widehat{p}_i; \cdot), \widehat{X}_j(\widehat{p}_j; \cdot)](G_1(-t_1; y)),$
        if $2 \le i < j = 3, \cdots, m,$

The right hand side in (13) is similar to that in (10) and the same argument as used above in (13) allow one to write

(14)        $\dfrac{\partial X_j}{\partial t_i}(p_j; y) = [X_i(p_i; \cdot), X_j(p_j; \cdot)](y)$     $2 \le i < j = 3, \cdots, m.$

and the proof is complete.

## 1.3  Gradient Systems Determined by a Lie Algebra

We noticed previously that any finite composition of flows $y(p) \overset{\triangle}{=} G_1(t_1) \circ \cdots \circ G_m(t_m)(x_0)$ can be associated with a gradient system

$$\frac{\partial y}{\partial t_1} = Y_1(y), \ \frac{\partial y}{\partial t_2} = X_2(t_1; y), \cdots, \ \frac{\partial y}{\partial t_m} = X_m(t_1, \cdots, t_{m-1}; y), \ y(0) = x_0,$$

and its local solution.

Both solution and gradient system are based essentially on the property that $\dfrac{\partial}{\partial t_1}, \cdots, \dfrac{\partial}{\partial_m}$ are commutative derivations in $\mathrm{Der}(R^m)$ and this must be reconsidered if $\dfrac{\partial}{\partial t_i}, i = 1, \cdots, m$, are replaced by some noncommuting

$$\vec{g}_i \in \mathrm{Der}(R^n), \ (\text{or } g_i \overset{\triangle}{=} \vec{g}_i \, I \in \mathbf{F}_n), \ i = 1, \cdots, m).$$

A simple system of linear hyperbolic equations is defined

(1) $\qquad \vec{g}_i(\varphi)(x) = 0 \ (\text{or } \langle \dfrac{\partial \varphi}{\partial x}(x), g_i(x) \rangle = 0), \ i = 1, \cdots, m),$

and a nontrivial solution $\varphi(x)$, $x \in V(x_0) \subseteq R^n$, with $\dfrac{\partial \varphi}{\partial x}(x_0) \neq 0$ must satisfy (1) on some integral manifold $M_{x_0}$ of the Lie algebra $L(g_1, \cdots, g_m)$ determined by the vector fields $g_i \in \mathbf{F}_n$.

The domain $M_{x_0}$ necessarily obeys to $\dim M_{x_0} = k < n$, otherwise a nontrivial solution for (1) does not exist.

A more general system of quasiliniar hyperbolic equations

(2) $\qquad \left\langle \dfrac{\partial \varphi}{\partial x}(x), g_i(x) \right\rangle = d_i(\varphi(x), x), \qquad d_i \in C^\infty(R^{n+1}, R) \qquad i = 1, \cdots, m,$

lead us to the same problem of defining integral manifolds of a Lie algebra if (2) is rewriten as a linear system

(3) $\qquad \left\langle \dfrac{\partial F}{\partial y}(y), Y_i(y) \right\rangle = 0, \ y \overset{\triangle}{=} (t, x) \in R^{n+1}, y_0 \overset{\triangle}{=} (0, x_0),$

$$i = 1, \cdots, m,$$

where

$$Y_i(y) \overset{\triangle}{=} (\ d_i(y), y_i(x)\ ) \in \mathbf{F}_{n+1}, \ i = 1, \cdots, m.$$

The domain $M_{y_0}$ of (3) will be an integral manifold of the corresponding Lie algebra $L(Y_1, \cdots, Y_m) \subseteq \mathbf{F}_{n+1}$ fulfilling $\dim M_{y_0} = k < n + 1$. In addition, some regularity condition of the solution $F(t, x)$ in (3) with respect to the variable $t$ will be necessary for obtain ting a nontrivial solution of (2). It is accomplished impossing a maximal rank condition on the original Lie algebra $L(g_1, \cdots, g_m)$; or provided a new maximal rank Lie algebra $L(q_1, \cdots, q_m)$ is associated and the homomorphism $\lambda$ is given explicitly. The homomorphism $\lambda$ between the two Lie algebra defines an integral manifold of $L(g_1, \cdots, g_m)$ and is defined usually by a solution in a gradient system (see applications in the next chapter).

All these can be summarized as follows. For a Lie algebra $L \subseteq \mathbf{F}_n$ denote $L(x)$ the linear subspace determined by the vector fields in $L$ taken at the fixed point $x \in R^n$.

**Theorem 1.**
*Let $L \subseteq \mathbf{F}_n$ be a Lie algebra and $x_0 \in R^n$ fixed. Assume that there exist $\{Y_1, \cdots, Y_m, \cdots, Y_M\} \subseteq L$ such that $y(p) \triangleq G_1(t_1) \circ \cdots \circ G_m(t_m)(x_0), p \triangleq (t_1, \cdots, t_m) \in D_m = \prod_{i=1}^{m}(-a_i, a_i)$, and $G_i(\tau)(x)$ the local flow generated by $Y_i$, fulfils the following*

$$i) \; \frac{\partial y}{\partial t_j}(p) = \sum_{k=1}^{M} \alpha_k^j(p) Y_k(y(p)) \text{ with } \alpha_k^j \in C^\infty(D_m) \text{ and } \frac{\partial y}{\partial t_1}(p), \cdots, \frac{\partial y}{\partial t_m}(p)$$

*are linearly independent $(\forall)\; p \in D_m$,*

$$ii) \; span \left\{ \frac{\partial y}{\partial t_1}(p), \cdots, \frac{\partial y}{\partial t_m}(p) \right\} = L(y(p)), \; (\forall)\; p \in D_m.$$

*Then there exist*

$$q_k \in C^\infty(D_m; R^m)$$

*and a homomorphism*

$$\lambda : L(q_1, \cdots, q_M) \to L \quad such \; that$$

*$c_1$) $\dim L(q_1, \cdots q_M)(p) = m, \;\; (\forall)\; p \in D_m,$*

*$c_2$) for any $y \in L$ there exist $q \in L(q_1, \cdots, q_M)$, such that*

$$\frac{\partial y}{\partial p}(p)q(p) = Y(y(p)), \; p \in D_m,$$

*and in particular*

$$\frac{\partial y}{\partial p}(p)q_i(p) = Y_i(y(p)), \; p \in D_m, \; i = 1, \cdots, M.$$

**Proof.**

By hypotheses the conditions in Theorem 2.3 are fulfilled and we associate a gradient system

$$(4) \quad \frac{\partial y}{\partial t_1} = Y_1(y), \frac{\partial y}{\partial t_2} = X_2(t_1; y), \cdots, \frac{\partial y}{\partial t_m} = X_m(t_1, \cdots, t_{m-1}; y), p \in D_m,$$

with Cauchy condition $y(0) = x_0$ for which $y(p)$ is the solution.

Using (i) we obtain $q_j(\cdot) \in C^\infty(D_m; R^m)$ fulfilling

$$(5) \quad \frac{\partial y}{\partial p}(p)q_j(p) = Y_j(p(y)), \; p \in D_m, \; j = 1, \cdots, M$$

Define the real Lie algebra $L(q_1, \cdots, q_M) \subset C^\infty(D_m; R^m)$. Taking directional derivatives and using the symmetry of $\dfrac{\partial^2 y_i}{\partial p^2}(p)$, $i = 1, \cdots, n$, from (5) we obtain $q \in L(q_1, \cdots, q_M)$ such that

$$(6) \quad \frac{\partial y}{\partial p}(p)q(p) = Y(y(p)), \; (\forall) \; p \in D_m,$$

if $Y \in L$ is an arbitrary product $P(Y_{i_1}, \cdots, Y_{i_M})$, $i_j \in \{1, \cdots, M\}$. Here

$$(7) \qquad q = P(q_{i_1}, \cdots, q_{i_M}), \text{ where } P(Y_{i_1}, \cdots, Y_{i_M}) \text{ defines } Y \in L$$

Using (6) and (7) we obtain $(c_2)$ valid and from the hypothesis (i) we write

$$(8) \quad \frac{\partial y}{\partial t_k}(p) = \sum_{j=1}^{M} a_j^k(p)Y_j(y(p)), \text{ with } a_j^k \in C^\infty(D_m), \; k = 1, \cdots, m$$

Define $\widehat{q}_s(p) = \displaystyle\sum_{j=1}^{M} a_j^s(p)q_j(p)$ for $s = 1, \cdots, m$, and using (8) in (5) we obtain

$$(9) \quad \frac{\partial y}{\partial p}(p)\widehat{q}_s(p) = \frac{\partial y}{\partial t_s}, \; (\forall) \; p \in D_m, \text{ for each } s \in \{1, \cdots, m\}$$

Using again (i) we obtain the unique solution $\widehat{q}_s(p) = e_s, s = 1, \cdots, m$, where $\{e_1, \cdots, e_m\} \subseteq R^m$ is the canonical basis. On the other hand $\widehat{q}_s(p) \in L(q_1, \cdots, q_M)(p)$ for each $p \in D_m$ and $s = 1, \cdots, m$, which shows that $\{e_1, \cdots, e_m\} \subseteq L(q_1, \cdots, q_M)(p)$ and $(c_1)$ follows. The proof is complete.

We have seen that a finite composition of flows determines a gradient system in $\mathbf{F}_n$ or Der $(R^n)$ provided the commutative derivations $\dfrac{\partial}{\partial t_i}$ $i = 1, \cdots, m$, are used. It may occur that in a Lie algebra $\Lambda$ it is not easy to define solutions for a differential equation but a definition of a gradient system is still possible. It is the case with a finitely generated over $R$ Lie algebra (or a finite dimensional Lie algebra).

### Definition

*We say that the real Lie algebra $\Lambda$ is finitely generated over $R(f.g.r.)$ if there exist $\{Y_1, \cdots, Y_M\} \subseteq \Lambda$ such that any $Y \in \Lambda$ can be written $Y = \displaystyle\sum_{j=1}^{M} a_j Y_j$, with some $a_j \in R$ depending on $Y$; $\{Y_1, \cdots, Y_M\}$ is called a system of generators.*

*We recall that for $Z \in \Lambda$ the linear application $\mathrm{ad} Z \in \mathrm{Der}(\Lambda)$ is defined by $\mathrm{ad} Z(Y) = [Z, Y]$, where $[\cdot, \cdot]$ is given by Lie algebra $\Lambda$.*

### Lemma 1.

*Let $\{Y_1, \cdots, Y_M\} \subseteq \Lambda$ be a system of generators for the Lie algebra $\Lambda$. Let $Z \in \Lambda$ and consider the formal series $\exp\ t\ \mathrm{ad}\ Z \triangleq I + \dfrac{t}{1!}\mathrm{ad}\ Z + \cdots \dfrac{t^k}{k!}\mathrm{ad}^k Z + \cdots$.*

*Define a matrix $(M \times M)B$ such that $\{\mathrm{ad}\ Z(Y_1), \cdots, \mathrm{ad}\ Z(Y_M)\} = \{Y_1, \cdots, Y_M\}B$. Then $(\exp\ t\ \mathrm{ad}\ Z)(X) \in \Lambda$ for each $X \in \Lambda$ and $t \in R$ fulfilling*

$c_1)$ $(\exp\ t\ \mathrm{ad}\ Z)(X) = \{Y_1, \cdots, Y_M\}\ (\exp\ t\ B)v$, *where* $v \in R^M$ *and* $X = \displaystyle\sum_{k=1}^{M} v^k Y_k$

$c_2)$ $(\exp\ t\ \mathrm{ad}\ Z)^{-1} = \exp\ -t\ \mathrm{ad}\ Z.$

### Proof.

The conclusion $(c_1)$ is a rewriting of

$$(\exp\ t\ \mathrm{ad}\ Z)(X) \triangleq X + \frac{t}{1!}\mathrm{ad} Z(X) + \frac{t^2}{2!}\mathrm{ad}^2 Z(X) + \cdots + \frac{t^k}{k!}\mathrm{ad}^k Z(X) + \cdots$$

and if we apply $(\exp\ -t\ \mathrm{ad}\ Z)$ to $(\exp\ t\ \mathrm{ad}\ Z)(X)$ we obtain $\{Y_1, \cdots, Y_M\}v$ $(= X)$ and $(c_2)$ is proved.

### Lemma 2.

With the same hypotheses as in Lemma 1 the analytical solution of the differential equation $\dfrac{dX}{dt}(t) = [Z, X(t)]$, $t \in R$, with Cauchy condition $X(0) = X \in \Lambda$ is unique and,

$$X(t) = (\exp t \text{ ad } Z)X = \{Y_1, \cdots, Y_M\} (\exp t B)v,$$

where $v \in R^M$ satisfies $\sum_{k=1}^{m} v^k Y_k = X$.

In addition, $(\exp t \text{ ad } Z)[X_1, X_2] = [X_1(t), X_2(t)]$ for any $X_1, X_2 \in \Lambda$, where $X_i(t) \triangleq (\exp t \text{ ad } Z)X_i$, $i = 1, 2$.

**Proof.**

By definition $\dfrac{dX}{dt}(t) = \{Y_1, \cdots, Y_M\}B(\exp t \ B)v = \text{ad } Z(X(t)) = [Z, X(t)]$, and $X(t)$, $t \in R$, is a solution.

For any other analytical solution.

$$X_1(t) = a_0 + \frac{t}{1!}a_1 + \cdots + \frac{t^k}{k!}a_k + \cdots, \quad a_j \in \Lambda,$$

we obtain

$$X = X_1(0) = a_0, \ a_1 = \frac{dX_1}{dt} = [Z, X] = \text{ad } Z(X), \cdots, a_k = \text{ad}^k Z(X), \cdots$$

$X_1(t)$ is rewriten as

$$X_1(t) = (\exp t \text{ ad } Z)(X) = X(t) \text{ for any } t \in R,$$

For the next conclusion we recall that it is valid for $X_i(t)$ as a formal series. On the other hand $X(t) \triangleq (\exp t \ adZ)(X)$, with $X \triangleq [X_1, X_2]$ is the unique analytical solution for $\dfrac{dX}{dt}(t) = [Z, X(t)]$, and Cauchy condition $X(0) = X$.

The function $Y(t) = [X_1(t), X_2(t)]$ fulfils $Y(0) = X$ and the above differential equation. Therefore $X(t) = Y(t)$.

The proof is complete.

Now, for a given system of generators $\{Y_1, \cdots, Y_M\} \subseteq \Lambda$, we associate the following mappings in $\Lambda$

(10)

$$X_1 = Y_1, X_2(t_1) \triangleq (\exp t_1 \text{ ad } Y_1)Y_2, \cdots, X_M(t_1, \cdots, t_{M-1})$$
$$= (\exp t_1 \text{ ad } Y_1) \cdots (\exp t_{M-1} adY_{M-1})Y_M \text{for } p \triangleq (t_1, \cdot, t_M) \in R^M, p_j$$
$$\triangleq (t_1, \cdots, t_{j-1})$$

A Baker–Campbell–Hausdorff type of formula is valid and it will give the meaning of a gradient system defined by the mappings in (10). Define $\widehat{X}_1 = Y_1$ and

$$
(11) \qquad
\begin{aligned}
\widehat{X}_{j+1}(p_{j+1}) =\ & \exp t_k \, \mathrm{ad}\, X_k(p_k) \cdots (\exp t_1 \, \mathrm{ad}\, X_1) \\
& (\exp t_{k+1} \, \mathrm{ad}\, Y_{k+1}) \cdots (\exp t_j \, \mathrm{ad}\, Y_j) Y_{j+1}
\end{aligned}
$$

for any $1 \le k \le j, j = 1, 2, \cdots, M - 1$, where $X_j(p_j)$ are defined in (10).

**Lemma 3.**

Let $\Lambda$ be a f.g.r. Lie algebra and $\{Y_1, \cdots, Y_M\} \subseteq \Lambda$ a system of generators. Define $X_j(p_j)$ and $\widehat{X}_j(p_j)$, $j = 1, \cdots, M$, as in (10) and (11) correspondingly. Then

$$
X_{j+1}(p_{j+1}) = \widehat{X}_{j+1}(p_{j+1}) \text{ for any } p \in R^M, j = 0, 1, \cdots, M - 1.
$$

**Proof.**

Using Lemma 1 we obtain $X_j(p_j) \in \Lambda$ $(\forall)$ $p \in R^M$, for any $j = 1, \cdots, M$. Denote by $X$ and $\widehat{X}$ the value of $X_{j+1}(p_{j+1})$ and $\widehat{X}_{j+1}(p_{j+1})$ for $t_k = 0$. By derivation with respect to $t_k$ we obtain

$$
(12) \qquad
\begin{aligned}
& \frac{\partial \widehat{X}_{j+1}}{\partial t_k}(p_{j+1}) \\
& = [X_k(p_k), \widehat{X}_{j+1}(p_{j+1})], \widehat{X}_{j+1}(t_1, \cdots, t_{k-1}, 0, t_{k+1}, \cdots, t_j) = \widehat{X}
\end{aligned}
$$

and

$$
(13) \qquad
\begin{aligned}
\frac{\partial X_{j+1}}{\partial t_k}(p_{j+1}) =\ & (\exp t_1 \, \mathrm{ad}\, Y_1) \cdots (\exp t_{k-1} \, \mathrm{ad}\, Y k_{-1}) \\
& [Y_k, (\exp t_k \, \mathrm{ad}\, Y_k) \cdots (\exp t_j \, \mathrm{ad}\, Y_j) Y_{j+1}] \\
& X_{j+1}(t_1, \cdots t_{k-1}, 0, t_{k+1}, \cdots, t_j) = X
\end{aligned}
$$

Using Lemma 2 we rewrite (13) as

$$
(14) \qquad
\begin{aligned}
\frac{\partial X_{j+1}}{\partial t_k}(p_{j+1}) =\ & [X_k(p_k), X_{j+1}(p_{j+1})], \\
& X_{j+1}(t_1, \cdot, t_{k-1}, 0, t_{k+1}, \cdot, t_j) = X.
\end{aligned}
$$

Comparing (12) and (14) we obtain the conclusion, provided $\widehat{X} = X$ (see Lemma 2).

By definition, $X$ and $\widehat{X}$ are similar to $X_j(p_j)$ and $\widehat{X}_j(p_j)$, respectively, containing only $j$ elements.

The proof is completed by an induction argument with respect to $j$. Now we are in position to formulate

**Theorem 2.**

Let $\Lambda$ be a f.g.r. Lie algebra and $\{Y_1, \cdots, Y_M\} \subseteq \Lambda$ a fixed system of generators. Then $X_j(p_j)$, $j = 1, \cdots, M$, defined in (10) is a gradient system in $\Lambda$ i.e.

$$\frac{\partial X_j}{\partial t_i}(p_j) = [X_i(p_i), X_j(p_j)],$$

for $1 \leq i < j = 2, \cdots, M$.

**Proof.**

By hypothesis the conditions in Lemma 3 are fulfilled. The derivative $\frac{\partial X_j}{\partial t_i}(p_j)$ is computed using the equality $X_{j+1}(p_{j+1}) = \widehat{X}_{j+1}(p_{j+1})$, for $j = 0, 1, \cdots, M-1$ where $\widehat{X}_j(p_j)$ is defined in Lemma 3.

Taking $k = i$ we obtain

$$\begin{aligned}
\frac{\partial X_{j+1}}{\partial t_i}(p_{j+1}) &= \frac{\partial \widehat{X}_{j+1}}{\partial t_i}(p_{j+1}) \\
&= [X_i(p_i), \widehat{X}_{j+1}(p_{j+1})] \\
&= [X_i(p_i), X_{j+1}(p_{j+1})]
\end{aligned}$$

and the proof is complete.

**Remark**

One may wonder why we have used a generator system $\{Y_1, \cdots, Y_M\} \subseteq \Lambda$ for a finite–dimensional linear space $\Lambda$.

A reason for this comes from the simpler form we can obtain for the mappings in (10) defining a gradient system. In the case where $\wedge$ is a nilpotent algebra then we can take a nilpotent system of generators $\{Y_1, \cdots, Y_M\}$, and $(\exp t_j \, \text{ad} \, Y_j)\{Y_1, \cdots, Y_M\} = \{Y_1, \cdots, Y_M\}(\exp t_j B_j)$, where the matrix $(M \times M)B_j$ is a nilpotent one (see $B_j^N = \odot$ for some natural $N$).

# Chapter 2

# Representation of a Gradient System

We have mentioned that a global gradient system can be associated with any system of generators provided the real Lie algebra $\Lambda$ is finite–dimensional (see theorem 1.3.2). As any linear application $T : \Lambda \to \Lambda$ can be represented by a matrix with respect to some basis in $\Lambda$ it is generally true that a gradient system associated with a system of generators can also be represented by a matrix with respect to the fixed system of generators. Here we shall confine ourselves to consider only a finite–dimensional Lie algebra $\Lambda$.

## 2.1   Finite–Dimensional Lie Algebra

For a fixed system of generators $\{Y_1, \cdots, Y_M\} \subseteq \Lambda$ in a finite–dimensional Lie algebra $\Lambda$ we associate the corresponding gradient system

(1)   $$X_1 = Y_1,\ X_2(t_1) = (\exp\ t_1\ \mathrm{ad}\ Y_1)Y_2, \cdots, X_M(t_1, \cdots, t_{M-1}) = = (\exp\ t_1 \mathrm{ad}Y_1) \cdots (\exp\ t_{M-1} \mathrm{ad}Y_{M-1})Y_M$$

for $p \overset{\triangle}{=} (t_1, \cdots, t_M) \in R^M,\ \ p_j \overset{\triangle}{=} (t_1, \cdots, t_{j-1})$,
     Let the matrix $(M \times M)B_j$ be fixed such that

(2)        $$\mathrm{ad}\ Y_j\{Y_1, \cdots, Y_M\} = \{Y_1, \cdots, Y_M\}B_j,\ \ j = 1, \cdots, M$$

Define the analytical matrix $(M \times M)A(p),\ p \in R^M$ by

(3)   $$A(p) = \{e_1,\ (\exp\ t_1 B_1)e_2, \cdots,\ (\exp\ t_1 B_1) \cdots (\exp\ t_{M-1} B_{M-1})e_M\}$$

where $e_1, \cdots, e_M \in R^M$ is the canonical basis.

**Theorem 1.**

Let an real finite–dimensional Lie algebra $\Lambda$ be given and consider $\{Y_1, \cdots, Y_M\}$
$\subseteq \Lambda$ a system of generators.

Then the associated gradient system in (1) can be represented as

$$\{X_1, X_2(t_1), \cdots, X_M(t_1, \cdots, t_{M-1})\} = \{Y_1, \cdots, Y_M\}A(p),$$

with $A(0) = I_M$ (identity), where the analytical $(M \times M)$ matrix $A(p)$ is given in (3).

**Proof**

Define the matrix $(M \times M)B_j$ such that $B_j e_j = 0$ and

(4)     $\{(\mathrm{ad} Y_j)Y_1, \cdots, (\mathrm{ad} Y_j)Y_M\} = \{Y_1, \cdots, Y_M\}B_j, \; j = 1, \cdots, M$

By a direct computation (see Lemma 1.3.1) we obtain

(5)     $(\exp t \, \mathrm{ad} \, Y_j)Y_{j+1} = \{Y_1, \cdots, Y_M\{ \exp t \, B_j)e_{j+1}, \; j = 1, \cdots, M-1,$

and

$$(\exp t \, \mathrm{ad} \, Y_j)\{Y_1, \cdots, Y_M\} = \{Y_1, \cdots, Y_M\} (\exp t \, B_j)$$

where $e_1, \cdots, e_M \in R^M$ is the canonical base.

Write $p_{j+1} = (t_1, \cdots, t_j)$ and $X_{j+1}(p_{j+1})$, $j = 1, 2, \cdots, M-1$, for the mappings defined in (1) with $p \overset{\triangle}{=} (t_1, \cdots, t_M) \in R^M$.

Using (5) in (1) it is easily seen that $X_{j+1}(p_{j+1})$ can be represented as follows:

(6)
$$X_{j+1}(p_{j+1}) = \{Y_1, \cdots, Y_M\}(\exp t_1 B_1) \cdots (\exp t_j B_j)e_{j+1}, j = 1, \cdots, M-1,$$

and

$$X_1 = Y_1 = \{Y_1, \cdots Y_M\}e_1.$$

Now, define the analytical matrix $A(p), p \in R^M$, as in (3) and we obtain

(7)     $\{X_1, X_2(t_1), \cdots, X_M(t_1, \cdots, t_{M-1})\} = \{Y_1, \cdots, Y_M\}A(p),$

with $A(0) = I_M$ (identity)

The proof is complete.

**Remark 1.**

*The matrix $A(p)$ associated with a gradient system in Theorem 1 fulfills $A(0) = I_M$ and by the continuity property it will be a nonsingular one for any $p \in D_M \overset{\triangle}{=} \prod_{i=1}^{M}(-a_i, a_i)$, where $a_i > 0$, $i = 1, \cdots, M$.*

It will be proved that the matrix $A(p)$ is nonsingular for any $p \in R^M$. To achieve this goal is necessary to see how the matrix changes if we take another system of generators. Choose an arbitrary $p^* \in D_M = \prod_{i=1}^{M}(-a_i, a_i)$ and by definition $A(p^*)$ is a nonsingular $(M \times M)$ matrix.

Define

$$(8) \qquad Y_1^* = Y_1, \ Y_2^* = X_2(t_1^*), \cdots, \quad Y_M^* = X_M(t_1^*, \cdots, t_{M-1}^*)$$

and using Theorem 1 we obtain a new system of generators

$$(9) \quad \{Y_1^*, \cdots, Y_M^*\} = \{Y_1, \cdots, Y_M\}A(p^*) \text{ for } p^* \in D_M = \prod_{i=1}^{M}(-a_i, a_i) \text{ fixed}$$

Using the new system of generators $\{Y_1^*, \cdots, Y_M^*\} \subseteq \Lambda$ define the corresponding gradient system

$$(10) \quad \begin{aligned} X_1^* &= Y_1^*, X_2^*(s_1) = (\exp s_1 \text{ ad } Y_1^*)Y_2^*, \cdots, X_M^*(s_1, \cdots, s_{M-1}) \\ &= (\exp s_1 \text{ ad } Y_1^*) \cdots (\exp s_{M-1} \text{ ad } Y_{M-1}^*)Y_M^* \end{aligned}$$

The algebraic representation of the gradient system in (10) could be obtained provided a $(M \times M)$ matrix $B_j^*$ is defined such that

$$(11) \qquad \text{ad } Y_j^*\{Y_1^*, \cdots, Y_M^*\} = \{Y_1^*, \cdots, Y_M^*\}B_j^*, \ j = 1, \cdots, M.$$

Using (9) we obtain

$$(12) \qquad Y_j^* = \{Y_1, \cdots, Y_M\}A(p^*)e_j = \sum_{i=1}^{M}(A(p^*)e_j)^i Y_i$$

and

$$(13) \qquad \text{ad } Y_j^*\{Y_1^*, \cdots, Y_M^*\} = \text{ad}Y_j^*\{Y_1, \cdots, Y_M\}A(p^*), \ j = 1, \cdots, M.$$

Using (12) in (13) it follows that

$$(14) \qquad \begin{aligned} &\text{ad } Y_j^*\{Y_1^*, \cdots, Y_M^*\} \\ &= \left(\sum_{i=1}^{M}(A(p^*)e_j)\right)^i \text{ad } Y_i\{Y_1, \cdots, Y_M\})A(p^*), \ j = 1, \cdots, M \end{aligned}$$

and replacing $\mathrm{ad}Y_i\{Y_1, \cdots, Y_M\}$ by $\{Y_1, \cdots, Y_M\}B_i$, where $B_i$ are defined in Theorem 1 we obtain

(15)

$$\mathrm{ad}Y_j^*\{Y_1^*, \cdots, Y_M^*\} = \{Y_1, \cdots, Y_M\} \left( \sum_{i=1}^{M} (A(p^*)e_j)^i B_i \right) A(p^*), j = 1, \cdots, M$$

Now, use again (9), where $A^{-1}(p^*)$ exists, and write

(16)              $$\mathrm{ad}Y_j^*\{Y_1^*, \cdots, Y_M^*\} = \{Y_1^*, \cdots, Y_M^*\} B_j^*,$$

where

(17)       $$B_j^* \triangleq A^{-1}(p^*) \left( \sum_{i=1}^{M} (A(p^*)e_j)^i B_i \right) A(p^*), j = 1, \cdots, M$$

The above given computation is summarized as follows

**Lemma 1.**

Let the conditions in Theorem 1 be fulfilled and $p^* \in D_M = \prod_{i=1}^{M}(-a_i, a_i)$ be fixed such that the matrix $A(p^*)$ is nonsingular. Define a new system of generators $\{Y_1^*, \cdots, Y_M^*\}$ and a corresponding gradient system as in (9) and (10) respectively. Then the following representation is valid

(c)    $X_{j+1}^*(s_1, \cdots, s_j) = \{Y_1^*, \cdots, Y_M^*\} (\exp s_1 B_1^*) \cdots (\exp s_j B_j^*)e_{j+1},$

$j = 1, \cdots, M - 1,$
where $B_j^*$ are defined in (17) and $e_1, \cdots, e_M \in R^M$ is the canonical basis.

**Proof.**
    The system $\{Y_1^*, \cdots, Y_M^*\}$ of generators and the corresponding gradient system in (10) fulfil the hypotheses in Theorem 1. Using (16) we obtain the $(M \times M)$ matrices $B_j^*$ such that

$$\{\mathrm{ad}\ Y_j^*(Y_1^*), \cdots, \mathrm{ad}\ Y_j^*(Y_M^*)\} = \{Y_1^*, \cdots, Y_M^*\}B_j^*, \ j = 1, \cdots, M.$$

where $B_j^*$ is defined in (17).
    Applying theorem 1 we obtain (c) and the proof is complete.
    In the same conditions as in Lemma 1 denote

(18)          $$A = A(p^*), \ C_j = \sum_{i=1}^{M}(Ae_j)^i B_i, \ j = 1, \cdots, M,$$

(19)     $Z_j(s_1, \cdots, s_j) = \exp(s_1 C_1) \cdots (\exp s_j C_j)$, $j = 1, \cdots, M-1$.


**Lemma 2.**
*Let the conditions in Lemma 1 be fulfilled. Then the gradient system in (10) has the following representation*

$(c_1)$ $\{X_1^*, X_2^*(s_1), \cdots, X_M^*(s_1, \cdots, s_{M-1})\} = \{Y_1^*, \cdots, Y_M^*\} A^*(s)$, $s \in R^M$,

*where the analytical $(M \times M)$ matrix $A^*(s)$ fulfils $A^*(0) = I_M$ and*
$(c_2)$
$A^*(s) =$
$\{e_1, A^{-1} Z_1(s_1) Ae_2, A^{-1} Z_2(s_1, s_2) Ae_3, \cdots, A^{-1} Z_{M-1}(s_1, \cdots, s_{M-1}) Ae_M\}$

*with $Z_j(s_1, \cdots, s_j)$ defined in (19) and A in (18).*
**Proof.**
By hypothesis, the conclusion (c) in Lemma 1 is fulfilled. The matrices
$A \overset{\triangle}{=} A(p^*)$ and $A^{-1}$ define $B_j^*$ by

(20)         $B_j^* = A^{-1} C_j A$, where $C_j$ isgivenin(18), $j = 1, \cdots, M$.

Using (20) we rewrite $\exp s_j B_j^* = A^{-1}(\exp s_j C_j)A$ and

(21)
$X_{j+1}^*(s_1, \cdots, s_j)$
$= \{Y_1^* \cdots, Y_M^*\} A^{-1}((\exp s_1 C_1) \cdots (\exp s_j C_j)) Ae_{j+1}$
$= \{Y_1^*, \cdots, Y_M^*\} A^{-1} Z_j(s_1, \cdots, s_j) Ae_{j+1}$, $j = 1, \cdots, M-1$,

where $Z_j(s_1, \cdots, s_j)$ is given in (19).
The conclusions $(c_1)$ and $(c_2)$ follows directly from (21) and the proof is complete.
Though the matrix $A^*(s)$ in Lemma 2 depends on $A^{-1}$ it is the purpose of the next lemma to state that the local nonsingularity of $A^*(s)$, $s \in D_M = \prod_{i=1}^{M}(-a_i, a_i)$ does not depend on $A^{-1}$. The proof of the next lemma is given in appendix (see Lemma a.1).


**Lemma 3.**
*Assume that the hypotheses in Lemma 2 are fulfilled and consider the matrix $A^*(s)$, $s \in R^M$, defined in $(c_2)$. Denote $Z_j^a(s) = A^{-1} Z_j(s_1, \cdots, s_j)A$, for $s \in R^M, j = 1, \cdots, M-1$, and*

$$M^a(s) \overset{\triangle}{=} (e_1, Z_1^a(s)e_2, \cdots, Z_{M-1}^a(s)e_M).$$

*Then*

$c_1$)                   $\det M^a(s) \geq d_1(s) \cdots d_{M-1}(s) + 0(s)$ if $d_i(s) \geq 0$,

*where*

$$d_i(s) \overset{\triangle}{=} \min \left\{ r : r \in \text{spectrum of } \frac{Z_i(s) + Z_i^T(s)}{2} \right\}$$

*(see $d_i(0) = 1$) and $| 0(s) | \leq C \max\{\|Z_i(s) - I\|, \ i = 1, \cdots, M - 1\}$ with a constant $C$ not depending on $A$.*

We notice that for a parameter $p \in R^M$ such that $p = p^* + s$, where $s, p \in D_M$, we may obtain the nonsingularity of the matrix $A(p)$ in Theorem 1 provided $A(p) = A(p^*) \cdot A^*(s)$, which means that the gradient system in Theorem 1 must equal the gradient system considered in Lemma 1.

**Lemma 4.**

*Assume the hypotheses in Theorem 1 are fulfilled and let $p^* \in D_M = \prod_1^M (-a_i, a_i)$ be fixed such that $A(p^*)$ is nonsingular. For an arbitrary $p \in R^M$ write $s_i \overset{\triangle}{=} t_i - t_i^*$, $i = 1, \cdots, M$, where $p = (t_1, \cdots, t_M)$, $p^* = (t_1^*, \cdots, t_M^*)$. Then*

c) $X_{j+1}^*(s_1, \cdots, s_j) = X_{j+1}(t_1, \cdots, t_j)$, $j = 1, \cdots, M - 1$

*where $X_{j+1}$ is defined in Theorem 1 and $X_{j+1}^*$ in Lemma 1.*

**Proof.**

The conclusion (c) is obtained provided each $X_{j+1}^*(s_1, \cdots, s_j)$ is rewriten and a Baker–Campbell–Hausdorff formula is used (see Lemma 1.3.3).

By definition for $s_1 = t_1 - t_1^*$,

$$X_2^*(s_1) = (\exp(t_1 - t_1^*)\text{ad}Y_1)X_2(t_1^*)$$

(22)                   $= (\exp t_1 \text{ad} Y_1)(\exp -t_1^* \text{ad} Y_1) X_2(t_1^*)$

$$= (\exp t_1 \text{ad} Y_1)Y_2 \overset{\triangle}{=} X_2(t_1)$$

and each $X_{j+1}(t_1^*, \cdots, t_j^*) \overset{\triangle}{=} Y_{j+1}^*$ can be rewriten (see Lemma 1.3.3)

(23)
$$Y_{j+1}^* = (\exp t_j^* \text{ad } X_j(p_j^*)) \cdots (\exp t_2^* \text{ad } X_2(t_1^*))(\exp t_1^* \text{ad } Y_1)Y_{j+1}$$

$$= (\exp t_j^* \text{ad } Y_j^*) \cdots (\exp t_2^* \text{ad } Y_2^*)(\exp t_1^* \text{ad } Y_1^*)Y_{j+1}, j = 1, \cdots, M - 1.$$

Using (23) for $j = 2$ we obtain

$$
\begin{aligned}
X_3^*(s_1, s_2) &= (\exp s_1 \mathrm{ad} Y_1^*)(\exp s_2 \mathrm{ad} Y_2^*) Y_3^* \\
&= (\exp s_1 \mathrm{ad} Y_1)(\exp t_2 \mathrm{ad} X_2(t_1^*))(\exp t_1^* \mathrm{ad} Y_1) Y_3 \\
&= (\exp s_1 \mathrm{ad} Y_1)(\exp t_1^* \mathrm{ad} Y_1)(\exp t_2 \mathrm{ad} Y_2) Y_3 = \\
&= (\exp t_1 \mathrm{ad} e Y_1)(\exp t_2 \mathrm{ad} Y_2) Y_3 \overset{\triangle}{=} X_3(t_1, t_2)
\end{aligned}
$$

(24)

if $s_1 = t_1 - t_1^*, \quad s_2 = t_2 - t_2^*$

Using the same argument as above we obtain

(25)  $\qquad X_{j+1}^*(s_1, \cdots, s_j) = X_{j+1}(t_1, \cdots, t_j), \quad j = 1, \cdots, M - 1$

provided $t_i = s_i + t_i^*, \ i = 1, \cdots, M - 1$.

The proof is complete.

**Theorem 2.**

*Assume the hypotheses in Theorem 1 are fulfilled and the analytical ($M \times$ $M$) matrix $A(p)$ is defined. Then $\det A(p) \neq 0$ for any $p \in R^M$.*

**Proof.**

Let $\widehat{p} \in R^M$ be arbitrarily fixed and consider the interval $[0, \widehat{p}]$. Using that $A(p)$, $p \in D_M \overset{\triangle}{=} \prod\limits_{i=1}^{M} (-a_i, a_i)$, is nonsingular, we shall show that it will remain nonsingular on the whole interval $[0, \widehat{p}]$. We choose a $p^* \in [0, \widehat{p}] \cap D_M$ with $\|p^*\| = \gamma > 0$ and define a new system of generators $\{Y_j^*, \cdots, Y_M^*\}$ and the corresponding gradient system, as in Lemma 1 . By hypothesis the conditions in Lemma 2 are fulfilled and we obtain the algebraic representation of the new gradient system:

(26)  $\{X_1^*, X_2^*(s_1), \cdots, X_M^*(s_1, \cdots, s_{M-1})\} = \{Y_1^*, \cdots, Y_M^*\} A^*(s), \ s \in R^M$

where

(27)  $\{Y_1^*, \cdots, Y_M^*\} = \{Y_1, \cdots, Y_M\} A(p^*)$ with $A(p^*)$ nonsingular

On the other hand, using Theorem 1 we obtain

(28)  $\{X_1, X_2(t_1), \cdots, X_{M-1}(t_1, \cdots, t_{M-1})\} = \{Y_1, \cdots, Y_M\} A(p),$

for any $p \in R^M$. From Lemma 4 we obtain

(29)  $\qquad X_{j+1}^*(s_1, \cdots, s_j) = X_{j+1}(t_1, \cdots, t_j), \ j = 1, \cdots, M - 1,$

provided
$$s_i = t_i - t_i^*, \ p^* \overset{\triangle}{=} (t_1^*, \cdots, t_M^*), \ p = (t_1, \cdots, t_M),$$

where $X_{j+1}^*$ and $X_{j+1}$ fulfil (26) and correspondingly (28).

Using (27) and (28) in (29) we obtain

$$
\begin{aligned}
(30) \quad \{X_1, X_2(t_1), \cdots, X_{M-1}(t_1, \cdots, t_{M-1}\} &= \{Y_1, \cdots, Y_M\} A(p^*) \cdot A^*(s) \\
&= \{Y_1, \cdots, Y_M\} A(p)
\end{aligned}
$$

provided
$$s = (s_1, \cdots, s_M) \text{ and } s_i = t_i - t_i^*, \ i = 1, \cdots, M.$$

Now, Lemma 3 shows that $A^*(s), s \in D_M = \prod_{i=1}^{M}(-a_i, a_i)$, is a nonsingular matrix and choose $s_1 \in D_M \cap [0, \widehat{p}]$, such that $\|s_1\| = \gamma = \|p^*\|$. Then for $p_1 = p^* + s_1 \in [0, \widehat{p}]$ and $\|p_1\| = 2\gamma$ we obtain (see (30))

$$(31) \qquad A(p_1) = A(p^*)A^*(s_1),$$

where $A(p), p \in R^M$, is defined in Theorem 1 and $A(p^*), A^*(s_1)$ are both nonsingular matrices. Using the above algorithm we obtain $p_N = \widehat{p}$ such that

$$A(\widehat{p}) = A(p^*)A^*(s_1) \cdots A^*(s_N), \text{ where } p_{i+1} = p_i + s_{i+1}, i = 0, \cdots, N-1,$$

and the nonsingular matrices in the right hand side ensure that $A(\widehat{p})$ is a nonsingular one.

The proof is complete.

**Remark 3.**

*According to the proof given in Theorem 2, we are using the algebraic representation of the new gradient system in Lemma 1 with respect to the new system of generators $\{Y_1^*, \cdots, Y_M^*\}$ and finally Lemma 4 is applied for the new system of generators expressed with respect to the original one $\{Y_1, \cdots, Y_M\}$. Exploiting the gradient system structure of $Y_{j+1}^* \overset{\triangle}{=} X_{j+1}(t_1^*, \cdots, t_j^*), j = 0, 1, \cdots, M-1$, we may write directly this dependence but in a more implicit form.*

*Namely,*

$$
\begin{aligned}
(a) \quad \mathrm{ad}Y_1^*\{ \ Y_1^*, \cdots, Y_M^*\} &= \mathrm{ad}Y_1\{Y_1, X_2(t_1^*), \cdots, X_M(t_1^*, \cdots, t_{M-1}^*)\} \\
&= \frac{\partial}{\partial t_1}\{Y_1, X_2(t_1), \cdots, X_M(t_1, \cdots, t_{M-1})\}_{t_i = t_i^*} \\
&= \{Y_1, \cdots, Y_M\}\left\{0, \frac{\partial}{\partial t_1^*}A(p^*)e_2, \cdots, \frac{\partial}{\partial t_1^*}A(p^*)e_M\right\}
\end{aligned}
$$

Similarly for $j = 2, \cdots, M$, we obtain

(b)

$$\text{ad } Y_j^* \{Y_1^*, \cdots, Y_M^*\}$$

$$= \{-\text{ad } Y_1^*(Y_j^*), \cdots, -\text{ad } Y_{j-1}^* Y_j^*,$$

$$\frac{\partial X_{j+1}}{\partial t_j^*}(t_1^*, \cdots, t_j^*), \cdots, \frac{\partial X_M}{\partial t_j^*}(t_1^*, \cdots, t_{M-1}^*)\}$$

$$= \{Y_1, \cdots, Y_M\} \left\{ -\frac{\partial}{\partial t_1^*}(A(p^*)e_j), \cdots, -\frac{\partial}{\partial t_{j-1}^*}(A(p^*)e_j), \right.$$

$$\left. \frac{\partial}{\partial t_j^*}(A(p^*)e_{j+1}), \cdots, \frac{\partial}{\partial t_j^*}(A(p^*)e_M) \right\}$$

## 2.2 The Maximal Rank Lie Algebra

We shall start with an exercise. Consider a finite–dimensional Lie algebra $\Lambda$ and the natural mapping ad : $\Lambda \to \text{Der}(\Lambda)$ defined by $\text{ad} v(w) = [v, w]$ for any $v, w \in \Lambda$, where $[\cdot, \cdot]$ defines the Lie algebra. The linear space Der $(\Lambda)$ is a Lie algebra under the Lie bracket $[b_1, b_2]_1 \triangleq b_1 \circ b_2 - b_2 \circ b_1$, and the mapping ad is a homomorphism between the two Lie algebra $\Lambda$ and Der $(\Lambda)$.

On the other hand, taking a basis $\{Y_1, \cdots, Y_m\} \subseteq \Lambda$ we may and do associate a ($m \times m$ matrix of reals $B$ for each $\text{ad} v, v \in \Lambda$, and, in particular, let $B_i$ be associated with $\text{ad} Y_i, i = 1, \cdots, m$. Denote $L(B_1, \cdots, B_M) \subseteq \mathcal{L}(R^m; R^m)$ the Lie algebra determined by $B_1, \cdots, B_m$ under the Lie bracket $[B_i, B_j] = B_i B_j - B_j B_i$. Finally, we obtain a homomorphism $h : \Lambda \to L(B_1, \cdots, B_m)$ but $\dim L(B_1, \cdots B_m) \leq m$ is the main obstruction for recovering the elements in $\Lambda$ using matrices from $L(B_1, \cdots, B_m)$. It will not be the case if the new Lie algebra is constructed using a gradient system in $\Lambda$. We shall consider that the finite–dimensional Lie algebra $\Lambda$ is determined by $Y_1, \cdots, Y_m \in \Lambda, Y_i \neq 0, i.e. \Lambda = L(Y_1, \cdots, Y_m)$.

**Definition 1.**

*We say that $S \triangleq \{Y_1, \cdots, Y_{m+1}, \cdots Y_M\} \subseteq \Lambda$ is an $m-$ minimal system of generators if $S$ is a system of generators for $\Lambda$ and each $Y_j, j = m+1, \cdots, M$, cannot be written as a linear combination of $\{Y_1, \cdots, Y_m, \cdots, Y_{j-1}\}$; from a system of generators $\{Y_1, \cdots, Y_m, Y_{m+1}, \cdots, Y_M\}$ is always possible to obtain an m-minimal system $S$.*

Denote $C^\omega(R^M; R^m)$ the space consisting of all analytical entire function $q : R^M \to R^M$ and write $L(Y_1, \cdots, Y_M)$ for the real Lie algebra determined by $Y_1, \cdots, Y_M$ in a Lie algebra $\Lambda$.

**Theorem 1.**

Let $\{Y_1, \cdots, Y_m\} \subseteq \mathbf{F}_n$ $(\mathrm{Der}(R^n))$ be such that $L(Y_1, \cdots, Y_m)$ is f.g.r. and consider $S \overset{\triangle}{=} \{Y_1, \cdots Y_{m+1}, \cdots, Y_M\} \subseteq L(Y_1, \cdots, Y_m)$ an m-minimal system of generators. Let $X_j(p)$, $j = 1, \cdots, M, p \in R^M$, be the gradient system associated with $S$ in Theorem 1.1. Then there exist $q_i \in C^\omega(R^M; R^M)$, $i = 1, \cdots, m$, and a homomorphism $\lambda : L(Y_1, \cdots, Y_m) \to L(q_1, \cdots q_m)$ such that

$(c_1)$ $\qquad\qquad \dim L(q_1, \cdots, q_m)(p) = M$ for any $p \in R^M$,

$(c_2)$ $\qquad \lambda(P_k(Y_1, \cdots, Y_m)) = P_k(q_1, \cdots, q_m) \in L(q_1, \cdots, q_m)$

for any $P_k(Y_1, \cdots, Y_m) \overset{\triangle}{=} \mathrm{ad} Y_{i_1} \circ \cdots \circ \mathrm{ad}\, Y_{i_{k-1}}(Y_{i_k})$;

$(c_3)$ $\qquad\qquad \displaystyle\sum_{j=1}^{M} X_j(p) q^j(p) = Y$ if $\lambda(Y) = q \overset{\triangle}{=} (q^1, \cdots, q^M)$.

**Proof.**

By hypotheses, the conditions in theoremes 1.1. and 1.2 are fulfilled and let $A(p)$ be the nonsingular analytical matrix expressing the gradient system

$(1)$ $\qquad \{Y_1, X_2(t_1), \cdots, X_M(p_M)\} = \{Y_1, \cdots, Y_M\} A(p),$

$\qquad\qquad \det A(p) \neq 0\ (\forall)\ p \in R^M.$

Define $q_i \in C^\omega(R^M; R^M)$ such that

$(2)$ $\qquad A(p) q_i(p) = e_i,\quad i = 1, \cdots, m,$ where $e_1, \cdots, e_M \in R^M$

is the canonical basis. Mulplying with $q_i(p)$ in (1) and using (2) we obtain

$(3)$ $\qquad \{Y_1, X_2(t_1), \cdots, X_M(t_1, \cdots, t_{M-1})\} q_i(p) = Y_i,\ i = 1, \cdots, m$

Write $L(q_1, \cdots, q_m)$ for the Lie algebra determined by the vector fields $q_i \in C^\omega(R^M; R^M)$

The homomorphism $\lambda : L(Y_1, \cdots, Y_m) \to L(q_1, \cdots, q_m)$ is defined using the local solution in the following gradient system

$(4)$ $\qquad \dfrac{\partial y}{\partial t_1} = Y_1(y),\quad \dfrac{\partial y}{\partial t_2} = X_2(t_1; y),\quad \cdots,\quad \dfrac{\partial y}{\partial t_M} = X_M(t_1, \cdots, t_{M-1}; y)$

with $y(p_0) = x_0$, arbitrarily fixed $(p_0, x_0) \in R^M \times R^n$, where the vector fields $X_j$ in (4) define the gradient system in (1).

Let $y(p; x)$, $p \in V_1(p_0)$, $x \in V_2(x_0)$, be the local solution of (4) fulfilling $y(p_0; x) = x \in V_2(x_0)$. There holds

(5)
$$\frac{\partial y}{\partial t_j}(p; x) = X_j(p_j; y(p; x)), \quad j = 1, \cdots, M,$$

and (3) can be rewriten along $y(p; x)$ as

(6)
$$\sum_{j=1}^{M} \frac{\partial y}{\partial t_j}(p; x) \, q_i^j(p) = Y_i(y(p; x)), \quad i = 1, \cdots, m,$$

for any $p \in V_1(p_0)$, $x \in V_2(x_0)$.

Taking directional derivatives with respect to $q_k(p)$, $k \in \{1, \cdots, m\}$, in (6), and using the symmetry of the matrices

$$\frac{\partial^2 y^s}{\partial p^2}(p; x), \quad s = 1, \cdots, n,$$

we obtain

(7)
$$\sum_{j=1}^{M} \frac{\partial y}{\partial t_j}(p; x)[q_{i_1}, q_{i_2}]^{(j)}(p) = [Y_{i_1}, Y_{i_2}](y(p; x))$$

for any $i_1, i_2 \in \{1, \cdots, m\}$, $p \in V_1(p_0)$, $x \in V_2(x_0)$, where $[q_{i_1}, q_{i_2}]^{(j)}(p)$ denote the "$j$" component of a vector in $R^M$.

By the iteration of the above algorithm we obtain easily

(8)
$$\sum_{j=1}^{M} X_j(p; y(p; x)) \, P_k(q_1, \cdots, q_m)^j(p) = P_k(Y_1, \cdots, Y_m)(y(p; x))$$

for any $p \in V_1(p_0)$, $x \in V_2(x_0)$, where $P_k(v_1, \cdots, v_m) \overset{\triangle}{=} \mathrm{adv}_{i_1} \circ \cdots \circ \mathrm{adv}_{i_{k-1}}(v_{i_k})$ is the Lie product of $v_{i_1} \cdots v_{i_k}$.

Taking $p = p_0$, $x = x_0$ in (8) we obtain $y(p_0; x_0) = x_0$ and the correspondence $\lambda(P_k(Y_1, \cdots, Y_M)) = P_k(q_1, \cdots, q_m)$ is obtained from the following

(9)
$$\sum_{j=1}^{M} X_j(p) P_k(q_1, \cdots, q_m)^j = P_k(Y_1, \cdots, Y_m), \quad \text{where}$$

$P_k(v_1, \cdots, v_m)$ is defined as in (8). The mapping $\lambda$ fulfilling (9) is extended naturally and (2), (3) are fulfilled. In particular, each $Y_j$, $j = m+1, \cdots, M$, has the form of a product $P_k(Y_1, \cdots, Y_m)$ and from (9) we obtain

(10)
$$\sum_{j=1}^{M} X_j(p) q_s^j(p) = Y_s, \quad s = m+1, \cdots, M.$$

where $q_s \in L(q_1, \cdots, q_m)$ and $Y_s$ in the system of generators $\{Y_1, \cdots, Y_M\}$ are defined by the same product.

Now, using (1) in (10) we obtain

$$(11) \qquad \{Y_1, \cdots, Y_M\} A(p) q_s(p) = Y_s, \quad s = m+1, \cdots, M.$$

It is the $m$-minimality of the system $S = \{Y_1, \cdots, Y_m, Y_{m+1}, \cdots, Y_M\}$ which implies $A(p) q_s(p) = e_s$ as the unique solution from (11) for each $s = m+1, \cdots, M$, where $e_1, \cdots, e_M \in R^M$ is the canonical base.

By definition, $q_s \in L(q_1, \cdots, q_m)$ for any $s \in \{1, \cdots, M\}$ satisfies

$$(12) \qquad A(p) q_s(p) = e_s, \ s = 1, \cdots, M, \text{ for each } p \in R^M.$$

and we obtain $\dim L(q_1, \cdots, q_m)(p) = M$ for any $p \in R^M$.

The proof is complete.

## 2.3   Integral Manifolds

An integral manifold is determined by some integrals of vector fields in a Lie algebra $\Lambda$ and it is why a $\Lambda \subseteq \mathbf{F}_n$ (or $\mathrm{Der}(R^n)$) will be used.

The basic ingredients for an integral manifold to be defined have been described in Theorem 1.3.1 and the general hypotheses therein are fulfilled if we consider a finite–dimensional Lie algebra $\Lambda \subseteq \mathbf{F}_n$.

**Theorem 1.**

*Assume that a basis $\{Y_1, \cdots, Y_N\} \subseteq \Lambda$ and $x_0 \in R^n$ are fixed in a finite-dimensional Lie algebra $\Lambda$ such that $Y_j(x_0) = 0$, $j = k+1, \cdots, N$, and $Y_1(x_0), \cdots, Y_k(x_0)$ are linearly independent in $R^n$, where $k = \dim \Lambda(x_0)$. Define*

$$y(\widehat{p}) = G_1(t_1) \circ \cdots \circ G_k(t_k)(x_0) \text{ for } \widehat{p} \overset{\triangle}{=} (t_1, \cdots, t_k) \in D_k \overset{\triangle}{=} \prod_1^k (-a_i, a_i)$$

*where $G_i(t)(x)$ is the local flow generated by $Y_i$. Then $M_{x_0} \overset{\triangle}{=} \{x \in R^n : x = y(\widehat{p}), \widehat{p} \in D_k\}$ is a $C^\infty$ $k$-dimensional integral manifold of $\Lambda$ i.e.,*

*(c$_1$)* $\dfrac{\partial y}{\partial t_1}(\widehat{p}), \cdots, \dfrac{\partial y}{\partial t_k}(\widehat{p})$ *are linearly independent in $R^n$ for each $\widehat{p} \in D_k$,*

*(c$_2$)* $\dfrac{\partial y}{\partial t_i}(\widehat{p}) \in \Lambda(y(\widehat{p}))$ *and span* $\left\{\dfrac{\partial y}{\partial t_1}(\widehat{p}), \cdots, \dfrac{\partial y}{\partial t_k}(\widehat{p})\right\} = \Lambda(y(\widehat{p})), \widehat{p} \in D_k$

**Proof.**

The conclusions in the theorem restate the geometric properties of the set $M_{x_0}$, i.e. $\dim M_{x_0} = k$ and the tangent space $T_y M_{x_0}$ equals the linear space $\Lambda(y)$ provided $y = y(\widehat{p})$, for some $\widehat{p} \in D_k$. According to the fixed basis $\{Y_1, \cdots, Y_N\} \subseteq \Lambda$ and using Theorem 1.1 we obtain a gradient system
(1)

$$\frac{\partial y}{\partial t_1} = Y_1(y), \qquad \frac{\partial y}{\partial t_2} = X_2(t_1; y), \qquad \cdots, \qquad \frac{\partial y}{\partial t_k} = X_k(t_1, \cdots, t_{k-1}; y),$$

$$\cdots \qquad , \qquad \frac{\partial y}{\partial t_N} = \; = X_N(t_1, \cdots, t_{N-1}; y)$$

for which

$$y = G(p) = G_1(t_1) \circ \cdots \circ G_N(t_N)(x_0), \quad p \overset{\triangle}{=} (t_1, \cdot, t_N), -a_i < t_i < a_i,$$

is the solution with $y(0) = x_0$.

The algebraic representation of the gradient system shows that $X_j(p_j) \in \Lambda$, $j = 1, \cdots, N$, and in particular

(2)  $X_j(p_j; G(p)) \in \Lambda(G(p))$, $j = 1, \cdots, N$, forany $p \in D_N = \prod_{i=1}^N (-a_i, a_i)$

We notice that $y(\widehat{p}) = G(p)$, for any $p \in D_N$ (see $G_j(t_j)(x_0) = x_0$, $j = k+1, \cdot, N$) and from (1) and (2)

(3)  $\dfrac{\partial y}{\partial t_i}(\widehat{p}) \in \Lambda(y(\widehat{p}))$, forany $\widehat{p} \in D_k$, $i = 1, \cdots, k$

The same algebraic representation in Theorem 1.1 is used to have $(c_2)$ fulfilled . Namely,

(4) $\quad \left\{ \dfrac{\partial y}{\partial t_1}(\widehat{p}), \cdots \dfrac{\partial y}{\partial t_k}(\widehat{p}), 0, \cdots, 0 \right\} = \{Y_1, \cdots, Y_N\}(y(\widehat{p})) \, A(\widehat{p}) \text{ for } \widehat{p} \in D_k$

where the $(N \times N)$ matrix $A(\widehat{p})$ fulfils $\det A(\widehat{p}) \neq 0$ for any $\widehat{p} \in D_k$ (see Theorem 2.2).

Using that $\{Y_1, \cdots, Y_N\}$ is a basis for $\Lambda$, from (4) we see easily

(5) $\quad \operatorname{span} \left\{ \dfrac{\partial y}{\partial t_1}(\widehat{p}), \cdots, \dfrac{\partial y}{\partial t_k}(\widehat{p}) \right\} = \Lambda(y(\widehat{p})), \text{ forany } \widehat{p} \in D_k.$

and $(c_2)$ is proved

To prove that $\dfrac{\partial y}{\partial t_1}(\widehat{p}), \cdots, \dfrac{\partial y}{\partial t_k}(\widehat{p})$ are linearly independent it is enough to show that

(6) $\qquad \widehat{Y}_i = \left[ \dfrac{\partial G}{\partial x}(\widehat{p}; x_0) \right]^{-1} \dfrac{\partial y}{\partial t_i}(\widehat{p}), i = 1, \cdots, k,$

are linearly independent where

$$G(\widehat{p}; x) \stackrel{\triangle}{=} G_1(t_1) \circ \cdots \circ G_k(t_k)(x), \ \widehat{p} \in D_k, \ x \in V(x_0).$$

Multiplying by $\left( \dfrac{\partial G}{\partial x}(\widehat{p}; x_0) \right)^{-1}$ in (4) we obtain

(7) $\quad \{\widehat{Y}_1, \cdot, \widehat{Y}_k, 0, \cdots, 0\} = \{X_1(\widehat{p}; y(\widehat{p})), \cdots, X_N(\widehat{p}; y(\widehat{p}))\} A(\widehat{p}), \widehat{p} \in D_k,$

where

$$X_i(\widehat{p}; y(\widehat{p})) \stackrel{\triangle}{=} \left( \dfrac{\partial G}{\partial x}(\widehat{p}; x_0) \right)^{-1} Y_i(y(\widehat{p})), i = 1, \cdots, N.$$

It is easily seen that

(8) $\quad \left( \dfrac{\partial G}{\partial x}(\widehat{p}; x_0) \right)^{-1} = \left( \dfrac{\partial G_k}{\partial x}(t_k; y_k) \right)^{-1} \times \cdot \times \left( \dfrac{\partial G_1}{\partial x}(t_1; y_1) \right)^{-1},$

where

$$y_1 \stackrel{\triangle}{=} G_2(t_2) \circ \cdots \circ G_k(t_k)(x_0), \cdots, y_k = x_0, \ \widehat{p} = (t_1, \cdots, t_k)$$

Write $H_j(t; y_j) = \left( \dfrac{\partial G_j}{\partial x}(t; y_j) \right)^{-1}$, $t \in [0, t_j], j = 1, \cdots, k$, and there holds

(9) $\qquad \dfrac{dH_j}{dt} = -H_j \dfrac{\partial Y_j}{\partial x}(G_j(t; y_j)), \ H_j(0) = I_n$

Let the $(N \times N)$ matrix $B_j$ be such that

(10) $\quad \mathrm{ad} Y_j \{Y_1, \cdots, Y_N\} = \{Y_1, \cdots, Y_N\} B_j, \ j = 1, \cdots, N,$

Then define the nonsingular $(N \times N)$ matrix $Z_j(t), t \in [0, t_j]$,

(11) $\quad \dfrac{dZ_j}{dt} = -Z_j B_j, Z_j(0) = I_N, i.e. Z_j(t) = (\exp -B_j t), j = 1, \cdots, N,$

and by a direct computation we obtain

(12) $\quad \begin{aligned} \{X_1(\widehat{p}; y(\widehat{p})), &\cdots, X_N(\widehat{p}; y(\widehat{p}))\} A(\widehat{p}) \\ &= \{Y_1, \cdots, Y_M\}(x_0) Z_k(t_k) \times \cdots \times Z_1(t_1) A(\widehat{p}), \end{aligned}$

for any $\widehat{p} \in D_k = \displaystyle\prod_1^k (-a_i, a_i),$

where $X_i(\widehat{p}; y(\widehat{p}))$ and $A(\widehat{p})$ are given in (7).

Now using (11) (see $Y_j(x_0) = 0, \ j = k+1, \cdots, N$) and (7) we obtain that

$$\mathrm{span}\{\widehat{Y}_1, \cdots, \widehat{Y}_k\} = \mathrm{span} \ \{Y_1(x_0), \cdots, Y_k(x_0)\}$$

and by definition span $\{Y_1(x_0), \cdot, Y_k(x_0)\}$ is a $k$-dimensional subspace in $R^n$. Therefore, $\widehat{Y}_1, \cdot, \widehat{Y}_k$ are linearly independent and from (6) we obtain the condition $(c_1)$ fulfilled.

The proof is complete.

**Remark 1.**

*The integral manifold $M_{x_0}$ defined in Theorem 1 depends on the flows $G_i(t)(x)$, generated by the vector fields in the basis $\{Y_1, \cdots, Y_N\} \subseteq \Lambda$ (see $\widehat{p} \in D_k = \displaystyle\prod_1^k (-a_i, a_i)$). If the vector fields $Y_i$ are complete, i.e., $Y_i$ generates a global flow $G_i(t)(x), t \in R, x \in R^n$, then the associated integral manifold $M_{x_0} \stackrel{\triangle}{=} \{y \in R; y = y(\widehat{p}), \widehat{p} \in R^k\}$ will fulfill the conclusions $(c_1), (c_2)$, for any $\widehat{p} \in R^k$.*

**Remark 2.**

*Even for a finite–dimensional Lie algebra $\Lambda \subseteq \mathbf{F}_n$, $\dim \Lambda(x)$ is not constant when $x \in R^n$ varies and for each fixed $x_0 \in R^n$ we obtain a specific integral manifold $M_{x_0} \subseteq R^n$ with $\dim M_{x_0} = \dim \Lambda(x_0)$.*

*Theorem 1 shows that using the solution $y(\widehat{p})$ we may recover the full Lie algebra $\Lambda$ along with given integral manifold $M_{x_0}$. In this respect Theorem 2.1 gives us more information and the original Lie algebra $\Lambda$ is recovered by the homomorphism $\lambda$.*

## 2.4   Some applications

The maximal rank Lie algebra and integral manifolds associated with a finite–dimensional Lie algebra have direct implications in solving stabilization problems of affine control systems and to obtain nontrivial solutions for a system of hyperbolic equations.

To begin with a system of hyperbolic equations

$$(1) \qquad \left\langle \frac{\partial S}{\partial x}(x),\ g_i(x) \right\rangle = 0,\ i = 1, \cdots, m,$$

where $g_i \in \mathbf{F}_n(\vec{g}_i \in \mathrm{Der}(R^n))$ are given such that the real Lie algebra $\Lambda \overset{\triangle}{=} L(g_1, \cdots, g_m)$ determined by $g_i$ is finite–dimensional.

To obtain a nontrivial solution $S(x), x \in V(x_0) \subseteq R^n$ for (1) we must know that the following condition

$$(*) \qquad \dim L(g_1, \cdots, g_m)(x_0) = k < n$$

is fulfilled. Then take a basis $\{Y_1, \cdot, Y_N\} \subseteq \Lambda$ as in Theorem 3.1 and let

$$y(\widehat{p}) = G_1(t_1) \circ \cdots \circ G_k(t_k)(x_0),\ \widehat{p} \in D_k \overset{\triangle}{=} \prod_1^k (-a_i, a_i)$$

be defined accordingly. Associate the corresponding $k$-dimensional integral manifold $Mx_0$.

The domain where the system (1) must be fulfilled is taken as $M_{x_0}$ and for a $C^\infty(V(x_0))$ scalar function $S$ satisfying (1) along to $x = y(\widehat{p})$ , $\widehat{p} \in D_k$, we obtain

$$(2) \qquad \left\langle \frac{\partial S}{\partial x}(x),\ g(x) \right\rangle = 0 \text{ for any } \mathrm{g} \in \mathrm{L}(\mathrm{g}_1, \cdot, \mathrm{g}_m),\ \mathrm{x} \in \mathrm{M}_{x_0}.$$

The systems (1) and (2) are equivalent on the domain $M_{x_0}$ and $x = y(\widehat{p})$, $\widehat{p} \in D_k$, is the solution in a gradient system (see Theorem 3.1)

$$(3) \qquad \frac{\partial y}{\partial t_1} = Y_1(y),\ \frac{\partial y}{\partial t_2} = X_2(t_1; y), \cdots, \frac{\partial y}{\partial t_k} = X_k(t_1, \cdot, t_{k-1}; y),$$

where $X_j(\widehat{p}; y)$, $j = 1, \cdots, k$, are smooth for $y$ in a neighbourhood $V(y(\widehat{p})) \subseteq R^n$.

For a fixed $\widehat{p}_0 \in D_k$ write $y_0 = y(\widehat{p}_0)$ and consider the solution $G(r; x)$, $r \in V_1(\widehat{p}_0) \subseteq D_k$, $x \in V_2(y_0)$, in (3) with Cauchy condition $G(\widehat{p}_0; x) = x$.

On the other hand $G(\widehat{p}_0; y_0) = y_0 = y(\widehat{p}_0)$ and both $G(r; y_0)$, $r \in V_1(\widehat{p}_0)$, and $y(r)$, $r \in V_1(\widehat{p}_0)$ are solutions of the same gradient system. Therefore

(4) $$y(\widehat{p}) = G(\widehat{p}; y_0), \ (\forall)\ \widehat{p} \in V_1(\widehat{p}_0) \subseteq D_k$$

and

(5) $$\frac{\partial y}{\partial t_i}(\widehat{p}) = \frac{\partial G}{\partial t_i}(\widehat{p}; y_0) \ \ i = 1, \cdots, k, \text{ for any } \widehat{p} \in V_1(\widehat{p}_0).$$

Using again Theorem 3.1 we obtain that

(6) $$\frac{\partial G}{\partial t_i}(\widehat{p}; y_0), i = 1, \cdots, k, \text{ are linearly independent for } \widehat{p} \in V_1(\widehat{p}_0)$$

By definition $\frac{\partial G}{\partial x}(\widehat{p}_0; y_0) = I_n$ (identity), and without losing generality we may admit that

(7) $$B = \left\{ \frac{\partial G}{\partial t_1}(\widehat{p}_0; y_0), \cdots, \frac{\partial G}{\partial t_k}(\widehat{p}; y_0), \frac{\partial G}{\partial x_{k+1}}(\widehat{p}_0; y_0), \cdots, \frac{\partial G}{\partial x_n}(\widehat{p}_0; y_0) \right\}$$

is a basis in $R^n$.

Here we note that the chosen components $\{x_{k+1}, \cdots, x_n\}$ do not depend on the varying $\widehat{p} \in D_k$, and it ensures the definition of a global solution $S(x)$ for $x = y(\widehat{p})$, when $\widehat{p}$ is restricted to a compact set $K \subseteq D_k$. More precisely, write $z \overset{\triangle}{=} (\widehat{p}, x_{k+1}, \cdot, x_n)$, $F(z) = G(\widehat{p}; y_{10}, y_{k_0}, x_{k+1}, \cdots, x_n)$ and find the solution $z(x)$, $x \in \mathcal{U}(y_0)$, of the following

(8) $$F(z) = x \text{ with } x_0 = y_0 \text{ and } z_0 = (\widehat{p}_0, y_{0k+1}, \cdot, y_{0n})$$

In particular, for $x = y(\widehat{p})$, $\widehat{p} \in V_1(\widehat{p}_0)$, we obtain

(9) $$z(y(\widehat{p})) = (\widehat{p}, y_{0k+1}, \cdot, y_{0n}), \text{ i.e. } x_j(y(\widehat{p})) = x_j(y_0), j = k+1, \cdot, n,$$

for any $\widehat{p} \in V_1(\widehat{p}_0) \subseteq D_k$, and

(10) $$B(x) = \frac{\partial z}{\partial x}(x) = \left\{ \frac{\partial t_1}{\partial x}(x), \cdots, \frac{\partial t_k}{\partial x}(x), \frac{\partial x_{k+1}}{\partial x}(x), \cdots, \frac{\partial x_n}{\partial x}(x) \right\}$$

is a basis in $R^n$ for any $x \in \mathcal{U}(y_0)$.

The properties (9) and (10) are fulfilled for each $y_0$ in a compact set $C = \{y(\widehat{p}, \widehat{p} \in K \subseteq D_k\}$. Now, starting from $\widehat{p} = 0$ with Cauchy condition $G(0; x) = x$ for $x \in V(x_0)$ we obtain smooth mappings $x_{k+1}(x), \cdots, x_n(x)$ in a neighbourhood of the compact set $C$ such that

(11) $$x_j(y(\widehat{p})) = x_{0j}, \text{ forany } \widehat{p} \in K, \ j = k+1, \cdots, n$$

The equations (9) can be rewriten

(12)    $\left\langle \dfrac{\partial x_j}{\partial x}(y(\widehat{p})), \dfrac{\partial y}{\partial t_i}(\widehat{p}) \right\rangle = 0, \ (\forall) \ \widehat{p} \in K, \ j = k+1, \cdots, n$

and using Theorem 3.1 we obtain

(13)    $\left\langle \dfrac{\partial x_j}{\partial x}(y(\widehat{p})), g(y(\widehat{p})) \right\rangle = 0 \ (\forall) \ g \in L(g_1, \cdot, g_m), \ \widehat{p} \in K,$

for each $j = k+1, \cdots, n$, where $K \subseteq D_k$, is an arbitrary fixed compact with int K $\neq \varphi$.

Finally, we obtain $(n-k)$ nontrivial and global solutions for a system of hyperbolic equations

(14)    $\begin{array}{l} \left\langle \frac{\partial S}{\partial x}(x), g_i(x) \right\rangle = 0, \\ i = 1, \cdots, m, \ (\forall) \ x \in C \overset{\triangle}{=} \{y(\widehat{p}) : \ \widehat{p} \in K \subseteq D_k\} \end{array}$

provided dim $L(g_1, \cdots, g_m)(x_0) = k < n$, where the compact $0 \in K \subseteq D_k \overset{\triangle}{=}$ $\prod\limits_{i=1}^{k}(-a_i, a_i)$ is fixed arbitrarily.

Moreover, the mentioned solutions are defined using an itterative algorithm with the initial functions found from

(15)    $F(z) = x$ for $x \in V(x_0)$, $z \overset{\triangle}{=} (\widehat{p}, x_{k+1}, \cdots, x_n) \in \mathbf{R}^n$

where $F(z) = G(\widehat{p}; \widehat{x}_0, x_{k+1}, x_n)$, $\widehat{x}_0 \in R^k$, and $G(\widehat{p}; x)$ is the local solution in the gradient system (3) with Cauchy condition

$$y(0) = x \in V(x_0)$$

The above given computations are summarized in the following

**Proposition 1.**

*Let the Lie algebra $\Lambda \overset{\triangle}{=} L(g_1, \cdots, g_m)$ be finite- dimensional, where $g_i \in \mathbf{F}_n$ and assume that dim $\Lambda(x_0) = k < n$ for a fixed $x_0 \in R^n$. Define $y(\widehat{p}) = G_1(t_1) \circ \cdots, \circ G_k(t_k)(x_0), \widehat{p} = (t_1, \cdots, t_k) \in D_k = \prod(-a_i, a_i)$ where $G_i(t)(x)$ is the local flow generated by a $Y_i \in \Lambda$ and $\{Y_1, \cdots, Y_N\} \subseteq \Lambda$ is a basis with $Y_j(x_0) = 0, j = k+1, \cdots, n$, and $Y_1(x_0), \cdot, Y_k(x_0)$ linearly independent in $R^n$. Then, for each compact set $K \subseteq D_k, 0 \in$ int K, there exists $(n-k)$ smooth nontrivial solutions $S_j(x)$ defined on a neighbourhood of $C \overset{\triangle}{=} \{x \in R^n : x = y(\widehat{p}), \widehat{p} \in K\}$ fulfilling (1) for $x \in C$ (see (14)).*

In addition, if $\dim \Lambda(x) = n-1$, $(\forall)$ $x \in V(x_0)$ then the solution $S(x)$, $x \in$ $\mathcal{U}$ fulfils (1) for any $x \in \mathcal{U} \cap V(x_0)$.

**Remark 1.**

*The mentioned $(n - k)$ solutions are found using an implicit function theorem and a local solution in the gradient system (3) (see (15)).*

The Proposition 1 pointed out the implication of integral manifolds in solving a system of hyperbolic equations.

Next we shall consider a stabilization (asymptotic stabilization) problem for affine control systems

$$(16) \qquad \frac{dx}{dt} = \sum_{1}^{m} u_i g_i(x), \quad x \in R^n$$

where $g_i : R^n \to R^n$ is a smooth mapping $(g_i \in \mathbf{F}_n)$.

The real parameter $u_i$ must be defined as a function of $x$ such that the solution in (16) fulfils $\lim_{t \uparrow \infty} x(t; x_0) = 0$ for any Cauchy $x(0) = x_0$ data in a fixed bounded domain $D \subseteq R^n$.

The meaning of stabilization in the lack of full rank condition of $\Lambda \overset{\triangle}{=} L(g_1, \cdots, g_m)$ replaces the origin in $R^n$ by the set

$$(17) \qquad V = \{x \in R^n : \langle x, Y_j(x) \rangle = 0, \; j = 1, \cdots, M\}$$

where $\{Y_1, \cdots, Y_M\} \subseteq \Lambda$ is a fixed system of generators for $\Lambda$. In other words, starting with $x_0 \in R^n, \|x_0\| \leq C_0$, the $\omega$-limit set $\Omega_+(x_0)$ of the bounded solution in (16) must obey to

$$(18) \qquad \Omega_+(x_0) \subseteq V \cap \{x : \|x\| \leq C_0\},$$

where

$$\Omega_+(x_0) \overset{\triangle}{=} \{x \in R^n : x = \lim_{n \to \infty} x(t_n; ; x_0), \text{ for some } \{t_n\} \uparrow \infty\}$$

The target set $V$ measures the degeneracy of the finite–dimensional Lie algebra $\Lambda$ and assuming that for each $x \neq 0$ a $g \in \Lambda$ will exist such that $\langle x, g(x) \rangle \neq 0$ then $V = \{0\}$ and asymptotic stability follows. The control system (16) was studied intensively by well developed technics and among them a stabilizing periodic control $u(t, x_0)$ in the variable $t \geq 0$ and $C^\infty$ in $x_0 \in R^n$ solves the problem under a full rank condition $\dim \Lambda(x) = n$ for any $x \neq 0$. It is the full rank Lie algebra associated with $\Lambda$ which allow one to analyse the above problem in the lack of full rank condition on $\Lambda$. In what follows we shall use a fixed basis $B = \{Y_1, \cdots, Y_M\} \subseteq \Lambda \overset{\triangle}{=} L(g_1, \cdot, g_m)$ such that each $Y \in B$ is a product of $g_{i_1}, \cdot, g_{i_k}, i_j \in \{1, \cdots, m\}$, i.e., $Y = \mathrm{ad}g_{i_1} \circ \mathrm{ad}g_{i_2} \circ \cdots \circ \mathrm{ad}g_{i_{k-1}}(g_{i_k})$ for some natural number $k$. Define $C^\infty$ functions in $x_0 \in R^n$

(19)     $$v_j(x_0) = -\langle x_0, Y_j(x_0) \rangle, \ j = 1, \cdots, M,$$

$$a(x_0) = \|v(x_0)\|^2 \overset{\triangle}{=} \sum_{j=1}^{M} (v_j(x_0))^2, \ \widetilde{v}_j(x_0) = a^{l_j}(x_0)v_j(x_0)$$

where $l_j$ is the corresponding natural number $k$ associated with $Y_j$, $j = 1, \cdots, M$

Let $g_{i_j}$ be the vector field in $\{g_1, \cdots, g_m\}$ appearing in the first position of $Y_j$ (see $g_{i_1}$ in $Y$)

Let $T > 0$ be fixed and consider $u_{ij}(t), u_i(t) : [0, \infty) \to R$ some periodic functions of class $C^1$ with common period $T, j = 1, \cdots, M$, $i = 1, \cdots, m$, fulfilling

(20)     $$u_i(t) = -u_i(T - t), \ t \in [0, T], \ i = 1, \cdots, m.$$

By a periodically defined solution $\widetilde{x}(t; x_0)$, $t \geq 0$, of

(21)     $$\frac{dx}{dt} = a(x_0) \left( \sum_{j=1}^{M} u_{ij}(t)\widetilde{v}_j(x_0)g_{ij}(x) + \sum_{i=1}^{m} u_i(t)g_i(x) \right), \ x(0) = x_0$$

we mean that solution which starting with $x(0) = x_0$ is constructed on each interval $[kT, (k+1)T]$ as the solution $\widetilde{x}(t; x_k)$ of

(22)     $$\frac{dx}{dt} = a(x_k) \left( \sum_{j=1}^{M} u_{i_j}(t)\widetilde{v}_j(x_k)g_{ij}(x) + \sum_{i=1}^{m} u_i(t)g_i(x) \right)$$

with $x(kT) = x_k$, and $x_{k+1} \overset{\triangle}{=} \widetilde{x}((k+1)T); x_k, \ k = 0, 1, 2, \ldots.$

Let $\Omega_+(x_0)$ be the $w$-limit set for a periodically defined solution $\widetilde{x}(t; x_0), t \geq 0$, in (21), i.e.,

$$\Omega_+(x_0) = \{x \in R^n : x = \lim_{k \to \infty} \widetilde{x}(t_k; x_0), \text{ for some } \{t_k\} \uparrow \infty\}$$

Using the local flow $G_j(t)(x)$ generated by $Y_j$, we define a local diffeomorphism on $R^n$ $G(p; x) \overset{\triangle}{=} G_1(t_1) \circ \cdots, \circ G_M(t_M)(x)$, for $p \in W(0) \subseteq R^M, x \in U(0) \subseteq R^n$ which is the solution in a gradient system associated with $\{Y_1, \cdots, Y_M\}$.

**Proposition 2.**

Assume that $\Lambda = L(g_1, \cdots, g_m)$ is finite-dimensional and let $\{Y_1, \cdot, Y_M\} \subseteq \Lambda$ be a fixed basis. Define $a(x_0)$ and $\tilde{v}_j(x_0)$ as in (9) and let $C_0 > 0$ be fixed.

Then there exist periodic controls $u_{i_j}(t)$, $u_i(t)$ fulfilling (20) such that the periodically defined solution $\tilde{x}(t; x_0)$, $t \geq 0$, in (21) with $\|x_0\| \leq C_0$ fulfils $\Omega_+(x_0) \subset V \cap \{x : \|x\| \leq C_0\}$. In addition, the target set $V$ is an invariant one for any solution of (11) (see $x_0 \in V \leftrightarrow a(x_0) = 0$).

The proof will be given in Appendix (see a.2) but the following may help to see the meaning of the associated full rank Lie algebra $L(q_1, \cdot, q_m)$,

**Remark 2.**

A bilinear control system $\dfrac{dx}{dt} = \displaystyle\sum_{i=1}^{m} u_i A_i x$, $A_i \in L(R^n; R^n)$, determine a finite-dimensional Lie algebra $\Lambda = L(g_1, \cdot, g_m)(g_i(x) = A_i x)$ and Proposition 2 tells us that it can be stabilized with the target set

$$V = \{x \in R^n : \langle x, B_j x \rangle = 0, \ j = 1, \cdots, M\}$$

where $\{B_1, \cdots, B_M\} \subseteq L(A_1, \cdots, A_M)$ is a basis

**Remark 3.**

The Lie algebra $\Lambda = L(g_1, \cdots, g_m)$ is in homomorphism correspondence with a new Lie algebra $L(q_1, \cdots, q_m)$ of analytic vector fields fulfilling a full rank condition $\dim L(q_1, \cdots, q_m)(p) = M$, $p \in R^M$ (see Theorem 2.1). It allows one to replace the original control system (16) by a controllable one

$$\frac{dp}{dt} = \sum_{i=1}^{m} u_i q_i(p), \ p \in R^M, \ p(0) = 0, \ t \geq 0$$

and the homomorphism $\lambda : \Lambda \to L(q_1, \cdots, q_m)$ is defined (independently of $x_k \in R^n$) by $G(p; x_k)$ in (14) and $\{x_k\}$ is given in (12).

For the controllable system we may and do consider approximations such as in Chow's Theorem which determine corresponding approximations in (16) (see the proof of Proposition 2).

We shall complete this part of applications with a few remarks regarding the implication of control systems in so called nonholonomic constraints appearing as a nonintegrable system of phaffian forms. It is supposed to help us on deciding which coordinates are essential for the moving constrained mechanical system. In the case of holonomic constraints it can be done using an implicit function theorem which means that a differentiable manifold structure is given.

For nonholonomic constraints the main obstruction is that the derivative $\frac{dq}{dt}$ of the vector $q = (q_1, \cdots, q_d)$ describing parameters is not given explicitly by a function $b(q)$ as in a differential system.

By hypotheses we can solve the given system with respect to some

$$\frac{dq_j}{dt}, \ j = m+1, \cdots, d,$$

(23)

$$\frac{dx}{dt} = f(x) + \sum_{i=1}^{m} u_i g_i(x),$$

where

$$u_i \overset{\triangle}{=} \frac{dq_i}{dt}, \ i = 1, \cdots, m, \ x \overset{\triangle}{=} (q_1, \cdots, q_d),$$

and for the sake of simplicity take $f = 0$ and $g_i \in C^\infty(R^d; R^d)$. Now, the control parameter $u \overset{\triangle}{=} (u_1, \cdots, u_m)$ is taken in the space $L_\infty([0, T]; R^m) \overset{\triangle}{=} \mathcal{U}$ and from (23) we obtain a set of trajectories $X_{(x_0)} \overset{\triangle}{=} \{x^u(\cdot) : x^u(0) = x_0, \ u(\cdot) \in \mathcal{U}\}$ as solutions for the control system

(24) $$\frac{dx}{dt} = \sum_{1}^{m} u_i(t) g_i(x), \ x(0) = x_0 \in R^d, \ t \in [0, T], \ u(\cdot) \in \mathcal{U}$$

for each $x_0 \in R^d$ fixed.

An explicit description of the set $X_{x_0} \subseteq C([0, T]; R^d)$ is obtained, provided the slice $X_{x_0}(T) \overset{\triangle}{=} \{x^u(T) : x^u(\cdot) \in X_{x_0}\}$ appears as the image of a fixed function, say $y(p) \in R^d$ for $p \in W(0) \subseteq R^M$, which does not depend on $T > 0$.

In this respect, assuming that the Lie algebra $\Lambda = L(g_1, \cdots, g_m)$ is finite-dimensional and $B \overset{\triangle}{=} \{Y_1, \cdots, Y_M\} \subseteq \Lambda$ is a fixed basis of complete vector fields, we associate a full rank Lie algebra $L(q_1, \cdot, q_m)$ (see Theorem 2.1) and a new control system

(25) $$\frac{dp}{dt} = \sum_{1}^{m} u_i(t) q_i(p), \ p \in R^M, \ p(0) = 0, \ t \in [0, T], \ u(\cdot) \in \mathcal{U}.$$

For $u(\cdot) \in \mathcal{U}$ fixed and $p^u(\cdot)$ the solution in (25) we obtain a solution $x^u(\cdot) \in X(x_0)$ corresponding to the same control $u(\cdot) \in \mathcal{U}$. Moreover, $x^u(t) = y(p^u(t))$, $t \in [0, T]$, where

(26) $$y(p) \overset{\triangle}{=} G_1(t_1) \circ \cdots \circ G_M(t_M)(x_0), \ p = (t_1, \cdot, t_M) \in R^M,$$

and $G_i(t)(x)$ $t \in R$, $x \in R^d$, the flow generated by $Y_i \in B$. Using Chow's theorem for (25) we obtain that any $p \in R^M$ can be joined with the origin $0 \in R^M$ at time $T > 0$ by a trajectory of (25) corresponding to a $u(\cdot) \in \mathcal{U}$. It shows that $X_{x_0}(T) = \{y(p) : p \in R^M\}$ where $y(p)$, $p \in R^M$ is defined in (26).

In the case that the manifold structure of $X_{x_0}(T)$ is involved we restrict to a basis $B = \{Y_1, \cdots, Y_M\} \subseteq \Lambda$ such that $Y_j(x_0) = 0$, $j = k+1, \cdot, M$, and $Y_1(x_0), \cdots, Y_k(x_0)$ are linearly independent in $R^d$, where $\dim \Lambda(x_0) = k$. Define

$$(27) \qquad y(\widehat{p}) \overset{\triangle}{=} G_1(t_1) \circ \cdot \circ G_k(t_k)(x_0), \quad \widehat{p} = (t_1, \cdots, t_k),$$

Using the same arguments we obtain $X_{x_0}(T) = \{y(\widehat{p}) : \widehat{p} \in R^k\}$ as a $C^\infty$ $k$-dimensional integral manifold of $\Lambda$.

The above given computation can be summarized in the following

## Proposition 3.

*Assume that nonholonomic constraints can be rewriten as a control system in (24) with $g_i \in C^\infty(R^d, R^d)$ and the Lie algebra $L(g_1, \cdot, g_m) \overset{\triangle}{=} \Lambda$ is finite-dimensional with $\dim \Lambda(x_0) = k$ for a fixed $x_0 \in R^d$. Then for each $T > 0$ we obtain that $X_{x_0}(T) \overset{\triangle}{=} \{x^u(T) : x^u(\cdot) \in X_{x_0}\}$ is a $C^\infty$ $k$-dimensional integral manifold given by $X_{x_0}(T) = \{y(\widehat{p}) : p \in R^k\}$ where $y(\widehat{p}), \widehat{p} \in R^k$ is in (27).*

## Exercices

1) Let $\{g_1, \cdot, g_m\} \subseteq \mathbf{F}_n$ be given such that $L(g_1, \cdot, g_m) \overset{\triangle}{=} \Lambda$ is a nilpotent Lie algebra, i. e. there exists a natural $N$ with the property any Lie product $P(g_1, \cdot, g_m)$ containing more than $N$ factors equals zero.

Then $\Lambda$ is a finite–dimensional Lie algebra and a nilpotent system of generators $\{Y_1, \cdots, Y_M\} \subseteq \{g_1, \cdot, g_m\}$ can be defined for $\Lambda$, i.e.,

$$\mathrm{ad}Y_j\{Y_1, \cdot, Y_M\} \subseteq \mathrm{span}\ \{Y_{j+1}, \cdots, Y_M\}\ j = 1, \cdots, M$$

**Hint.** Start with $Y_1 = g_1$ and from the exponential mapping $(\exp \mathrm{ad}g_1)(g_2)$ take all the terms which are not vanishing

$$Y_2 = g_2, \ Y_3 = (\mathrm{ad}\ g_1)(g_2), \cdots, \ Y_{m_1} = (\mathrm{ad}\ g_1)^{N_1}(g_2).$$

Repeat the above procedure for $(\exp \mathrm{ad}\ Y_1)(\exp \mathrm{ad}\ Y_2)Y_3$ and adding the nonzero terms to $Y_1, \cdot, Y_{m_1}$ we obtain $\{Y_1, \cdots, Y_{m_1+m_2}\}$.

Finally, we obtain a finite set $\{Y_1, \cdots Y_{N_1}\} \subseteq \Lambda$ including all nonzero products of $g_1, g_2$. Then use $g_3$ and enlarge the system $\{Y_1, \cdot, Y_{N_1}\}$ by the same algorithm. It will lead us to the result.

2) The same hypotheses as for the gradient system associated with the nilpotent system of generators defined in exercise 1. Then the matrices $(M \times M)B_j$ are nilpotent, i.e., $B_j^N = \odot$ and we obtain a triangular matrix $A(p)$ for which the elements are polynomial functions of $p = (t_1, \cdots, t_M)$.

### Bibliographical notes

The algebraic representation of a gradient system in a finite–dimensional Lie algebra is adapted from Vârsan [12].

The application for stabilizing control systems is inspired by Coron [2] and adapted from Vârsan [13].

# Chapter 3

# Finitely Generated over Orbits Lie Algebras and Algebraic Representation of the Gradient System

We have seen that a finite–dimensional Lie algebra $\Lambda \subseteq \mathbf{F}_n$(or Der $(\mathbf{R}^n)$) can be associated with a full rank Lie algebra of analytical vector fields. Generally, the new Lie algebra is no longer finite–dimensional but it can be characterized using a global system of generators provided the space $R$ of the coefficients is replaced by analytical functions $C^\omega(R^M; R)$. That is to say, the new Lie algebra $L(q_1, \cdot, q_m)$ can be determined by a system $\{q_1, \cdot, q_m, Q_{m+1}, \cdot, Q_M\} \subseteq L(q_1, \cdot, q_m)$ in involution over the ring of analytical functions. It is the task of the finite generation over orbits of a Lie algebra to preserve both the infinite dimensionality property and the analysis of the associated gradient systems.

## 3.1 Lie algebras finitely generated over orbits (f.g.o) and gradient systems

As one may expect, the gradient system associated with a finite set of vector fields in a Lie algebra $\Lambda \subseteq \mathbf{F}_n$ (or Der $(\mathbf{R}^n)$) is defined locally and depends essentially on the orbit-solution of it. It is to be noted that by an orbit of $\Lambda$ we mean only a finite composition of local flows.

**Definition 1.**

Let $\Lambda \subseteq \mathbf{F}_n$ be a Lie algebra and $x_0 \in R^n$ fixed. By an orbit of the origin $x_0$ of $\Lambda$ we mean a function

$$G(p; x_0) \triangleq$$
$$G_1(t_1) \circ \cdots \circ G_k(t_k)(x_0), p \triangleq$$
$$(t_1, \cdot, t_k) \in D_k \triangleq \prod_{i=1}^{k}(-a_i, a_i),$$

where $G_i(t)(x)$, $t \in (-a_i, a_i)$, $x \in V(x_0)$ is the local flow generat by some $Y_i \in \Lambda$.

**Definition 2.**

We say that $\Lambda \subseteq \mathbf{F}_n$ is finitely generated with respect to the orbits of the origin $x_0 \in R^n(f.g.o; x_0)$ if $\{Y_1, \cdot, Y_M\} \subseteq \Lambda$ will exist such that any $Y \in \Lambda$ along to an arbitrary orbit $G(p; x_0), p \in D_k$, can be written

$$Y(G(p; x_0)) = \sum_{j=1}^{M} a_j(p) Y_j(G(p; x_0))$$

with $a_j \in C^\infty(\Omega_k)$ depending on $Y$ and $G(p; x_0), p \in D_k$; $\{Y_1, \cdot, Y_M\}$ is called a system of generators.

**Remark.**

It is easily seen that $\{g_1, \cdot, g_m\} \subseteq \mathbf{F}_n$ in involution, i.e., $[g_i, g_j](x) = \sum_{k=1}^{m} a_k(x) g_k(x)$ with $a_k \in C^\infty(R^n)$, determine a Lie algebra $\Lambda(g_1, \cdot, g_m)$ which is finitely generated over orbits. In particular, any f.g.r. Lie algebra is a f.g.o. Lie algebra with a fixed, system of generators independently of the origin $x_0$.

In what follows the origin $x_0 \in R^n$ of orbits is fixed, and for a system of generators $\{Y_1, \cdots Y_M\}$ in a $(f.g.0; x_0)$ Lie algebra $\Lambda$ we consider the general gradient system associated to it. That is to say, let $G_i(t)(x)$, $t \in (-a_i, a_i)$, $x \in V(x_0)$ be the local flow generated by $Y_i$ and write

$$H_i(t; y) \triangleq \left(\frac{\partial G_i}{\partial y}(t; y)\right)^{-1}, \quad y_{i+1} \triangleq G_i(-t_i; y_i), y_1 \triangleq y, i = 1, \cdot, M-1$$

Then define the vector fields

$$
\begin{aligned}
X_1(y) &: Y(y), \\
X_2(t_1; y) &= H_1(-t_1; y_1) Y_2(G_1(-t_1; y_1)), \\
X_M(t_1, \cdot, t_{M-1}; y) &: H_1(-t_1; y_1) \\
&\times H_2(-t_2; y_2) \times \cdots \\
&\times H_{M-1}(-t_{M-1}; y_{M-1}) Y_M(y_M)
\end{aligned}
$$

(1)

where $y \in V(x_0)$ and $p \triangleq (t_1, \cdot, t_M) \in D_M = \prod_{i=1}^{M}(-a_i, a_i)$.

The vector fields in (1) determine a gradient system (see theorem 1.2.3.)

(2) $\qquad \dfrac{\partial y}{\partial t_1} = X_1(y), \ \dfrac{\partial y}{\partial t_2} = X_2(t_1; y), \cdots , \dfrac{\partial y}{\partial t_M} = X_M(t_1, \cdot, t_{M-1}; y)$

for $p \in D_M$ and $y \in V(x_0)$, and the orbit of the origin $x_0 \in R^n$,

(3) $\qquad y(p) = G_1(t_1) \circ G_2(t_2) \circ \cdots \circ G_M(t_M)(x_0) \ \ p \triangleq (t_1, \cdot, t_M) \in D_M,$

satisfies (2)

**Lemma 1.**

Let $\Lambda \subseteq \mathbf{F}_n$ be a (f.g.o; $x_0$) Lie algebra and $x_0 \in R^n$ fixed. Consider the gradient system in (2) associated to a fixed system $\{Y_1, \cdots, Y_M\} \subseteq \Lambda$ of generators. Then the orbit in (3) is the solution of the gradient system (2) and there exist $(M \times M)$ nonsingular matrices $Z_j(t_j; t_j, \cdot, t_M), j = 1, \cdot, M-1$, such that the vector fields $X_j(p_j; y)$, in (2) for $y = y(p)$ in (3) fulfil

(c) $\ X_{j+1}(p_{j+1}; y(p)) \ = \{Y_1(y(p)), \cdots, Y_M(y(p))\} Z_1(t_1; t_1, \cdot, t_M) \times \cdots$
$\qquad\qquad\qquad\qquad\quad \times Z_j(t_j; t_j, \cdot, t_M) e_{j+1}$

where $e_1, \cdot, e_M \in R^M$ is the canonical basis and $Z_j$ is $C^\infty$ in $p \in D_M$ and fulfil linear differential equations (see (28)).

**Proof.**

By hypothesis the conditions in Theorem 1.2.3 are fulfilled and the orbit $y(p)$ in (3) is the solution of the gradient system in (2). Using the (f.g.0; $x_0$) property of $\Lambda$ a $C^\infty$ matrix $B_1(p)$ in $D_M$ is determined such that

(4) $\qquad ad \ Y_1\{Y_1, \cdot, Y_M\}(y(p)) = \{Y_1, \cdot, Y_M\}(y(p)) \ B_1(t_1, \cdot, t_M)$

Let $Z_1(t; t_1, \cdot, t_M)$, $t \in [0, t_1]$, be the matrix solution of

(5) $\qquad \dfrac{dZ_1}{dt} = Z_1 B_1(t_1 - t, t_2, \cdot, t_M), \ Z_1(0) = I_M(identity)$

The standard successive approximations method applied in (5) allows one to write the solution $Z_1(t; t_1, \cdot, t_M)$, $t \in [0, t_1]$, as a convergent Volterra series, and by a direct computation we obtain

(6) $\qquad X_2(t_1; y(p)) = \{Y_1, \cdot, Y_M\}(y(p)) Z_1(t_1; t_1, \cdot, t_M) e_2$

where the vector field $X_2(t_1; y)$ is defined in (1) and the nonsingular $(M \times M)$ matrix $Z_1$ fulfils (5).
More precisely, denote

(7)                       $N_1(t; y) = H_1(-t; y)\{Y_1, \cdot, Y_M\}(G_1(-t; y))$

and $X_2(t_1; y)$ can be rewriten

(8)                               $X_2(t_1; y) = N_1(t_1; y)e_2$

Then we notice that $N_1(t; y(p))$, $t \in [0, t_1]$, satisfies

(9)        $\dfrac{dN_1}{dt} = N_1 B_1(t_1 - t, t_2, \cdot, t_M)$, $N_1(0) = \{Y_1, \cdot, Y_M\}(y(p))$

where the matrix $B_1(p)$ is defined in (4) and obeys

(10)  $ad\, Y_1\{Y_1, \cdot, Y_M\}(G_1(-t; y)) = \{Y_1, \cdot, Y_M\}(G_1(-t; y))B_1(t_1 - t, t_2, \cdot, t_M)$

for $y = y(p)$
   Using the same standard iterative procedure we obtain

$$N_1^0(t) \overset{\triangle}{=} \{Y_1, \cdot, Y_M\}(y(p)) = N_1(0)$$

(11)     $N_1^{k+1}(t) = N_1(0) + \displaystyle\int_0^t N_1^k(s)B_1(t_1 - s, t_2, \cdot, t_M)ds, \ k = 0, 1, 2 \cdots$

and it is easily seen that

(12)                       $N_1^{k+1}(t) = N_1(0)Z_1^{k+1}(t; t_1, \cdot, t_M)$

where $\{Z_1^k(t; t_1, \cdot, t_M\}_{k \geq 0}$ defines the solution in (5) and

(13)        $\displaystyle\lim_{k \to \infty} Z_1^k(t; t_1, \cdot, t_M) = Z_1(t; t_1, \cdot, t_M)$ uniformly in $t \in [0, t_1]$.

   Therefore, using (12) and (13) in (11) we obtain that

(14)                       $N_1(t) \overset{\triangle}{=} \displaystyle\lim_{k \to \infty} N_1^k(t), \ t \in [0, t_1]$

is the solution in (9) fulfilling

(15)  $N_1(t; y) = \{Y_1, \cdot, Y_M\}(y) \, Z_1(t; t_1, \cdot, t_M), \ t \in [0, t_1]$, provided $y = y(p)$,

and (6) holds.

The next vector field $X_3(t_1, t_2; y)$ can be represented similarly. It is rewriten as

(16)     $X_3(t_1, t_2; y) = H_1(-t_1; y)\widehat{X}_3(t_2; y_2)$, where $y_1 = y(p) = y$,

$$y_2 \overset{\triangle}{=} G_1(-t_1; y_1) = G_2(t_2) \circ \cdots \circ G_M(t_M)(x_0)$$

and

(17)             $\widehat{X}_3(t_2; y) = H_2(-t_2; y_2)Y_3(G_2(-t_2; y_2))$

Firstly, we have represented $\widehat{X}_3$ as

(18)             $\widehat{X}_3(t_2; y_2) = \{Y_1, \cdot, Y_M\}(y_2)Z_2(t_2; t_2, t_M)e_3$

where the nonsingular and $C^\infty(M \times M)$ matrix $Z_2(t; t_2, \cdot, t_M)$, $t \in [0, t_2]$, is the unique solution of

(19)         $\dfrac{dZ_2}{dt} = Z_2 B_2(t_2 - t, t_3, \cdot, t_M)$,     $Z_2(0) = I_M$

Here the $C^\infty(M \times M)$ matrix $B_2(t_2, \cdot, t_M), p \in D_M$, is determined such that

(20)     $ad\, Y_2\{Y_1, \cdot, Y_M\}(y_2) = \{Y_1, \cdot, Y_M\}(y_2)B_2(t_2, \cdot, t_M)$, $p \in D_M$,

and denoting

(21)             $N_2(t; y_2) \overset{\triangle}{=} H_2(-t; y_2)\{Y_1, \cdot, Y_M\}(G_2(-t; y_2))$

we obtain

(22)     $\begin{aligned} \widehat{X}_3(t_2; y_2) &= N_2(t_2; y_2)e_3 \\ &= \{Y_1, \cdot, Y_M\}(y_2)Z_2(t_2; t_2, \cdot, t_M)e_3 \end{aligned}$

That is to say, $N_2(t; y_2)$, $t \in [0, t_2]$, as a solution in

(23)         $\dfrac{dN_2}{dt} = N_2 B_2(t_2 - t, t_3, \cdot, t_M)$, $N_2(0) = \{Y_1, \cdot, Y_M\}(y_2)$

can be represented using the standard itterative procedure and we obtain

(24)         $N_2(t; y_2) = \{Y_1, \cdot, Y_M\}(y_2)Z_2(t; t_2, \cdot, t_M\}$, $t \in [0, t_2]$,

where $Z_2$ is the solution in (19).

Now we use (22) in (16), and taking into account that $y_2 = G_1(-t_1; y_1)$ we rewrite

(25)     $H_1(-t_1; y_1)\{Y_1, \cdot, Y_M\}(G_1(-t_1; y_1)) = N_1(t_1; y_1)$ for $y_1 \overset{\triangle}{=} y(p)$.

where $N_1(t; y(p))$, $t \in [0, t_1]$ fulfils (15)

Therefore, the vector field $X_3(t_1, t_2; y)$ in (16) obeys

(26)     $X_3(t_1, t_2; y) = \{Y_1, \cdot, Y_M\}(y)Z_1(t_1; t_1, \cdot, t_M) \times Z_2(t_2; t_2, \cdot, t_M)e_3,$

for $y = y(p)$

An induction argument will complete the proof and each $X_j(p_j; y)$ in conclusion (c) obtains the form

(27)   $X_{j+1}(p_{j+1}; y) = \{Y_1, \cdot, Y_M\}Z_1(t_1; t_1, \cdot, t_M) \times \cdots \times Z_j(t_j; t_j, \cdot, t_M)e_{j+1},$

provided $y = y(p)$. Here $Z_j(t; t_j, \cdot, t_M)$, $t \in [0, tj]$, is the solution of the linear equation

(28)                 $\dfrac{dZ_j}{dt} = Z_j B_j(t_j - t, t_{j+1}, \cdot, t_M), \; Z_j(0) = I_M,$

where the $(M \times M)$ and $C^\infty$ matrix $B_j(t_j, \cdot, t_M)$ for $p \in D_M$ is determined such that

(29)        $\operatorname{ad} Y_j\{Y_1, \cdot, Y_M\}(y_j) = \{Y_1, \cdot, Y_M\}(y_j)(B_j(t_j, \cdots, t_M)$

$y_j \overset{\triangle}{=} G_j(t_j) \circ \cdots \circ G_M(t_M)(x_0), \; j = 1, \cdot, M.$
The proof is complete.

**Remark 1.**

It the case that each $Y_j$ generates a global flow $G_j(t)(x), t \in R, x \in R^n$, $j = 1, \cdot, M$, then conclusion (c) in Lemma 1 holds for any $p \in R^M$.

**Remark 2.**

Under the hypotheses in Lemma 1 write

$Z_j(p) \overset{\triangle}{=} Z_1(t_1; t_1, \cdot, t_M) \times \cdot \times Z_j(t_j; t_j, \cdot, t_M), p \overset{\triangle}{=} (t_1, \cdot, t_M) \in D_M, j = 1, \cdots, M-1$

,
where $Z_j(t_j; t_j, \cdot, t_M)$ are defined in Lemma 1, fulfilling the conclusion (c). Define an $(M \times M)C^\infty$ matrix by:

(a) $A(p) = (e_1, Z_1(p)e_2, \cdots, Z_{M-1}(p)e_M), \; p \in D_M$.

*Then the vector fields $X_j(p_j; \bullet)$ fulfil*

$(b)\{Y_1, X_2(t_1), \cdots, X_M(t_1, \cdot, t_{M-1})\}(y) =$
$\{Y_1, \cdots, Y_M\}(y)A(p)$ *provided* $y = y(p)$, *for each* $p \in D_M$, *where the*
$C^\infty$ *matrix* $A(p), p \in D_M$, *is given in (a) and obeys to* $A(0) = I_M$.

### Remark 3.

*As is stated in Lemma 1, the algebraic representation of the gradient system is not anymore a global one and in addition it depends on the local solution* $y(p), p \in D_M$. *There are given two types of local properties, one is conditioned by the existence of local flows* $G_i(t)(x)$ *(see* $p \in D_M$*) and the other expresses the nonsingularity of the matrix* $A(p)$ *(see Remark 2) for* $p \in V(0) \subseteq D_M$ *(see* $A(0) = I_M$*).*

*It is the goal of the following considerations to make sure that the only real restriction for a global nondegenerate representation is determined by the local existence of the flows* $G_i(t)(x)$.

## 3.2   Nonsingularity of the gradient system

Let $\Lambda \subseteq \mathbf{F}_n$ be a (f.g.o; $x_0$) Lie algebra and fix $x_0 \in R^n$ as the origin of orbits.

Using the same notations as in Lemma 1.1 we obtain a gradient system

$$\frac{\partial y}{\partial t_j} = X_j(p_j; y), \; j = 1, \cdots, M, \; p_j \overset{\triangle}{=} (t_1, \cdot, t_{j-1}),$$

and its local solution

$$y(p) = G_1(t_1) \circ \cdots \circ G_M(t_M)(x_0), \qquad p \in D_M = \prod_{i=1}^M (-a_i, a_i)$$

such that

$$(1) \qquad \begin{aligned} \frac{\partial y}{\partial p}(p) &= \{Y_1, X_2(t_1), \cdots, X_M(t_1, \cdot, t_{M-1})\} \\ &= \{Y_1, \cdots, Y_M\}(y(p))A(p), \; p \in D_M. \end{aligned}$$

where $A(0) = I_M$ (see Remark 1.2) and

$$(2) \qquad \det A(p) \neq 0 \text{ for any } p \in V(0) \subseteq D_M.$$

The following step is to show that for a fixed $p^* \in V(0)$ we obtain a new system of generators for $\Lambda$ provided the gradient system is used.

**Lemma 1.**

*Let the hypotheses in Lemma 1.1 be fulfilled and consider the vector fields*
$X_j(p_j; y), j = 1, \cdots, M,$ *of the gradient system fulfilling (1) and (2).*
*Then there exist an* $(M \times M)C^\infty$ *matrix*

$$A(p^*; p), p^* \in V(0) \subseteq D_M, \ p \in D_M$$

*such that* $A(0; p) = I_M,$

$$c_1) \ \{Y_1, X_2(t_1^*), \cdots, X_M(t_1^*, \cdot, t_{M-1}^*)\} = \{Y_1, \cdots, Y_M\}(y(p))A(p^*; p)$$

*and for an arbitrarily fixed compact* $K \subseteq D_M$ *there exists* $\mathcal{U}(0) \subseteq V(0)$ *such that*

$$c_2) \ \det A(p^*; p) \neq 0 \ (\forall) \ p^* \in \mathcal{U}(0), \ p \in K.$$

**Proof.**

Any $p \in D_M$ is rewriten $p = p^* + s$, where $p^* \in V(0)$ and $s \in R^M$.
Then the solution $y(p), p \in D_M$, will be

$$(3) \qquad y(p) = y(s_1, \cdot, s_M; x^*) \triangleq G_1(s_1) \circ G_2^*(s_2) \circ \cdots \circ G_M^*(s_M)(x^*)$$

where $G_i^*(t)(x)$ is the local flow generated by the locally defined vector field

$$(4) \qquad Y_i^*(y) \triangleq X_i(t_1^*, \cdot, t_{i-1}^*; y), Y_1^*(y) = Y_1(y), \ i = 2, \cdots, M,$$

Here $X_j(p_j; y)$ determine the gradient system and

$$x^* \triangleq G_1(t_1^* \circ G_2(t_2^*) \circ \cdots \circ G_M(t_M^*)(x_0), \qquad p^* = (t_1^*, \cdots, t_M^*) \in V(0).$$

The function $y(s; x^*) = y(p)$, defined in (3), is the local solution of a new
gradient system

$$(5) \qquad \frac{\partial y}{\partial s_i} = X_i^*(s_1, \cdot, s_{i-1}; y), \ i = 1, \cdots, M, \ s \in W(0) \subseteq D_M,$$

where $X_1^* = X_1 = Y_1$ and
(6)

$$X_2^*(s_1; y) \triangleq H_1(-s_1; y)Y_2^*(G_1(-s_1; y))$$

$$\vdots$$
$$X_M^*(s_1^*, \cdots, s_{M-1}; y) = H_1(-s_1; y) \times H_2^*(-s_2; y_2) \times \cdots$$
$$\times H_{M-1}^*(-s_{M-1}; y_{M-1})Y_M^*(G_{M-1}^*(-s_{M-1}; y_{M-1}))$$

Here

$$y_1 \triangleq y, \ y_{i+1} \triangleq G_i^*(-s_i; y_i), \ H_i^*(\tau; y) \triangleq \left(\frac{\partial G_i^*(\tau; y)}{\partial y}\right)^{-1}, \ i = 1, \cdots, M-1.$$

and

$$y_i = G_i^*(s_i) \circ \cdots \circ G_M^*(s_M)(x^*), \quad i = 2, \cdots M, \text{ hold.}$$

The algebraic representation of the locally defined vector fields in (4) depends on the system

$$\{Y_1, Y_2, \cdots, Y_M\}(y(p))$$

In this respect, for $y = y(s; x^*) = y(p)$ (see (3)) we rewrite

(7)     $$Y_1^* = Y_1, \quad Y_2^*(y) = X_2(t_1^*; y_1) Y_2(G_1(-t_1^*; y)), \cdots,$$

$$\begin{aligned} Y_M^*(y) &= X_M(t_1^*, \cdots, t_{M-1}^*; y) \\ &= H_1(-t_1^*; y_1) \times \cdots \times H_{M-1}(-t_{M-1}^*; y_{M-1}) Y_M(G_{M-1}(-t_{M-1}^*; y_{M-1})) \end{aligned}$$

where

$$y_1 \stackrel{\triangle}{=} y; \, y_{j+1} = G_j(-t_j^*; y_j), \quad j = 1, \cdots, M-1,$$

and

$$y_j = G_j(t_j^*) \circ \cdots \circ G_M(t_M^*)(x_0), \quad j = 2, \cdots, M.$$

It is to be noted that $p^* \neq p$ and the algebraic representation for (7) will depend on $p^*$ and $p$ as well. Nevertheless, the same algorithm for Lemma 1.1 can be applied and we recall it for $Y_2^*(y)$.

Write

(8)     $$N_1(t; p) \stackrel{\triangle}{=} H_1(-t; y(p))\{Y_1, \cdots, Y_M\}(G_1(-t; y(p))), \quad t \in [0, t_1^*]$$

and choosing a matrix $(M \times M) B_1(p), p \in D_M$ such that

(9)     $$ad\, Y_1\{Y_1, \cdots, Y_M\}(y(p)) = \{Y_1, \cdots, Y_M\}(y(p)) B_1(p),$$

we obtain that the matrix $N_1(t, p)$ in (8) fulfils

(10)    $$N_1(t_1^*; p) = \{Y_1, \cdot, Y_M\}(y(p)) + \int_0^{t_1^*} N_1(t; p) B_1(t_1 - t, t_2, \cdot, t_M) dt$$

Now, let $Z_1(t; p), t \in [0, t_1^*]$, be the matrix solution of

(11)    $$\frac{dZ_1}{dt} = Z_1 B_1(t_1 - t, t_2, \cdot, t_M), \quad Z_1(0) = I_M \, (B_1 e_1 = 0, Z_1 e_1 = e_1)$$

and $N_1$ fulfilling (10) is represented as

(12)    $$N_1(t_1^*; p) = \{Y_1, \cdot, Y_M\}(y(p)) Z_1(t_1^*; p)$$

On the other hand, using (12) we obtain

$$Y_2^*(y) \overset{\triangle}{=} X_2(t_1^*; y)$$

(13)
$$\overset{\triangle}{=} N_1(t_1^*; p)e_2$$

$$= \{Y_1, \cdot, Y_M\}(y(p))Z_1(t_1^*; p)e_2$$

where $Z_1(t; p), t \in [0, t_1^*]$, is the solution in (11). It is easily seen that following the scheme in Lemma 1.1. we obtain any $Y_j^*(y)$, $j = 2, \cdot, M$, in (7) represented as

(14)  $$Y_3^*(y) \overset{\triangle}{=} X_3(t_1^*, t_2^*; y) \overset{\triangle}{=} H_1(-t_1^*; y_1) \times H_2(-t_2^*; y_2)Y_3(G_2(-t_2^*; y_2))$$
$$= \{Y_1, \cdot, Y_M\}(y(p)Z_1(t_1^*; p) \times Z_2(t_2^*; t_1^*, p)e_3 \quad \text{for } y = y(p)$$

and
(15)

$$Y_{j+1}^*(y) \overset{\triangle}{=} X_{j+1}(t_1^*, \cdots, t_j^*; y)$$
$$= \{Y_1, \cdots, Y_M\}(y(p)Z_1(t_1^*; p) \times \cdots \times Z_j(t_j^*; t_1^*, \cdots, t_{j-1}^*, p)e_{j+1}$$

for $y = y(p)$, $j = 1, \cdots, M-1$, where $e_1, \cdots, e_M \in R^M$ is the canonical base. Here the matrix $Z_j(t; t_1^*, \cdot, t_{j-1}^*, p)$, $t \in [0, t_j^*]$, is the solution of the equation

(16)  $$\frac{dZ_j}{dt} = Z_j B_j(t, t_1^*, \cdots, t_{j-1}^*, p), \ Z_j(0) = I_M \ (B_j e_j = 0, Z_j e_j = e_j)$$

where $B_j(p), p \in D_M$, of class $C^\infty$ is chosen such that

$$ad \ \ Y_j\{Y_1, \cdots, Y_M\}(G_j(-t) \circ G_{j-1}(-t_{j-1}^*) \circ \cdots \circ G_1(-t_1^*)(y))$$

(17)
$$= \{Y_1, \cdots, Y_M\}(G_j(-t) \circ G_{j-1}(-t_{j-1}^*)\circ$$

$$\cdots \circ G_1(-t_1^*)(y)B_j(t, t_1^*, \cdot, t_{j-1}^*, p),$$

provided $y = y(p)$, $t \in [0, t_1^*]$.
    For the sake of simplicity write

(18)
$$Z_j(p_j^*; p) \overset{\triangle}{=} Z_1(t_1^*; p) \times \cdots \times Z_j(t_j^*; t_1^*, \cdot, t_{j-1}^*, p),$$

$$p_j^* \overset{\triangle}{=} (t_1^*, \cdots, t_j^*), j = 1, \cdot, M - 1$$

and define the matrix $(M \times M)A(p^*; p)$

(19)      $$A(p^*; p) = \{e_1, Z_1(p_1^*; p)e_2, \cdots, Z_{M-1}(p_{M-1}^*; p)e_M\}$$

It holds $A(0; p) = I_M$ (see $Z_j(0) = I_M$) and the smooth matrix $A(p^*; p), p^* \in V(0), p \in D_M$ fulfill the hypotheses in Lemma a.1 uniformly with respect to $p \in K \subseteq D_M$, and arbitrarily fixed compact set $K$.

Aplying Lemma a.1 we obtain $\mathcal{U}(0) \subseteq V(0) \subseteq D_M$ depending on $K \subseteq D_M$ fixed such that $A(p^*; p)$, obeys to

(20)     $\det A(p^*; p) \neq 0 \ (\forall) \ p^* \in \mathcal{U}(0) \subseteq D_M$ and $p \in K \subseteq D_M$.

Now, the locally defined vector fields in (15) fulfil

(21)     $\{Y_1, Y_2^*, \cdot, Y_M^*\}(y(p)) = \{Y_1, \cdot, Y_M\}(y(p)) \ A(p^*; p)$

where the smooth matrix $A(p^*; p)$ satisfies (20).
The proof is complete.

### Remark 1.
*Lemma 1 shows that the locally defined vector fields $\{Y_1, Y_2^*, \cdot, Y_M^*\}(y)$ can be used as a new system of generators for the new gradient system in (6) provided we restrict ourselves to the orbit*

$$\{y(p); p \in K \subseteq D_M\} \text{ and for } p^* \in \mathcal{U}(0).$$

*The gradient system defined in (1) with Cauchy data $y(p^*) = x^*$ admits $y(p), p \in D_M$, as the unique solution and therefore $\{Y_1, Y_2^*, \cdot, Y_M^*\}(y)$ could be used as a system of generators for (1) also, provided $y = y(p), p \in K$.*

### Lemma 2.
*Asumme that the hypotheses in Lemma 1 are fulfilled and consider the nonsingular matrix $A(p^*; p), p^* \in \mathcal{U}(0) \subseteq D_M, p \in K \subseteq D_M$ satisfying $(c_1)$ and $(c_2)$. Then there exist a nonsingular matrix $A^*(s; p)$ such that gradient system in (1) and (6) obey to the following*

(c)
$$\begin{aligned}
&\{Y_1, X_2(t_1), \ \cdot, X_M(t_1, \cdot, t_{M-1})\}(y(p)) \\
&= \{Y_1, X_2^*(s_1), \cdots, X_M^*(s_1, \cdots, s_{M-1})\}(y(p)) \\
&= \{Y_1, Y_2^*, \cdots, Y_M^*\}(y(p))A^*(s; p) \quad \text{for} \quad p \\
&= p^* + s \in K, \ s \in U(0),
\end{aligned}$$

*where $Y_{j+1}^* = X_{j+1}(t_1^*, \cdot, t_j^*)$, $j = 1, \cdots, M - 1$, and $p^* = (t_1^*, \cdot, t_M^*) \in \mathcal{U}(0)$ is fixed.*

### Proof.
By definition

$$\begin{aligned}
\tfrac{\partial y}{\partial t_j}(p) = X_j(p_j; y(p)) &= \tfrac{\partial y}{\partial s_j}(s; x^*) \\
&= X_j^*(s_1, \cdot, s_{j-1}; y(s; x^*)), \qquad j = 1, \cdot, M,
\end{aligned}$$

provided $p = p^* + s$ and $y(s; x^*) \triangleq y(p)$ (see (1) and (6)).

To obtain the conclusion we may and do consider only the gradient system in (6). Namely,

$$
\begin{aligned}
X_2^*(s_1; y) &= X_2^*(0; y) + \int_0^{s_1} \frac{dX_2^*}{dt}(t; y)dt \\
&= Y_2^*(y) + \int_0^{s_1} H_1(-t; y)[Y_1, Y_2^*](G(-t; y))dt \\
&= Y_2^*(y) + \int_0^{s_1} H_1(-t; y)\frac{\partial X_2}{\partial t_1}(t_1^*; G(-t; y))dt
\end{aligned}
$$

$(22)$

where the gradient structure $\dfrac{\partial X_2}{\partial t_1}(t_1^*; x) = [Y_1, X_2(t_1^*)](x)$ of the system (1) is used.

Write $N_1^*(t; p) \triangleq H_1(-t; y)\{Y_1, Y_2^*, \cdot, Y_M^*\}(G(-t; y))$, $t \in [0, s], y = y(p)$, and (22) means

$(23)$ $$X_2^*(s_1; y) = N_1^*(s_1; p)e_2$$

for $y = y(p)$, where

$$
\begin{aligned}
N_1^*(s_1; p) &= \{Y_1, Y_2^*, \cdot, Y_M^*\}(y) + \\
&\quad \int_0^{s_1} H_1(-t; y)ad\, Y_1\{Y_1, Y_2^*, \cdot, Y_M^*\}(G(-t; y))dt
\end{aligned}
$$

$(24)$

for $y = y(p)$.

Now, using the gradient system (1) we rewrite

$$
ad\, Y_1(Y_j^*)(x) = \frac{\partial X_j}{\partial t_1^*}(t_1^*, \cdot, t_{j-1}^*; x), j = 1, \cdots, M,
$$

and (24) will change accordingly

$$
\begin{aligned}
N_1^*(s_1; p) &= \{Y_1, Y_2^*, \cdot, Y_M^*\}(y) \\
&\quad + \int_0^{s_1} H_1(-t; y)\frac{\partial}{\partial t_1^*}\{Y_1, Y_2^*, \cdots, Y_M^*\}(G(-t; y))dt
\end{aligned}
$$

$(25)$

for $y = y(p)$

The explicit dependence on $p^* = (t_1^*, \cdot, t_M^*)$ of the vector fields $\{Y_1, Y_2^*, \cdot, Y_M^*\}$ is given in Lemma 1, i.e.,

(26)        $\{Y_1, Y_2^*, \cdot, Y_M^*\}(y) = \{Y_1, \cdots, Y_M\}(y)A(p^*; p)$        for $y = y(p)$,

where the $C^\infty$ matrix $A(p^*; p), p^* \in \mathcal{U}(0) \subseteq D_M, p \in K \subseteq D_M$, fulfils $\det A(p^*; p) \neq 0$.

Using (26) we obtain the expression

$$\frac{\partial}{\partial t_1}\{Y_1, Y_2^*, \cdots, Y_M^*\}(y)$$

(27)
$$= \{Y_1, \cdots, Y_M\}(y)\frac{\partial}{\partial t_1^*}A(p^*; p)$$
$$= \{Y_1, Y_2^*, \cdots, Y_M^*(y)A^{-1}(p^*; p)\frac{\partial}{\partial t_1^*}A(p^*; p)$$
$$= \{Y_1, Y_2^*, \cdot, Y_M^*\}B_1^*(p^*; t_1, \cdots, t_M),$$

provided $y = y(p)$ where

(28)        $$B_1^*(p^*; t_1, \cdots, t_M) \triangleq A^{-1}(p^*; p)\frac{\partial}{\partial t_1^*}A(p^*; p)$$

Substitute (27) into (25) and it is easily seen that

$$N_1^*(s_1; p) = \{Y_1, Y_2^*, \cdots, Y_M^*\}(y)$$

(29)
$$+ \int_0^{s_1} N_1^*(t; p)B_1^*(p^*; t_1 - t, t_2, \cdot, t_M)dt, \qquad y = y(p)$$

Now, using the standard iterative procedure we obtain

(30)        $$N_1^*(s_1; p) = \{Y_1, Y_2^*, \cdots, Y_M^*\}(y)Z_1^*(s_1; p)$$

if $y = y(p)$, $p = p^* + s \in K$ where $p^* \in \mathcal{U}(0) \subseteq D_M$ is fixed arbitrarily and $Z_1^*(t; p), t \in [0, s_1]$, is the matrix solution of

(31)        $$\frac{dZ_1}{dt} = Z_1 B_1^*(p^*; t_1 - t, t_2, \cdots, t_M), \ Z_1^*(0) = I_M$$

Here the $(M \times M)$ matrix $B_1^*$ is that defined in (28). It is $N_1^*(s_1; p)$ in (30) which allow one to write

(32)        $$X_2^*(s_1, y) = \{Y_1, Y_2^*, \cdot, Y_M^*\}(y)Z_1^*(s_1; p)e_2 \text{ if } y = y(p),$$

$p = p^* + s \in K \subseteq D_M$, where $X_2^*$ and $Z_1^*$ are defined in (23) and (31), respectively. Now, it is a routine matter to obtain each $X_{j+1}^* (s_1, \cdot, s_j; y)$

algebraically represented provided we notice that Lemma 1 allows one to represent the operation $ad\ Y_i^*\{Y_1, Y_2^*, \cdot, Y_M^*\}(x)$ explicitly as

$$(33)\qquad ad\ Y_i^*\{Y_1, Y_2^*, \cdot, Y_M^*\}(x) = \{Y_1, Y_2, \cdot, Y_M\}(x)C_i(p^*;p)\ \text{for}\ x = y(p)$$

and $p \in K \subseteq D_M$, where the matrix $C_i(p^*;p)$ is obtained from $A(p^*;p)$ (given in Lemma 1) by taking partial derivatives. More precisely, the gradient structure of the system (1) allows one to write

$$(34)\qquad \begin{aligned} [Y_i^*, Y_j^*](x) &\overset{\triangle}{=} [X_i(t_1^*, \cdot, t_{i-1}^*), X_j(t_1^*, \cdot, t_{j-1}^*)](x) \\ &= \frac{\partial X_j}{\partial t_i^*}(t_1^*, \cdot, t_{j-1}^*; x), i \le j, \end{aligned}$$

and

$$[Y_i^*; Y_j^*](x) = -\frac{\partial X_i}{\partial t_j^*}(t_1^*, \cdot, t_{j-1}^*; x)\ \text{if}\ i > j.$$

and using Lemma 1 we may and do perform the derivatives in the right hand side involving the given matrix $A(p^*;p)$. The matrix $C_i(p^*;p)$ in (33) receives an explicit form and using the new system of generators we rewrite (33) as

$$(35)\qquad ad\ Y_i^*\{Y_1, Y_2^*, \cdots, Y_M^*\}(y) = \{Y_1, Y_2^*, \cdot, Y_M^*\}(y)A^{-1}(p^*;p)C_i(p^*;p)$$

if $y = y(p), p^* \in \mathcal{U}$ is fixed and $p = p^* + s \in K$.

The matrices in (35) determine the nonsingular matrix solutions $Z_i^*(t;p), t \in [0, s_i]$ of

$$(36)\qquad \frac{dZ_i^*}{dt} = Z_i^* B_i^*(p^*; t_1, \cdots, t_{i-1}, t_i - t, t_{i+1}, \cdots, t_M), Z_i^*(0) = I_M$$

where

$$(37)\qquad B_i(p^*; t_1, \cdots, t_M) = A^{-1}(p^*;p)C_i(p^*;p),\ i = 1, \cdots, M - 1$$

Write

$$(38)\qquad \begin{aligned} Z_j^*(s_1, \cdot, s_j; p) &= Z_1^*(s_1; p) \times \cdots \times Z_j^*(s_j; p), \\ &j = 1, \cdots, M - 1 \end{aligned}$$

and define an $(M \times M)$ matrix

$$(39)\qquad A^*(s;p) \overset{\triangle}{=} \{e_1, Z_1^*(s_1; p)e_2, \cdots, Z_{M-1}^*(s_1, \cdots, s_{M-1}; p)e_M\}$$

where $Z_j^*(s_1, \cdot, s_j; p)$ are given in (38).

Then each vector field in (6) can be written

(40)          $X^*_{j+1}(s_1, \cdot, s_j; y) = \{Y_1, Y^*_2, \cdots, Y^*_M\}(y)Z^*_j(s_1, \cdot, s_j; p)e_{j+1}$

$j = 1, \cdots, M-1$, where $e_1, \cdots, e_M \in R^M$ is the canonical base and $y = y(p), p = p^* + s \in K \subseteq D_M$.

Finally, from (40) and (39) we obtain the conclusion (c), i.e.,

$$\{Y_1, X^*_2(s_1), \cdots, X^*_M(s_1, \cdots, s_{M-1})\}(y(p)) = \{Y_1, Y^*_2, \cdot, Y^*_M\}(y(p))A^*(s; p),$$

where the nonsingular matrix $A^*(s; p)$, $s \in \mathcal{U}(0) \subseteq D_M$, $p \in p^* + s$, is defined in (39) and fulfils the conditions in Lemma a.1.

The proof is complete.

We shall conclude this part specifying the nonsingularity of the gradient system in (1) with respect to a given system of generators.

**Theorem 1.**

Let $\Lambda \subseteq \mathbf{F}_n$, $x_0 \in R^n$ be fixed such that $\Lambda$ is (f.g.o; $x_0$). For a fixed system $\{Y_1, \cdots, Y_M\} \subseteq \Lambda$ of generators associate the orbit $y(p) = G_1(t_1) \circ \cdot \circ G_M(t_M)(x_0), p \in D_M \overset{\Delta}{=} \prod_1^M (-a_i, a_i)$, and the corresponding gradient system

(i)          $\dfrac{\partial y}{\partial t_1} = Y_1(y), \dfrac{\partial y}{\partial t_2} = X_2(t_1; y), \cdots, \dfrac{\partial y}{\partial t_M} = X_M(t_1, \cdot, t_{M-1}; y)$

$$p \overset{\Delta}{=} (t_1, \cdot, t_M) \in D_M, \quad y \in V(x_0) \subseteq R^n.$$

Then for each $p_0 \in D_M$ there exist $\mathcal{U}(0) \subseteq R^M$ and a nonsingular matrix $A(s; p_0), s \in \mathcal{U}(0)$, such that

(c)          $\{Y_1, X_2(t_1), \cdot, X_M(t_1, \cdot, t_{M-1})\}(y) = \{Y_1, \cdots, Y_M\}(y)A(s; p_0).$

provided $y = y(p), p = p_0 + s, \ s \in \mathcal{U}(0)$.

**Proof.**

Let $P \in D_M$ be arbitrarily fixed and the segment $[0, P] \subseteq D_M$ is covered with a finite set of open spheres $S_i(p_i) \overset{\Delta}{=} p_i + \mathcal{U}(0) \subseteq D_M$, $i = 1, \cdots, N$, where $p_{i+1} \in p_i + \mathcal{U}(0)$ and $p_N = P$. By hypothesis the conditions in Lemma 2 are fulfilled and a chain of nonsingular matrices $A^{(i)}(s; p), s \in \mathcal{U}(0), i = 1, \cdots, N$, are defined such that

(41)
$$\begin{aligned}
\frac{\partial y}{\partial p}(p) &\overset{\Delta}{=} \{Y_1, X_2(t_1), \cdots, X_M(t_1, \cdots, t_{M-1})\}(y) \\
&= \{Y_1, X^{(i)}_2(s_1), \cdots, X^{(i)}_1(s_1, \cdot, s_{M-1})\}(y) \\
&= \{Y_1, \cdot, Y_M\}(y)A^{(i)}(s; p)
\end{aligned}$$

if $y = y(p)$, $p = p_i + s$, $s \in \mathcal{U}(0)$.

We recall that $y(s_1, \cdot, s_M; x_i) \stackrel{\triangle}{=} y(p)$, $p = p_i + s$, $s \in \mathcal{U}(0), x_i \stackrel{\triangle}{=} y(p_i)$, is the solution of the gradient system

$$(42) \quad \frac{\partial y}{\partial s_1} = Y_1(y), \quad \frac{\partial y}{\partial s_2} = X_2^{(i)}(s_1; y), \cdots, \quad \frac{\partial y}{\partial s_M} = X_M^{(i)}(s_1, \cdots, s_{M-1}; y)$$

with Cauchy data $y_0 = x_i \stackrel{\triangle}{=} y(p_i)$.

Actually $y(s_1, \cdot, s_M; x_i) \stackrel{\triangle}{=} G_1(s_1 + t_1^i) \circ \cdots \circ G_M(s_M + t_M^i)(x_0)$, where $p_i \stackrel{\triangle}{=} (t_1^i, \cdots, t_M^i)$ and we obtain

$$(43) \quad \frac{\partial y}{\partial s_j}(s; x_i) = \frac{\partial y}{\partial t_j}(p), j = 2, \cdots, M, \text{ provided p} = \text{p}_i + \text{s}, \text{i} = 1, \cdots \text{N}.$$

In particular, for $p_N = P$, we obtain

$$(44) \quad \begin{aligned} \frac{\partial y}{\partial t_j}(p) &= X_j(p; y(p)) = X_j^N(s; y(s; x_N)) \\ &= \frac{\partial y}{\partial s_j}(s; x_N), \ j = 2, \cdots, M, \end{aligned}$$

provided $p = P + s$, $s \in \mathcal{U}(0)$, and

$$(45) \quad \begin{aligned} \{Y_1, X_2(\ t_1), \cdots, X_M(t_1, \cdots, t_{M-1})\}(y) \\ = \{Y_1, \cdot, Y_M\}(y) A^{(N)}(s; p), \ s \in \mathcal{U}(0) \end{aligned}$$

if $y = y(p)$, $p = P + s$, where the nonsingular and $C^\infty$ matrix $A^N(s; p), s \in \mathcal{U}(0)$, fulfils the conclusion.

The proof is complete.

**Remark 2.**

*Theorem 1 is a local version of the algebraic representation which is suitable for (f.g.o.) Lie algebras. Comparing the algebraic representation given for the same gradient system in Lemma 1.1 and Theorem 2.1 we are not able to obtain the conclusion that the original $A(p)$ (see Remark 1.2) is nonsingular for all $p \in D_M$.*

**Definition.**

*Let $x_0 \in R^n$ and $\Lambda \subseteq \mathbf{F}_n$ be a (f.g.o; $x_0$) Lie algebra. A system $\{Y_1, \cdots, Y_M\} \subseteq \Lambda$ of generators is called k-minimal if $\{Y_1(x_0), \cdot, Y_k(x_0)\} \subseteq R^n$ are linearly independent and*

$$Y_j(x_0) = 0, \ j = k+1, \cdots, M, \quad where \ k = \dim \Lambda(x_0).$$

**Theorem 2.**

Let $\Lambda \subseteq \mathbf{F}_n$ be a (f.g.o.; $x_0$) Lie algebra and $x_0 \in R^n$ fixed. Assume that a $k$-minimal system of generators $\{Y_1, \cdot, Y_M\} \subseteq \Lambda$ was fixed and define

(a)
$$y(\widehat{p}) = G_1(t_1) \circ \cdots \circ G_k(t_k)(x_0),$$
$$\widehat{p} \overset{\triangle}{=} (t_1, \cdots, t_k) \in D_k = \prod_{i=1}^{k}(-a_i, a_i),$$

where $k = \dim \Lambda(x_0)$ and $G_i(t)(x), t \in (-a_i, a_i)$, $x \in V(x_0)$ is the local flow generated by $Y_i$. Then $y(\widehat{p}), \widehat{p} \in D_k$, is the solution of the gradient system

(b)
$$\frac{\partial y}{\partial t_1} = Y_1(y), \qquad \frac{\partial y}{\partial t_2} = X_2(t_1; y), \cdots,$$
$$\frac{\partial y}{\partial t_M} = X_M(t_1, \cdots, t_{M-1}; y), y(0) = x_0$$

fulfilling

(c_1)
$$\frac{\partial y}{\partial t_j}(\widehat{p}) \in \Lambda(y(\widehat{p})), \ j = 1, \cdots, k, \ (\forall) \ \widehat{p} \in D_k;$$

(c_2)
$$\frac{\partial y}{\partial t_1}(\widehat{p}), \cdots, \frac{\partial y}{\partial t_k}(\widehat{p}) \text{ are linearly independent in } R^n \text{ and}$$

$$\text{span} \left\{ \frac{\partial y}{\partial t_1}(\widehat{p}), \cdots, \frac{\partial y}{\partial t_k}(\widehat{p}) \right\} = \Lambda(y(\widehat{p})) \ (\forall) \ \widehat{p} \in D_k.$$

**Proof.**

By hypothesis, the conditions in Theorem 1 are fulfilled and for a $k$-minimal system of generators we obtain that $y(p) \overset{\triangle}{=} G_1(t_1) \circ \cdots \circ G_M(t_M)(x_0)$ equals $y(\widehat{p})$ in (a) for any $p \in D_M$. In addition, $y(\widehat{p}), \widehat{p} \in D_k$, the solution for (b) means $X_{j+1}(t_1, \cdots, t_j; y(\widehat{p})) = 0$ for any $j = k, \cdots, M-1$. Let $\widehat{p}_0 \in D_k$ be fixed arbitrarily and write the conclusion (c) in Theorem 1 according to the solution $y = y(\widehat{p}), \widehat{p} \in D_k$. We obtain

(46)
$$\left\{ \frac{\partial y}{\partial t_1}(\widehat{p}_0), \cdots, \frac{\partial y}{\partial t_k}(\widehat{p}_0), 0, \cdots, 0 \right\} = \{Y_1, \cdots, Y_M\}(y(\widehat{p}_0)) A(0; \widehat{p}_0),$$
$$\det A(0; \widehat{p}_0) \neq 0.$$

and from (46) we obtain the conclusion (c_1). In addition,

$$\text{span}\{Y_1, \cdot, Y_M\}(y(\widehat{p}_0)) = \Lambda(y(\widehat{p}_0)) \text{ and } \det A(0; \widehat{p}_0) \neq 0$$

allow one to see that

(47)
$$\text{span} \left\{ \frac{\partial y}{\partial t_1}(\widehat{p}_0), \cdots, \frac{\partial y}{\partial t_k}(\widehat{p}_0) \right\} = \Lambda(y(\widehat{p}_0))$$

for an arbitrary fixed $\widehat{p}_0 \in D_k$.

It remains to show that $\dfrac{\partial y}{\partial t_1}(\widehat{p}_0), \cdots, \dfrac{\partial y}{\partial t_k}(\widehat{p}_0)$ are linearly independent and it is done by proving that

(48)
$$\begin{aligned}
\widehat{Y}_1 &= \left\{ \left( \frac{\partial G}{\partial x}(\widehat{p}_0; x_0) \right)^{-1} \frac{\partial y}{\partial t_1}(\widehat{p}_0), \cdots, \right. \\
\widehat{Y}_k &= \left( \frac{\partial G}{\partial x}(\widehat{p}_0; x_0) \right)^{-1} \frac{\partial y}{\partial t_k}(\widehat{p}_0)
\end{aligned}$$

are linearly independent, where

(49)   $G(\widehat{p}; x) \overset{\triangle}{=} G_1(t_1) \circ \cdots \circ G_k(t_k)(x), \ \widehat{p} = (t_1, \cdot, t_k) \in D_k, x \in V(x_0).$

By definition,

(50)   $\left( \dfrac{\partial G}{\partial x}(\widehat{p}_0; x_0) \right)^{-1} = \left( \dfrac{\partial G_k}{\partial x}(t_k; x_k) \right)^{-1} \times \cdots \times \left( \dfrac{\partial G_1}{\partial x}(t_1; x_1) \right)^{-1},$

where $x_1 \overset{\triangle}{=} G_2(t_2) \circ \cdots \circ G_k(t_k)(x_0), \cdots, x_k \overset{\triangle}{=} x_0, \widehat{p}_0 \overset{\triangle}{=} (t_1, \cdots, t_k)$

On the other hand, using (46) we obtain

(51)   $\{\widehat{Y}_1, \cdots, \widehat{Y}_k, 0, \cdots, 0\} = \left( \dfrac{\partial G}{\partial x}(\widehat{p}_0; x_0) \right)^{-1} \{Y_1, \cdot, Y_M\}(y(\widehat{p}_0)) A(0; \widehat{p}_0)$

and as one may expect the right hand side can be expressed

(52)
$$\begin{aligned}
\left( \tfrac{\partial G}{\partial x}(\widehat{p}_0; x_0) \right)^{-1} & \{Y_1, \cdots, Y_M\}(y(\widehat{p}_0)) \\
&= \{Y_1, \cdot, Y_M\}(x_0) Z(t_1, \cdots, t_k) A(0; \widehat{p}_0)
\end{aligned}$$

where the $(M \times M)$ matrix $Z$ fulfils $\det Z(t_1, \cdots, t_k) \neq 0$. More precisely, $H_j(t; x_j) \overset{\triangle}{=} \left( \dfrac{\partial G}{\partial x}(t; x_j) \right)^{-1}, t \in [0, t_j]$, fulfils

(53)   $\dfrac{dH_j}{dt} = -H_j \dfrac{\partial Y_j}{\partial x}(G_j(t; x_j)), H_j(0) = I_n, \ j = 1, \cdots, k,$

and it allows a direct computation of the matrix $Z(t_1, \cdots, t_k)$ as

(54)   $Z(t_1, \cdots t_k) \overset{\triangle}{=} Z_k(t_k) Z_{k-1}(t_{k-1}; t_k) \times \cdots \times Z_1(t_1; t_2, \cdots, t_k)$

where $Z_j(t; \cdot), t \in [0, t_j]$ is the solution of

(55)   $\dfrac{dZ_j}{dt} = -Z_j B_j(t, t_{j+1}, \cdots, t_k), \ Z_j(0) = I_M$

Here, the $C^\infty$ matrix $B_j(t, t_{j+1}, \cdot, t_k)$ is determined such that

(56) $\quad ad\, Y_j\{Y_1, \cdot, Y_M\}(G_j(t; x_j)) = \{Y_1, \cdots, Y_M\}(G_j(t; x_j))B_j(t; t_{j+1}, \cdot, t_k)$

where

$$x_j \triangleq G_{j+1}(t_{j+1}) \circ \cdots \circ G_k(t_k)(x_0)$$

Now, using (51) and (52), we obtain

(57) $\quad$ span $\{\widehat{Y}_1, \cdots, \widehat{Y}_k\} =$ span $\{Y_1(x_0), \cdots, Y_k(x_0), 0, \cdots, 0\}$

and $\widehat{Y}_1, \cdots, \widehat{Y}_k$ are linearly independent.

It shows that the conclusion $(c_2)$ holds and the proof is complete.

Theorem 2 is the keypoint of a homomorphism between a (f.g.o; $x_0$) Lie algebra $\Lambda$ and a maximal rank new Lie algebra which is stated in the next.

**Theorem 3.**

*Let $x_0 \in R^n$ and $\Lambda \subseteq \mathbf{F}_n$ be a (f.g.o; $x_0$) Lie algebra. Consider $\{Y_1, \cdots, Y_M\} \subseteq \Lambda$ a k-minimal system of generators where $\dim \Lambda(x_0) = k$, and define*

$$y(\widehat{p}) = G_1(t_1) \circ \cdots \circ G_k(t_k)(x_0), \widehat{p} = (t_1, \cdots, t_k) \in D_k \triangleq \prod_1^k (-a_i; a_i)$$

*as in Theorem 2. Then there exist $Q_s(\cdot) \in C^\infty(D_k; R^k)$ $s = 1, \cdots, M$, and a homomorphism $\lambda : \Lambda \to L(Q_1, \cdots, Q_M)$ such that*

$c_1)\lambda(Y_s) = Q_s$ $s = 1, \cdots, M$,

$c_2) \dim L(Q_1, \cdots, Q_M)(\widehat{p}) = k$, $(\forall)\, \widehat{p} \in D_k$, *and each* $Q \in L(Q_1, \cdot, Q_M)$ *can be written*

$$Q(\widehat{p}) = \sum_1^M a_i(\widehat{p})Q_i(\widehat{p}),$$

*where $a_i(\cdot) \in C^\infty(D_k)$.*

**Proof.**

By hypothesis, the conditions in Theorem 2 are fulfilled and let $Q_s(\cdot) \in C^\infty(D_k; R^k)$ be such that

(58) $\qquad \sum_{j=1}^k \dfrac{\partial y}{\partial t_j}(\widehat{p})Q_s^j(\widehat{p}) = Y_s(y(\widehat{p})),\quad s = 1, \cdots, M.$

Define the Lie algebra $L(Q_1, \cdots, Q_M) \subseteq C^\infty(D_k; R^k)$ over $R$ and using the symmetry of the matrices $\left(\dfrac{\partial^2 y_\sigma}{\partial t_i \partial t_j}(\hat{p})\right)_{i,j}$, $\sigma = 1, \cdots, n$, from (58) we obtain that any Lie bracket $Y$ of $\{Y_1, \cdots Y_M\}$ along to $y(\hat{p})$ has a corresponding element $Q \in L(Q_1, \cdot, Q_M)$ such that

$$(59) \qquad \sum_{j=1}^{k} \frac{\partial y}{\partial t_j}(\hat{p}) Q(\hat{p}) = Y(y(\hat{p})) \ (\forall) \ \hat{p} \in D_k$$

Therefore, the homomorphism $\lambda : \Lambda \to L(Q_1, \cdots, Q_M)$ is defined by the unique solution in (59) for each $Y \in \Lambda$ and the conclusion $(c_1)$ holds.

On the other hand, for the fixed orbit $y(\hat{p}), \hat{p} \in D_k$, and $\dfrac{\partial y}{\partial t_j}(\hat{p}) \in \Lambda(y(\hat{p}))$, $j = 1, \cdot, k$ (see Th. 2) we determine $a_j^i \in C^\infty(D_k; R^k)$ such that

$$(60) \qquad \sum_{i=1}^{M} a_j^i(\hat{p}) Y_i(y(\hat{p})) = \frac{\partial y}{\partial t_j}(\hat{p}), \ j = 1, \cdots, k.$$

Combining (58) and (60) we obtain

$$(61) \qquad \sum_{j=1}^{M} \frac{\partial y}{\partial t_j}(\hat{p}) \left(\sum_{i=1}^{M} a_s^i(\hat{p}) Q_i^j(\hat{p})\right) = \frac{\partial y}{\partial t_s}(\hat{p}), s = 1, \cdots, k$$

and $(c_2)$ from Theorem 2 allow one to conclude

$$(62) \qquad \sum_{i=1}^{M} a_s^i(\hat{p}) Q_i(\hat{p}) = e_s, \ s = 1, \cdots, k$$

where $\{e_1, \cdot, e_k\} \subseteq R^k$ is the canonical basis.
It is easily seen that (62) means

$$(63) \qquad \operatorname{span} \{Q_1(\hat{p}), \cdots, Q_M(\hat{p})\} = R^k \text{ foreach } \hat{p} \in D_k$$

and according to (62) an arbitrary $Q \in L(Q_1, \cdot, Q_M)$ is written as

$$(64) \qquad Q(\hat{p}) = \sum_{j=1}^{M} a_s(\hat{p}) Q_s(\hat{p}), \ \hat{p} \in D_k,$$

where $a_s(\cdot) \in C^\infty(D_k)$
The equations (63) and (64) prove that the conclusion $(c_2)$ is fulfilled and the proof is complete.

**Remark 3.**

*It might be useful to know how the new Lie algebra $L(Q_1, \cdot, Q_M)$ could be determined in Theorem 3 provided the original Lie algebra $\Lambda$ is determined by some $g_i \in \mathbf{F}_n$, $i = 1, \cdot, m$, i.e. $\Lambda = L(g_1, \cdots, g_m)$. In this case, we may rewrite $L(Q_1, \cdots, Q_M)$ as $L(q_1, \cdots, q_m)$, where $q_i(\cdot) \in C^\infty(D_k; R^k)$, $i = 1, \cdots, m$, are found such that*

$$\sum_{j=1}^{k} \frac{\partial y}{\partial t_j}(\widehat{p}(q_i^j(\widehat{p})) = g_i(y(\widehat{p})), \quad (\forall)\; \widehat{p} \in D_k, i = 1, \cdots, m.$$

## 3.3 Some Applications

Some considerations regarding a system of linear hyperbolic equations have been made in connection with a finite dimensional Lie algebra (see Proposition 2.1) and the domain of the system was found as an integral manifold. A system of linear hyperbolic equations will be considered in a more general setting of a (f.g.o; $x_0$) Lie algebra and the basic tools of analysis are contained in Theorems 2.2 and 2.3. A linear system is defined

(1) $$\left\langle \frac{\partial S}{\partial x}(x), g_i(x) \right\rangle = l_i(x),\; i = 1, \cdots, m$$

where $g_i \in C^\infty(R^n; R^n)$, $l_i \in C^\infty(R^n)$.

A nontrivial solution $S(x), x \in V(x_0) \subseteq R^n$, is a nonzero scalar $C^\infty$ function fulfilling (1) on a domain $M_{x_0} \cap V(x_0)$, where $M_{x_0}$ appears as an integral manifold of the Lie algebra $L(g_1, \cdot, g_m)$.

A special case is

(2) $$\dim L(g_1, \cdot, g_m)(x_0) = n, \quad (l_1(x_0), \cdots, l_m(x_0)) \neq 0$$

and the system (1) fulfilling (2) is called nonsingular at $x_0$.

To make sure that a solution of (1) exists a compatibility condition on $\{g_1, \cdot, g_m\}$ and $\{l_1, \cdot, l_m\}$ is necessary. In the classical Frobenius theorem $g_i = e_i$, $x \in R^m$, where $e_1, \cdot, e_m \in R^m$ is the canonical basis, and the integrability of (1) is determined by $\frac{\partial l_i}{\partial x_j}(x) = \frac{\partial l_j}{\partial x_i}(x), i, j \in \{1, \cdots, m\}$. For noncommuting $g_i \in \mathbf{F}_n$ define $\widetilde{g}_i(x) = \begin{pmatrix} l_i(x) \\ g_i(x) \end{pmatrix} \in R^{n+1}$ and consider the Lie algebra

$$L(\widetilde{g}_1, \cdot, \widetilde{g}_m) \subseteq \mathbf{F}_{n+1} \stackrel{\triangle}{=} C^\infty(R^{n+1}; R^{n+1})$$

(c) **Compatibility condition**.

Let $x_0 \in R^n$ be fixed and assume that $L(\tilde{g}_1, \cdot, \tilde{g}_m)$ is a (f.g.o; $x_0$) Lie algebra such that $\begin{pmatrix} \alpha \\ 0 \end{pmatrix} \notin L(\tilde{g}_1, \cdot, \tilde{g}_m)(x_0)$ if $\alpha \neq 0$.

**Proposition 1.**

Let $g_i \in \mathbf{F}_n, l_i \in C^\infty(R^n)$, be given $i = 1, \cdots, m$, and $x_0 \in R^n$ is fixed. Assume that (1) is nonsingular at $x_0$ and $L(\tilde{g}_1, \cdot, \tilde{g}_m)$ is fulfilling the compatibility condition. Then there exists a nontrivial solution $S(x)$, $x \in V(x_0)$, of (1).

**Proof.** The assumed system of generators $\{Z_1, \cdot, Z_M\} \subseteq L(\tilde{g}_1, \cdot, \tilde{g}_m)$ is arranged to be an $n$-minimal one, i.e. $Z_1(x_0), \cdots, Z_n(x_0)$ are linearly independent in $R^{n+1}$ and $Z_j(x_0) = 0$, $j = n+1, \cdots, M$. Here the compatibility condition ensures that it is enough to rearrange $\{Y_1, \cdots, Y_M\} \subseteq L(g_1, \cdots, g_m)$ such that $Y_1(x_0), \cdot, Y_n(x_0)$ are linearly independent in $R^n$ and $Y_j(x_0) = 0$, $j = n+1, \cdot, M$.

Write $z = (t, x) \in R^{n+1}$, $z_0 = (0, x_0)$ and define

$$
(3) \qquad
\begin{aligned}
z(\hat{p}) &= \tilde{G}_1(t_1) \circ \cdots \circ \tilde{G}_n(t_n)(z_0), \\
\hat{p} &\triangleq (t_1, \cdot, t_n) \in D_n \triangleq \textstyle\prod_1^n (-a_i, a_i)
\end{aligned}
$$

where $\tilde{G}_i(t)(z)$, $t \in (-a_i, a_i)$, $z \in \mathcal{U}(z_0) \subseteq R^{n+1}$, is the local flow generated by $Z_i$.

On the other hand $\dim L(\tilde{g}_1, \cdots, \tilde{g}_m)(z_0) = \dim L(g_1, \cdot, g_m)(x_0) = n < n+1$ and consider the linear homogeneous system

$$
(4) \qquad \left\langle \frac{\partial F}{\partial z}(z), \tilde{g}_i(z) \right\rangle = 0, \quad i = 1, \cdots, m
$$

Using Theorem 2.2 we obtain that

$$
(5) \qquad M_{z_0} = \{z \in R^{n+1}; z = z(\hat{p}), \hat{p} \in D_n\}
$$

is an integral $n$-dimensional manifold of $L(\tilde{g}_1, \cdot, \tilde{g}_m)$ for $z(\hat{p})$ in (3), and (4) is equivalent to

$$
(6) \qquad \left\langle \frac{\partial F}{\partial z}(z), \tilde{g}(z) \right\rangle = 0 \text{ for any } \tilde{g} \in \mathrm{L}(\tilde{g}_1, \cdots, \tilde{g}_m),
$$

provided $z \in M_{z_0}$.

To obtain a nontrivial solution $F(z)$, $z \in \mathcal{U}(z_0)$, for (4) we need to know that $M_{z_0}$ in (5) can be rewriten as

$$
(7) \qquad M_{z_0} \cap \mathcal{U}(z_0) = \{z \in \mathcal{U}(z_0) : F(z) = F(z_0)\}
$$

where $F$ is a $C^\infty$ nonzero scalar function.

Starting with the integral manifold $M_{z_0}$ in (5), where $z(\widehat{p}), \widehat{p} \in D_n$, is the solution in a nonsingular gradient system, the same computations as in Proposition 2.4.1 will allow one to have (7) fulfilled. That is to say, the system (4) (or (6)) has a nontrivial solution $F(t,x), (t,x) \in \mathcal{U}(z_0)$, fulfilling

(8) $$\frac{\partial F}{\partial t}(t,x)l_i(x) + \left\langle \frac{\partial F}{\partial x}(t,x), g_i(x) \right\rangle = 0, \ i = 1, \cdot, m,$$

for any $(t,x) \in M_{z_0} \cap \mathcal{U}(z_0) \subseteq R^{n+1}$, or

(9) $$\left\langle \frac{\partial F}{\partial z}(z), \widetilde{g}(z) \right\rangle = 0 \ (\forall) \ z \in M_{z_0} \cap \mathcal{U}(z_0), \ \widetilde{g} \in L(\widetilde{g}_1, \cdot, \widetilde{g}_m)$$

By hypotheses $\dim L(g_1, \cdots, g_m)(x_0) = n$ and from (9) we obtain

(10) $$\frac{\partial F}{\partial t}(0,x_0) \stackrel{\triangle}{=} \frac{\partial F}{\partial t}(z_0) \neq 0$$

Therefore, a unique $C^\infty$ solution $t(x), x \in V(x_0) \subseteq R^n$, can be found from

(11) $$F(t,x) = F(0,x_0)$$

and it holds

(12) $$t(x_0) = 0, \ \frac{\partial t}{\partial x}(x) = \frac{-\dfrac{\partial F}{\partial x}(t(x),x)}{\dfrac{\partial F}{\partial t}(t(x),x)}, \qquad x \in V(x_0).$$

Combining (8) and (12) we obtain that $t(x), x \in V(x_0) \subseteq R^n$ is a nontrivial solution for

(13) $$\left\langle \frac{\partial t}{\partial x}(x), g_i(x) \right\rangle = l_i(x), \ (\forall) \ x \in V(x_0), \ i = 1, \cdots, m$$

and the proof is complete.

### Comment

The condition $\dim L(g_1, \cdot, g_m)(x_0) = n$ is essential to have the Proposition 1 proved.

The general case $\dim L(g_1, \cdot, g_m)(x_0) = k \leq n$ is the goal of the following and the homomorphism correspondence in Theorem 3.3 is the main tool allowing one to handle with $k < n$.

($H_1$) Assume that $g_i \in \mathbf{F}_n, x_0 \in R^n$ are given such that the Lie algebra $L(g_1, \cdots, g_m)$ is (f.g.o; $x_0$) and $\dim L(g_1, \cdot, g_m)(x_0) = k < n$.

The hypothesis ($H_1$) allow one to apply Theorem 3.3 and choosing a $k$-minimal system $\{Y_1, \cdot, Y_M\} \subseteq L(g_1, \cdot, g_m)$ of generators we define

$$(14) \qquad y(\widehat{p}) = G_1(t_1) \circ \cdot \circ G_k(t_k)(x_0), \widehat{p} = (t_1, \cdot, t_k) \in D_k \stackrel{\triangle}{=} \prod_1^k (-a_i, a_i)$$

where $G_i(t)(x), t \in (-a_i, a_i)$, $x \in V(x_0)$, is the local flow generated by $Y_i$

Now, the system (1) is replaced by

$$(15) \qquad \left\langle \frac{\partial \mathcal{U}}{\partial \widehat{p}}(\widehat{p}), q_i(\widehat{p}) \right\rangle = f_i(\widehat{p}), \ i = 1, \cdots, m,$$

where $f_i(\widehat{p}) \stackrel{\triangle}{=} l_i(y(\widehat{p})), \widehat{p} \in D_k$, and $q_i(\cdot) \in C^\infty(D_k; R^k)$ are found as in Theorem 3.3 (see Remark 3.3) fulfilling

$$(16) \qquad \sum_{j=1}^k \frac{\partial y}{\partial t_j}(\widehat{p}) \cdot q_i^j(\widehat{p}) = g_i(y(\widehat{p})), \ (\forall) \ \widehat{p} \in D_k, \ i = 1, \cdot, m.$$

In addition, it is known that

$$(17) \qquad \dim L(q_1, \cdot, q_m)(\widehat{p}) = k \text{ for any } \widehat{p} \in D_k.$$

and (15) is a nonsingular system at $\widehat{p}_0 = 0$.

The new system (15) is solved provided a compatibility condition is imposed.

### Proposition 2.

*Let $g_i \in \mathbf{F}_n, l_i \in C^\infty(R^n)$ be given, $i = 1, \cdots, m$, and $x_0 \in R^n$ is fixed. Assume that ($H_1$) is fulfilled for $L(g_1, \cdot, g_m)$ and the compatibility condition is fulfilled for $L(\widetilde{g}_1, \cdot, \widetilde{g}_m)$. Then, there exists a nontrivial solution for (1), i.e. a smooth $S(x), x \in V(x_0)$ and an integral manifold $M_{x_0}$ of $L(g_1, \cdot, g_m)$ with $\dim M_{x_0} = \dim L(g_1, \cdot, g_m)(x_0)$ will exists such that*

$$\left\langle \frac{\partial S}{\partial x}(x), g_i(x) \right\rangle = l_i(x), i = 1, \cdot, m, \ (\forall) \ x \in M_{x_0} \cap V(x_0).$$

### Proof.

The compatibility condition for $L(\widetilde{g}_1, \cdots, \widetilde{g}_m)$ and ($H_1$) for $L(g_1, \cdots, g_m)$ means that a $k$-minimal systems of generators $\{Z_1, \cdot, Z_M\} \subseteq L(\widetilde{g}_1, \cdot, \widetilde{g}_m)$ will exist such that

$$Z_j(x) = \begin{pmatrix} L_j(x) \\ Y_j(x) \end{pmatrix}$$

fulfils $Z_j(x_0) = 0, j = k+1, \cdot, M$, and $Y_1(x_0), \cdot, Y_k(x_0)$ are linearly independent in $R^n$. Define

$$z(\widehat{p}) = \widehat{G}_1(t_1) \circ \cdots \circ \widehat{G}_k(t_k)(z_0) \in R^{n+1} \quad z_0 = (0, x_0),$$

$$\widehat{p} = (t_1, \cdot, t_k) \in D_k \stackrel{\triangle}{=} \prod_1^k(-a_i, a_i)$$

where $\widehat{G}_i(t)(z), t \in (-a_i, a_i), z \in \mathcal{U}(z_0) \subseteq R^{n+1}$ is the local flow generated by $Z_i$.

Using Theorem 2.3 (see Remark 2.3) for $\Lambda = L(\widetilde{g}_1, \cdot, \widetilde{g}_m)$ and $\dim \Lambda(z_0) = k$ we obtain $q_i \in C^\infty(D_k; R^k)$, $i = 1, \cdots, m$, such that

(18)
$$\sum_{j=1}^k \frac{\partial z}{\partial t_j}(\widehat{p}) q_i^j(\widehat{p}) = \widetilde{g}_i(z(\widehat{p})), \quad i = 1, \cdot, m, \ \widehat{p} \in D_k.$$

(19)
$$\dim L(q_1, \cdots, q_m)(\widehat{p}) = k \quad (\forall)\widehat{p} \in D_k.$$

Denote $y(\widehat{p}) = pr_{R^n} z(\widehat{p})$, $\widehat{p} \in D_k$, i.e. $y(\widehat{p})$ is defined by the last $n$-components of $z(\widehat{p})(y(\widehat{p}) = G_1(t_1) \circ \cdots \circ G_k(t_k)(x_0))$ and (18) can be rewriten

(20)
$$\sum_{j=1}^k \frac{\partial y}{\partial t_j}(\widehat{p}) q_i^j(\widehat{p}) = g_i(y(\widehat{p})), \quad i = 1, \cdots, m,$$

(21)
$$\sum_{j=1}^k \frac{\partial \tau}{\partial t_j}(\widehat{p}) q_i^j(\widehat{p}) \stackrel{\triangle}{=} \left\langle \frac{\partial \tau}{\partial \widehat{p}}(\widehat{p}), q_i(\widehat{p}) \right\rangle$$
$$= l_i(y(\widehat{p})), i = 1, \cdots, m$$

where $\tau(\widehat{p})$ denotes the first component of $z(\widehat{p})$, $z \stackrel{\triangle}{=} \begin{pmatrix} \tau \\ y \end{pmatrix}$. The equations (21) show that the system (15) has a nontrivial solution $\tau(\widehat{p}), \widehat{p} \in D_k$, and (21), (20) allow one to obtain a nontrivial solution for (1) provided $\widehat{p} = \varphi(x), x \in V(x_0)$, with $\varphi(y(\widehat{p})) = \widehat{p}$ will be determined. Write

$$G(y) = G(\widehat{p}; \widehat{x}) \stackrel{\triangle}{=} G_1(t_1) \circ \cdots \circ G_k(t_k)(\widehat{x}, x_{10}, \cdot, x_{k0}),$$
$$\widehat{x} \stackrel{\triangle}{=} (x_{k+1}, \cdot, x_n), \quad y = (\widehat{p}, \widehat{x})$$

and it is easily seen that $\dfrac{\partial G}{\partial y}(0, \widehat{x}_0) \neq 0$, where $y_0 = (0, \widehat{x}_0)$, $\widehat{x}_0 \stackrel{\triangle}{=} (x_{0k+1}, \cdot, x_{0n})$

The solution $y(x)$, $x \in V(x_0) \subseteq R^n$ is found such that $y(x) = (\varphi(x), \widehat{x}(x))$ and

(22) $$G(\varphi(x), \widehat{x}(x)) = x, \quad x \in V(x_0), \quad \varphi(x) \in D_k$$

Using $\tau(\widehat{p}), \widehat{p} \in D_k$, defined in (21) and $\widehat{p} = \varphi(x)$ fulfilling (22) we obtain

(23) $$\varphi(y(\widehat{p})) = \widehat{p}, \text{ and } S(x) \stackrel{\triangle}{=} \tau(\varphi(x)), x \in V(x_0).$$

A direct computation proves that $S(x), x \in V(x_0)$ defined in (23) is a nontrivial solution for (1).

More precisely, from (23) we obtain

(24)
$$\sum_{j=1}^{n} \frac{\partial \varphi^i}{\partial x_j}(y(\widehat{p})\frac{\partial y^j}{\partial t_i}(\widehat{p}) = 1,$$
$$\sum_{j=1}^{n} \frac{\partial \varphi^i}{\partial x_j}(y(\widehat{p}))\frac{\partial y^j}{\partial t_s}(\widehat{p}) = 0, \ s \neq i$$

and using (19) it follows

(25)
$$\left\langle \frac{\partial S}{\partial x}(y(\widehat{p})), g_i(y(\widehat{p})) \right\rangle \stackrel{\triangle}{=} \left\langle \frac{\partial \tau}{\partial \widehat{p}}(\widehat{p})\frac{\partial \varphi}{\partial x}(y)(\widehat{p}), g_i(y(\widehat{p})) \right\rangle$$
$$= \left\langle \frac{\partial \tau}{\partial \widehat{p}}(\widehat{p})\frac{\partial \varphi}{\partial x}(y(\widehat{p})), \frac{\partial y}{\partial \widehat{p}}(\widehat{p})q_i(\widehat{p}) \right\rangle$$
$$= \left\langle \frac{\partial \tau}{\partial \widehat{p}}(\widehat{p}), q_i(\widehat{p}) \right\rangle$$
$$= l_i(y(\widehat{p})).$$

for $\widehat{p} \in D_k$ with $y(\widehat{p}) \in V(x_0) \subseteq R^n$.

Therefore, (25) means

(26) $$\left\langle \frac{\partial S}{\partial x}(x), g_i(x) \right\rangle = l_i(x), \ (\forall) \ x \in V(x_0) \cap M_{x_0}, i = 1, \cdots, m,$$

where the $k$-dimensional integral manifold $M_{x_0} \stackrel{\triangle}{=} \{x \in R^n : x = y(\widehat{p}), \widehat{p} \in D_k\}$.

The proof is complete.

**Bibliographical notes**
(see Th.2, Th.3)


The main results are adapted from Vârsan [13] and the analysis has been done in a more complete form. The concept of finite generation over orbits (f.g.o.) is usually used in the geometry of foliations associated with distribution with singularities (see Hermann [6]).

# Chapter 4

# Applications

Systems of semilinear hyperbolic equations and their solutions are considered focussing the analysis on the domain-integral manifold as the main part of the solution.

Here a similarity with a Frobenius gradient system is easily seen provided the comutative derivations $\dfrac{\partial}{\partial x_1}, \cdots, \dfrac{\partial}{\partial x_m}$ are replaced by some non-commuting vector fields, $g_1, \cdot, g_m \in \text{Der}(\mathbb{R}^n)$ and the Frobenius complete integrability conditions take the form of a compatibility condition on a Lie algebra.

Then, integral representation of the solution for stochastic differential equations will point out the complexity of the input (Wiener process)–output (solution) application via the gradient system of deterministic vector fields.

In particular, the dependence of the solution on the Cauchy data $x(0) = x \in \mathbb{R}^n$ is separated from the influence of the input and an approximation of the original input process will generate a corresponding approximation for the stochastic flow. An anticipating drift in the stochastic equation may and do involve a carefull analysis.

## 4.1  Systems of Semiliniar Equations

A scalar $C^\infty$ function $t(x), x \in V(x_0) \in \mathbb{R}^n$ is a nontrivial solution for the system

$$(1) \qquad \left\langle \frac{\partial t}{\partial x}(x), g_i(x) \right\rangle = f_i(t(x), x), \ i = 1, \cdots, m,$$

if there exists an integral manifold $M_{x_0} \subseteq \mathbb{R}^n$ of $L(g_1, \cdot, g_m)$ such that (1) in fulfilled for all $x \in M_{x_0} \cap V(x_0)$ and $\dfrac{\partial t}{\partial x}(x_0) \neq 0$, where $g_i \in \mathbf{F}_n$ (orDer $(\mathbb{R}^n)$

and $f_i \in C^\infty(R \times R^n; R)$, $i = 1, \cdots m$.

A special case

$$(2) \qquad \dim L(g_1, \cdot, g_m)(x_0) = n \text{ for } x_0 \in R^n \text{ fixed,}$$

where $L(g_1, \cdot, g_m)$ is the Lie algebra determined by $\{g_1, \cdot, g_m\}$, allow one to obtain an $n$-dimensional $M_{x_0}$ of $L(g_1, \cdot, g_m)$ and the domain $M_{x_0} \cap V(x_0)$ of the system (1) will be a neighbourhood of $x_0, V(x_0) \subseteq R^n$.

The semilinear system (1) is associated with a linear one

$$(3) \qquad \left\langle \frac{\partial F}{\partial z}(z), \widetilde{g}_i(z) \right\rangle = 0, \ i = 1, \cdot, m, \ z_0 = (0, x_0)$$

where

$$z \stackrel{\triangle}{=} (t, x) \in R^{n+1}, \widetilde{g}_i(z) \stackrel{\triangle}{=} \left( \begin{array}{c} f_i(z) \\ g_i(x) \end{array} \right), \ i = 1, \cdots, m.$$

and a nontrivial solution $F(t, x)$ for (3) will determine a nontrivial solution for (1) provided

$$(4) \quad \dim L(\widetilde{g}_1, \cdots, \widetilde{g}_m)(z_0) = n, \ (\text{see } L(\widetilde{g}_1, \cdots, \widetilde{g}_m) \subseteq \mathbf{F}_{n+1} (\text{Der } (R^{n+1}))$$

By the definition of the nontrivial solution $\mathbf{F}(z)$ in (3) we obtain an integral manifold $M_{z_0}$ of $L(\widetilde{g}_1, \cdots, \widetilde{g}_m)$ such that (3) is fulfilled for all $z \in M_{z_0} \cap W(z_0)$ with $W(z_0) \subseteq R^{n+1}$. It shows that under the existence of a nontrivial solution of (3) we may associate an equivalent system to (3)

$$(5) \qquad \left\langle \frac{\partial F}{\partial z}(z), \widetilde{g}(z) \right\rangle = 0, \ (\forall) \ \widetilde{g} \in L(\widetilde{g}_1, \cdot, \widetilde{g}_m), z \in M_{z_0} \cap W(z_0).$$

and the degeneracy of the Lie algebra $L(\widetilde{g}_1, \cdot, \widetilde{g}_m)$ arround to $z = z_0$ contained in (5) appears as necessary.

### c) Compatibility condition

$L(\widetilde{g}_1, \cdot, \widetilde{g}_m)$ is finitely generated with respect to orbits starting from $z_0$ (f.g.o; $z_o$) and $\left( \begin{array}{c} \alpha \\ 0 \end{array} \right) \notin L(\widetilde{g}_1, \cdot, \widetilde{g}_m)(z_0)$ if $\alpha \neq 0, \alpha \in R$.

### Theorem 1.

Assume that $g_i \in \mathbf{F}_n, f_i \in C^\infty(R^{n+1}; R), x_0 \in R^n, i = 1, \cdot, m$, are given such that $(f_i(0, x_0), \cdots, f_m(0, x_0)) \neq 0, \dim L(g_1, \cdots g_m)(x_0) = n$ and compatibility condition (c) are fulfilled. Then (1) has a nontrivial solution.

**Proof.** By hypotheses (see (c)) we may choose a system of generators $\{Z_1, \cdot, Z_M\} \subseteq L(\widetilde{g}_1, \cdots, \widetilde{g}_m)$, $Z_j(z) = \begin{pmatrix} L_j(z) \\ Y_j(x) \end{pmatrix}$ such that $Y_1(x_0), \cdots, Y_m(x_0)$ are linearly independent in $R^n$ and $Z_j(z_0) = 0, j = n+1, \cdots, M$. That is to say $\{Z_1, \cdot, Z_M\}$ is an $n$-minimal system of generators for $L(\widetilde{g}_1, \cdots, \widetilde{g}_m)$ and $\dim L(\widetilde{g}_1, \cdots, \widetilde{g}_M)(z_0) = n$. The conditions in Theorem 3.2.3 are fulfilled and the homomorphism $\lambda : L(\widetilde{g}_1, \cdot, \widetilde{g}_m) \to L(q_1, \cdot, q_m)$ is defined by

$$z(p) = \widetilde{G}_1(t_1) \circ \cdots \circ \widetilde{G}_n(t_n)(z_0), \ p = (t_1, \cdot, t_n) \in D_n \stackrel{\triangle}{=} \prod_1^n (-a_i, a_i),$$

such that $q_i \in C^\infty(D_n; R^n)$ and

(6)
$$\frac{\partial z}{\partial p}(p)q_i(p) = \widetilde{g}_i(z(p)), \ i = 1, \cdots, m.$$

$$\frac{\partial z}{\partial p}(p)Q_s(p) = Z_s(z(p)), \ s = 1, \cdots, M, \ p \in D_n$$

Here $Q_j \in L(q_1, \cdots, q_m)$ and

(7)
$$\text{span} \{Q_1(p), \cdots, Q_M(p)\} = L(q_1, \cdots, q_m)(p) \ p \in D_n,$$

$$\dim L(q_1, \cdots, q_m)(p) = n, \ (\forall) \ p \in D_n$$

By definition $z = (t, x), \widetilde{g}_i(z) = \begin{pmatrix} f_i(z) \\ g_i(x) \end{pmatrix}$ and from (6) for $z(p) = (\tau(p), y(p))$, $p \in D_n$, we obtain

(8)
$$\left\langle \frac{\partial \tau}{\partial p}(p), q_i(p) \right\rangle = f_i(\tau(p), y(p)), \ i = 1, \cdots, m,$$
$$\frac{\partial y}{\partial p}(p)q_i(p) = g_i(y(p)),$$

and

(9)
$$\frac{\partial y}{\partial p}(p)Q_s(p) = Y_s(y(p)), \ s = 1, \cdots, M,$$

Here $y(p) = G_1(t_1) \circ \cdots \circ G_n(t_n)(x_0)$ and $G_j(t)(x)$ is the local flow generated by $Y_j$ in $Z_j = \begin{pmatrix} L_j \\ Y_j \end{pmatrix}$. Now, from the system

(10)
$$y(p) = x, \ x \in V(x_0)$$

we find a $C^\infty$ function $p(x)$ (see $\frac{\partial y}{\partial t_1}(0) = Y_1(x_0), \cdot, \frac{\partial y}{\partial t_n}(0) = Y_n(x_0)$ are linearly independent in $R^n$) such that

$$(11) \qquad y(p(x)) \equiv x, \frac{\partial y}{\partial p}(p(x))\frac{\partial p}{\partial x}(x) = I_n, \; x \in V(x_0)$$

Write $t(x) = \tau(p(x)), x \in V(x_0)$, and a direct computation allow one to see

$$(12) \qquad \left\langle \frac{\partial t}{\partial x}(x), g_i(x) \right\rangle = \left\langle \frac{\partial \tau}{\partial p}(p(x)), q_i(p(x)) \right\rangle, \; x \in V(x_0)$$
$$q_i(p(x)) = \frac{\partial p}{\partial x}(x)g_i(x).$$

Using (12) in (8) we obtain

$$(13) \qquad \left\langle \frac{\partial t}{\partial x}(x), g_i(x) \right\rangle = f_i(t(x), x), \; x \in V(x_0), \; i = 1, \cdot, m.$$

and the proof is complete.

**Remark 1.**

The domain $V(x_0)$ for the system (1) (see (13)) is determined in (11) and depends on $\left\{ \frac{\partial y}{\partial t_1}(p), \cdots, \frac{\partial y}{\partial t_n}(p) \right\}$ are linearly independent in $R^n$. The Lie algebra $L(g_1, \cdots, g_m)$ is (f.g.o; $x_0$) (see (c)) and $\left\{ \frac{\partial y}{\partial t_1}(p), \partial y \cdots, \frac{\partial y}{\partial t_n}(p) \right\} \subseteq R^n$, are linearly independent for any $p \in D_n$ allowing one to obtain a maximal domain $\mathcal{U}(x_0) \subseteq V(x_0)$ where the system (1) is fulfilled In the general case

$$(14) \qquad \dim L(g_1, \cdot, g_m)(x_0) = k < n$$

we restate the result in Theorem 1.

**Theorem 2**

Assume that $g_i \in \mathbf{F}_n, f_i \in C^\infty(R^{n+1}; R), x_0 \in R^n, i = 1, \cdots, n$, are given such that $(f_1(0, x_0), \cdots, f_m(0, x_0)) \neq 0$, $\dim L\{g_1, g_m)(x_0)\} = k < n$, and compatibility condition (c) are fulfilled. Then (1) has a nontrivial solution.

**Proof.**

Using compatibility condition (c) we see easily that $L(g_1, \cdot, g_m)$ is (f.g.o; $x_0$) and $\dim L(\tilde{g}_1, \cdots, \tilde{g}_m)(z_0) = k = \dim L(g_1, \cdots, g_m)(x_0)$. Define a $k$-minimal system of generators $Z_j(z) = \begin{pmatrix} L_j(z) \\ Y_j(x) \end{pmatrix}, j = 1, \cdots, M$, for

$L(\tilde{g}_1, \cdots, \tilde{g}_m)$ and a $k$-dimensional manifold $M_{z_0}$ is obtained as

$$(15) \quad z(\hat{p}) = \tilde{G}_1(t_1) \circ \cdots \circ \tilde{G}_k(t_k)(z_0), \ \hat{p} \overset{\triangle}{=} (t_1, \cdots, t_k) \in D_k \overset{\triangle}{=} \prod_1^k (-a_i, a_i)$$

where $\tilde{G}_j(t)(z)$ is the local flow generated by $Z_j$.

The condition is Theorem 3.2.3 are fulfilled and the homomorphism $\lambda :$ $L(\tilde{g}_1, \cdots, \tilde{g}_m) \to L(q_1, \cdots q_m)$ is defined by $z(\hat{p}), \hat{p} \in D_k$, in (15), such that $q_i \in C^\infty(D_k; R^k)$ and

$$(16) \quad \begin{aligned} &\tfrac{\partial z}{\partial \hat{p}}(\hat{p}) q_i(\hat{p}) = \tilde{g}_i(z(\hat{p})), \ 1, \cdots, m, \\ &\dim L(q_1, \cdot, q_m)(\hat{p}) = k \ (\forall) \ \hat{p} \in D_k. \end{aligned}$$

By definition $z = (t; x), \tilde{g}_i(z) = \begin{pmatrix} f_i(z) \\ g_i(x) \end{pmatrix}$ and from (16) and $z(\hat{p}) \overset{\triangle}{=}$ $(\tau(\hat{p}), y(\hat{p}))$ we obtain

$$(17) \quad \begin{aligned} &\left\langle \tfrac{\partial \tau}{\partial \hat{p}}(\hat{p}), q_i(\hat{p}) \right\rangle = f_i(\tau(\hat{p}), y(\hat{p})), i = 1, \cdots, m, \\ &\tfrac{\partial y}{\partial \hat{p}}(\hat{p}) q_i(\hat{p}) = g_i(y(\hat{p})), \ \hat{p} \in D_k. \end{aligned}$$

Here $y(\hat{p}) = G_1(t_1) \circ \cdots \circ G_k(t_k)(x_0)$, $\hat{p} \in D_k$, and $G_j(t)(x)$ is the local flow generated by $Y_j$ in $Z_j = \begin{pmatrix} L_j \\ Y_j \end{pmatrix}$.

Now, $y(\hat{p}), \hat{p} \in D_k$, defines a $k$-dimensional integral manifold $M_{x_0}$ of $L(g_1, \cdots, g_m)$ and in particular

$$(18) \quad \frac{\partial y}{\partial t_i}(\hat{p}), \ i = 1, \cdots, k$$

are linearly independent in $R^n$, for any $\hat{p} \in D_k$. We are looking for a $C^\infty$ function $\hat{p} = v(x)$, $x \in V(x_0) \subseteq R^n$, such that

$$(19) \quad v(x) = \hat{p} \text{ provided } x = y(\hat{p}),$$

and $y(v(x)) = x$ for any $x \in V(x_0) \cap M_{x_0}$. In this respect, consider $G(\hat{p}; x) \overset{\triangle}{=}$ $G_1(t_1) \circ \cdots \circ G_k(t_k)(x)$, for $x \in V(x_0), \hat{p} \in D_k$ and notice that

$$(20) \quad \frac{\partial G}{\partial t_i}(\hat{p}; x_0) = \frac{\partial y}{\partial t_i}(\hat{p}), \ \frac{\partial G}{\partial x}(0; x) = I_n, \ i = 1, \cdot, k, \ (\forall) \ \hat{p} \in D_k.$$

For the sake of simplicity define $\hat{x} = (x_{k+1}, \cdot, x_n), w \overset{\triangle}{=} (\hat{p}, \hat{x}) \in R^n$, and consider that $F(w) \overset{\triangle}{=} G(\hat{p}; x_{01}, \cdots, x_{0k}, x_{k+1}, \cdot, x_n)$ fulfils

$$(21) \quad \begin{aligned} &\det \frac{\partial F}{\partial w}(w_0) \neq 0 \\ &\text{where} \\ &w_0 = (0, \hat{x}_0) == (0, x_{0k+1}, \cdot, x_{0n}) (\text{see}(18) + (20)). \end{aligned}$$

Then we find a unique $C^\infty$ function $w(x)$, $x \in V(x_0) \subseteq R^n$, fulfilling $w(x) = (v(x), u(x))$ (see (19))

$$(22) \quad F(w(x)) = x, \ x \in V(x_0), y(v(x)) = x \quad \text{provided } x \in V(x_0) \cap M_{x_0}$$

and

$$(23) \qquad v(y(\widehat{p})) = \widehat{p}, \ u(y(\widehat{p})) = \widehat{x}_0, \ (\forall) \ y(\widehat{p}) \in V(x_0)$$

Taking derivatives in (23) we obtain that $v \overset{\triangle}{=} (v^1, \cdots, v^k)$ fulfils

$$(24) \qquad\qquad \frac{\partial v_i}{\partial x}(y(\widehat{p})) \frac{\partial y}{\partial t_i}(\widehat{p}) = 1 \ (\forall) \ i \in \{1, \cdot, k\}$$

$$\frac{\partial v^j}{\partial x}(y(\widehat{p})) \frac{\partial y}{\partial t_i}(\widehat{p}) = 0, \ \text{if } i \neq j$$

Define $t(x) = \tau(v(x))$, $x \in V(x_0)$, where $\tau$ fulfils (17). Using (24) and (17) we get

$$(25) \qquad \left\langle \frac{\partial t}{\partial x}(x), g_i(x) \right\rangle = \left\langle \frac{\partial \tau}{\partial p}(v(x)) \frac{\partial v}{\partial x}(x), \ g_i(x) \right\rangle$$
$$i = 1, \cdots, m \ \text{ forany } x \in M_{x_0} \cap V(x_0)$$

and (17) is rewriten

$$(26) \qquad \left\langle \frac{\partial t}{\partial x}(x), g_i(x) \right\rangle = f_i(t(x), x), \quad i = 1, \cdot, m,$$
$$(\forall) \ x \in M_{x_0} \cap V(x_0)$$

The proof is complete.

**Remark 2.**
*A nontrivial $C^\infty$-solution $u(x)$, $x \in V(0) \subseteq R^n$, for a semiliniar elliptic equation*

$$\sum_{i=1}^{n} \frac{\partial^2 u}{\partial x_i^2} \overset{\triangle}{=} \Delta u = n f(u) f'(u) \ \left( f \in C^\infty(R; R), f'(u) = \frac{df}{du}(u) \right)$$

*is obtained applying Frobenius theorem to the system* $\dfrac{\partial u}{\partial x_i} = f(u), \ i = 1, \cdots, n, \ u(0) = u_0.$

*Applying Theorem 3.2.3 for some commuting $g_i \in \mathbf{F}_n$ (Der $(R^n)$), $i = 1, \cdots, m$, we can obtain a nontrivial solution $u(x)$, $x \in V(x_0) \cap M_{x_0}$, for the system*

(*)
$$\left\langle \frac{\partial u}{\partial x}(x), g_i(x) \right\rangle = f(u), \ i = 1, \cdots, m$$

*where $M_{x_0}$ is an integral manifold of $L(g_1, \cdots, g_m)$. Then the solution $u(x)$, $x \in V(x_0) \cap M_{x_0}$, of (*) will fulfil the following elliptic semiliniar equation*

(**)
$$\sum_{i,j=1}^{m} [(T_r \ g_i(x)g_j^T(x)\frac{\partial^2 u}{\partial x^2}(x))$$
$$+ \langle \frac{\partial u}{\partial x}(x), \frac{\partial g_i}{\partial x}(x)g_j(x)\rangle]$$
$$= m^2 f(u)f'(u) \ (\forall) \ x \in V(x_0) \cap M_{x_0}.$$

*where $f \in C^\infty(R; R)$ is arbitrarily fixed.*

# 4.2 Stochastic Differential Equations

We shall consider nonlinear stochastic differential equations of diffusion type where the input Wiener process appears in a multiplicative form.

An analytic and deterministic description of the input (Wiener process) output (solution) mapping is associated with a stochastic differential equations and as one may expect, the simplest case is when the input Wiener process can be neglected by a translation of the solution. It is not the case when the drift and diffusion part are defined by some noncommuting vector field, $g_0, g_1, \cdot, g_m$, and the Lie algebra $L(g_0, g_1, \cdots, g_m)$ plays the key role in describing the input–output mapping. In particular, the dependence of the stochastic flow on the Cauchy data $x(0) = x_0 \in R^n$ can be easily seen provided the Lie algebra $L(g_0, g_1, \cdots, g_m)$ is finitely generated.

A stochastic differential equation of diffusion type is defined

(1)
$$dx = g_0(x)dt + \sum_{i=1}^{m} g_i(x) \circ dw^i,$$
$$t \in [0, T], \ x(0) = x_0 \in R^n$$

where $w(t), t \geq 0$, is a standard $m$-dimensional. Wiener process over the probability space $\{\Omega, \mathcal{F}, P\}$ (see appendix a.3) and $\bullet$ in the stochastic part means Stratonovich–Fisk integral.

For the time being we accept that $g_i \in C^\infty(R^n; R^n) = F_n$, $i = 0, 1, \cdot, m$, and we know that (1) has a unique solution $x(t, \omega) : [0, T] \times \Omega \to R^n$ for each $x_0 \in R^n$ fixed which is a measurable process with continuous trajectories $x(\cdot, \omega) \in C([0, T]; R^n)$ for $\omega \in \Omega$.

In addition, the solution $x(t, \omega)$, is nonanticipating with respect to the $\sigma$-algebras $\mathcal{F}_t \subseteq \mathcal{F}$, $t \geq 0$, determined by $\mathcal{F}_t = \sigma(w(s) : 0 \leq s \leq t))$ i.e. $x(t, \bullet)$ is $\mathcal{F}_t$-measurable for each $t \in [0, T]$ (see a.3).

The unique solution in (1) is viewed as the limit in $L_2(\Omega; P)$ of a sequence of the nonanticipative solutions $\{x_\varepsilon(t, \omega))\}_{\varepsilon \downarrow 0}$ for the associated control system

$$(2) \qquad \frac{dx}{dt} = g_0(x) + \sum_{j=1}^{m} g_j(x)\frac{dv_\varepsilon^j}{dt}(t, \omega), \ x(0) = x_0, t \in [0, T]$$

where $v_\varepsilon(t, \omega)$ is the Langevin's approximation of the Wiener process $w(t)$ (see a.3).

A good integral representation of the solution in (1) is obtained provided the analytical description of the input $\left(\frac{dv_\varepsilon}{dt}(t)\right)$ output (solution $x_\varepsilon(t)$) mapping in (2) depends only on the given vector fields $g_i \in \mathbf{F}_n$. It is why the Lie algebra $L(g_0, g_1, \cdots, g_m)$ and gradient systems in $L(g_0, g_1, \cdots, g_m)$ are the main tools for describing the mapping and local solutions in (1) will be represented under the following assumptions.

## Assumption 1.

The Lie algebra $L(g_0, g_1, \cdots, g_m) \overset{\triangle}{=} \Lambda$ is finitely generated over $C^\infty(R^n)$ (or of the finite type) i.e. there exist $\{Y_1, \cdots, Y_M\} \subseteq \Lambda$ such that any $Y \in \Lambda$ can be written $Y(x) = \sum_{j=1}^{M} a_j(x)Y_j(x)$, with $a_j \in C^\infty(R^n)$ depending on $Y; \{Y_1, \cdots, Y_M\}$ is a system of generators.

For a fixed system of generators $\{Y_1, \cdots, Y_M\} \subseteq \Lambda$ consider the composition of flows

$$(3) \qquad G(p; x) = G_1(t_1) \circ \cdots \circ G_M(t_M)(x), \ p = (t_1, \cdots, t_M)$$

where $G_i(t)(x)$ is the local flow generated by $Y_i$. The function in (3) exists only locally and it is assumed that $G(p; x)$ for $p \in D_M = \prod_{1}^{M}(-a_i, a_i)$ and $x \in V(x_0)$ is well defined.

In the case that $Y_i$ is a complete vector field for $i = 1, \cdots, M$, then $D_M = R^M$, $V(x_0) = R^n$. The complexity of a diffusion type equation is preserved even for a finite dimensional Lie algebra $\Lambda$.

**Assumption 2.**

The Lie algebra $\Lambda = L(g_0, g_1, \cdots, g_m)$ is a finite dimensional space and there exists a basis $\{Y_1, \cdots, Y_M\} \subseteq \Lambda$ of complete vector fields.

In what follows $\{v_\varepsilon(t, w)\}_{\varepsilon > 0}$ is the Langevin's approximation of the Wiener process $w(t, w)$ (see a.3).

**Lemma 1.**

Let $g_i \in C^\infty(R^n; R^n)$ $i = 0, 1, \cdots, m$, be given such that the assumption 2 is fulfilled with $\dim \Lambda = M$.

Then there exist $q_i \in C^\omega(R^n; R^n)$, $i = 0, 1, \cdots, m$, such that

(a) $\quad \dfrac{\partial G}{\partial p}(p; x) q_i(p) = g_i(G(p; x))$, $i = 0, 1, \cdots, m$, $p \in R^M$, $x \in R^n$.

(b) $\qquad\qquad$ if $p_\varepsilon(t, w)$, $t \in [0, \mathrm{T}(w)]$, is the solutionof

$$\frac{dp}{dt} = q_0(p) + \sum_1^m q_i(p) \frac{dv_\varepsilon^i}{dt}(t, w), p(0) = 0, \text{ then}$$

$x_\varepsilon(t, w; x_0) = G(p_\varepsilon(t, w); x_0)$, $t \in [0, T(w)]$, is the solution in (2) for each $x_0$, where $G(p; x)$ is defined in (3).

**Proof.**

By hypotheses, the conditions in Theorem 2.2.1 are fulfilled and conclusion (a) is obtained provided we notice that $G(p; x)$ in (3) is the solution in a gradient system

$$\frac{\partial y}{\partial t_1} = Y_1(y), \frac{\partial y}{\partial t_2} = X_2(t_1; y), \cdots, \frac{\partial y}{\partial t_M} = X_M(t_1, \cdot, t_{M-1}; y).$$

The conclusion (b) is obtained from (a) by a straight computation. The proof is complete.

**Remark 1.**

The solution of (2) is represented in Lemma 1 using a globally defined function $G(p; x)$ and a locally defined solution $p_\varepsilon(t)$ with $p_\varepsilon(0) = 0$. It is easily seen that the mapping $x_0 \to x_\varepsilon(t, w; x_0)$ from $R^n$ to $R^n$ is a smooth one and completely determined by the smoothness of $G(p; x)$ with respect to $x \in R^n$.

As far as the representation for solutions in (1) is involved we shall keep fixed the function $G(p; x)$ in (3) and use a local solution for the following stochastic differential equation

$$(4) \qquad dp = q_0(p)dt + \sum_1^m q_i(p) \circ dw^i(t), \ p(0) = 0, \ t \geq 0$$

where $q_i \in C^\omega(R^M; R^M)$ are defined in Lemma 1. The vector fields $q_i$ in (4) are not global Lipschitz continuous functions and to define a local solution in (3) we need to replace $q_i$ by a $C^\infty$ and bounded function

$$(5) \qquad \widetilde{q}_i(p) = \widetilde{\alpha}(p)q_i(p), \ i = 0, 1, \cdots, m.$$

Here $\widetilde{\alpha}(p)$, $p \in R^M$, is a scalar $C^\infty$ function such that $\widetilde{\alpha}(p) = 1$ for $p \in S_N(0), \widetilde{\alpha}(p) = 0$ for $p \in R^M \setminus S_{2N}(0)$ and $0 \leq \widetilde{\alpha}(p) \leq 1$ for any $p \in R^M$, where $S_\rho(0) \subseteq R^M$ stands for a ball with the radius $\rho$, and $N$ is a fixed natural number. Let $\widetilde{p}(t)$, $t \geq 0$, be the solution in

$$(6) \qquad dp = \widetilde{q}_0(p)dt + \sum_1^m \widetilde{q}_i(p) \circ dw^i(t), \ p(0) = 0$$

and define a stopping time

$$(7) \qquad \widetilde{\tau}(\omega) = \inf\{t \geq 0; \|\widetilde{p}(t; \omega)\| \geq N\}$$

Then we obtain (see a.3)

$$(8) \qquad \widetilde{p}(t \wedge \widetilde{\tau}) = \int_0^{t \wedge \widetilde{\tau}} q_0(\widetilde{p}(s))ds + \sum_{i=1}^m \int_0^{t \wedge \widetilde{\tau}} q_i(\widetilde{p}(s)) \circ dw^i(s)$$

for any $t \geq 0$, where $\widetilde{\tau}$ is defined in (7).

Let $\widehat{C}(t, \omega)$, $t \geq 0$, be an $\{\mathcal{F}_t\}$ nonanticipating function defined as

$$(9) \qquad \widehat{C}(t, \omega) = \begin{cases} 1 & \text{if} \ \widetilde{\tau}(\omega) > t \\ 0 & \text{if} \ \widetilde{\tau}(\omega) < t \end{cases}$$

and the solution in (8)can be rewriten in a more suitable form

$$\tilde{p}(t \Lambda \tilde{\tau}) = \int_0^t \widehat{C}(s, w) q_0(\tilde{p}(s)) ds$$

$$+ \sum_{i=1}^m \int_0^t \widehat{C}(s, w) q_i(\tilde{p}(s)) \circ dw^i(s)$$

(10)

$$= \int_0^t \widehat{C}(s, w) q_0(\tilde{p}(s) \Lambda \tilde{\tau})) ds$$

$$+ \sum_1^m \int_0^t \widehat{C}(s, w) q_i(\tilde{p}(s \Lambda \tilde{\tau})) \circ dw^i(s)$$

That is to say, $\widehat{p}(t) \overset{\triangle}{=} \tilde{p}(t \Lambda \tilde{\tau})$, $t \in [0, T]$, fulfils

$$\widehat{p}(t) = \int_0^t \widehat{C}(s, w)[q_0(\widehat{p}(s))$$

(11)

$$+ \frac{1}{2} \sum_1^m \left( \frac{\partial q_i}{\partial p} q_i \right)(\widehat{p}(s))] ds$$

$$+ \int_0^t \widehat{C}(s, w) q_i(\widehat{p}(s)) dw^i(s),$$

where the Ito's stochastic integral is used, and $\widehat{C}(t, w)$ is defined in (9).

**Definition 1.**
We say that a nonanticipating process $x(t)$, $t \in [0, T]$, and a stopping time $\tau : \Omega \to [0, \infty)$ is a local solution in (1) if

$$x(t \Lambda \tau) = x_0 + \int_0^{t \Lambda \tau} g_0(x(s)) ds$$

($\alpha$)

$$+ \sum_1^m \int_0^{t \Lambda \tau} g_i(x(s)) \circ dw^i(s), \qquad t \in [0, T]$$

and

$$\|g_j(x(t))\| \le b, \ 0 \le t \le T \Lambda \tau(w).$$

*for some constant b > 0.*

**Lemma 2.**

*Assume that the hypotheses in Lemma 1 are fulfilled and for a fixed natural number $N$ define $\widetilde{p}(t), t \geq 0, \widetilde{\tau}$, the solution in (6) and respectively a stopping time in (7).*

*Then $\widetilde{x}(t; x_0) = G(\widetilde{p}(t); x_0)$ and $\widetilde{\tau}, t \in [0,T]$, is a local solution in (1).*

**Proof.**

By definition, $\widetilde{x}(t \wedge \widetilde{\tau}; \ x_0) = G(\widetilde{p}(t \wedge \widetilde{\tau}; \ x_0)$, $t \in [0,T]$, and using (7) we obtain $\|\widehat{p}(t \wedge \widetilde{\tau})\| \leq N$ for any $t \in [0,T], \omega \in \Omega$. On the other hand $\widehat{p}(t) = \widetilde{p}(t \wedge \tau)$, $t \in [0,T]$, fulfils the stochastic equation (11) and applying Ito's stochastic rule of differentiation we obtain that $\widehat{x}(t; x_0) \stackrel{\triangle}{=} \widetilde{x}(t \wedge \widetilde{\tau}; x_0)$, $t \in [0,T]$, obeys

$$
\begin{aligned}
\widehat{p}(t; x_0) \ &= x_0 \ + \int_0^t \frac{\partial G}{\partial p}(\widehat{p}(s); x_0) d\widehat{p}(s) \\
(12) \qquad &= x_0 \ + \int_0^t \widehat{C}(s) \frac{\partial G}{\partial p}(\widehat{p}(s); x_0) q_0(\widehat{p}(s)) ds \\
&\quad + \sum_{i=1}^m \int_0^t \widehat{C}(s) \left( \frac{\partial G}{\partial p}(\widehat{p}(s); x_0) q_i(\widehat{p}(s)) \right) \circ dw^i(s).
\end{aligned}
$$

Using Lemma 1 we notice that

$$
(13) \qquad \frac{\partial G}{\partial p}(\widehat{p}(s); x_0) q_j(\widehat{p}(s)) = g_j(G(\widehat{p}(s); x_0)), \ j = 0, 1, \cdots, m.
$$

and (12) can be rewriten

$$
\begin{aligned}
\widehat{x}(t; x_0) = \ &x_0 + \int_0^t \widehat{C}(s) g_0(\widehat{x}(s; x_0)) ds \\
(14) \qquad &+ \sum_{i=1}^m \int_0^t \widehat{C}(s) g_i(\widehat{x}(s; x_0)) \circ dw^i(s)
\end{aligned}
$$

or

$$
\begin{aligned}
\widetilde{x}(t \wedge \widetilde{\tau}; x_0) = \ &x_0 + \int_0^{t \wedge \widetilde{\tau}} g_0(\widetilde{x}(s; x_0)) ds \\
(15) \qquad &\sum_{i=1}^m \int_0^{t \wedge \widetilde{\tau}} g_i(\widetilde{x}(s; x_0)) \circ dw^i(s)
\end{aligned}
$$

(∀) $t \in [0, T]$ where $\widehat{C}(s, \omega)$ is defined in (9).

The proof is complete.

The representation of solutions in Lemma 2, allow one to aproximate a local solution in (1) by non–anticipating solutions of ordinary differential equations as follows. For a given non–anticipative and bounded process $u(t, \omega); [0, T] \times \Omega \to [0, 1]$ consider

(16)
$$\frac{dx}{dt} = u(t, \omega)g_0(x) + u(t, \omega) \sum_{i=1}^{m} g_i(x)\frac{dv_\varepsilon^i}{dt}(t, \omega),$$
$$t \in [0, T], x(0) = x_0.$$

where $v_\varepsilon(t, \omega)$ is the Langevin approximation of the Wiener process $w(t, \omega)$. Each pair $x(t; x_0)$, $u(t))$, $t \in [0, T]$, fulfilling (16) is called a solution of (16).

**Lemma 3.**

*Assume the hypotheses in Lemma 1 fulfilled and for a fixed natural $N$ define $(\tilde{p}(t), \tilde{\tau})$, $t \in [0; T]$, a local solution for (4) (see (7) and (8)). Then $\tilde{x}(t; x_0), \tilde{\tau})$, $t \in [0, T]$, is a local solution in (1), where $\tilde{x}(t; x_0) = G(\tilde{p}(t); x_0)$, and $G$ is given in (3). In addition, there exists $\{x_\varepsilon(t; x_0), u_\varepsilon(t) : t \in [0, T]\}_{\varepsilon > 0}$ solutions in (16) such that*

$$\lim_{\varepsilon \to 0} E\|x_\varepsilon(t; x_0) - \tilde{x}(t\Lambda\tilde{\tau}; x_0)\|^2 = 0, \ (\forall) \ t \in [0, T]$$

*uniformly with respect to $x_0$ in a bounded set.*

**Proof.**

Applying Lemma 2 we notice that $\{\tilde{x}(t; x_0), \tilde{\tau}\}$, $t \in [0, T]$, is a local solution of (1). On the other hand, by definition $\tilde{q}_j(p) = \tilde{\alpha}(p)q_j(p)$, $j = 0, 1, \cdots, m$, (see (5)) and $\tilde{p}(t)$, $t \in [0, T]$, is the solution in (6) which can be rewriten as (8), i.e. $(\tilde{p}(t), \tilde{\tau})$, $t \in [0, T]$, is a local solution for (4) fulfilling $\|\tilde{p}(t)\| \leq N$ (∀) $0 \leq t \leq T\Lambda\tilde{\tau}$. Now using (a.3) we approximate $\tilde{p}(t)$, $t \in [0, T]$, by solutions $\{p_\varepsilon(t), t \in [0, T], \varepsilon > 0\}$ of ordinary differential equations

(17)
$$\frac{dp}{dt} = \tilde{q}_0(p) + \sum_{i=1}^{m} \tilde{q}_i(p)\frac{dv_\varepsilon^i}{dt}(t), \ p(0) = 0, \ t \in [0, T].$$

and we obtain

(18)
$$\lim_{\varepsilon \to 0} E\|p_\varepsilon(t) - \tilde{p}(t)\|^2 = 0, \ \|p_\varepsilon(t)\| \leq 2N, \ t \in [0, T]$$
$$\lim_{\varepsilon \to 0} E\|p_\varepsilon(t\Lambda\tilde{\tau}) - \tilde{p}(t\Lambda\tilde{\tau})\|^2 = 0, \ t \in [0, T]$$

Using $\|\tilde{p}(t\Lambda\tilde{\tau})\| \leq N$ (∀) $t \in [0, T]$, we obtain that set $A_\varepsilon(t) \stackrel{\Delta}{=} \{\omega \in \Omega : \|p_\varepsilon(t\Lambda\tilde{\tau})\| > N\}$ fulfils

(19)
$$\lim_{\varepsilon \to 0} \mathbf{P} \, A_\varepsilon(t) = 0, \ t \in [0, T]$$

Denote $x_\varepsilon(t; x_0) = G(p_\varepsilon(t\Lambda\widehat{\varepsilon}); x_0)$, $t \in [0, T]$, and using (18) and $G$ is locally Lipschitz continuous with respect to $p$ we obtain the conclusion

$$(20) \qquad \lim_{\varepsilon \to 0} E\|x_\varepsilon(t; x_0) - \widetilde{x}(t\Lambda\widetilde{\tau}; x_0)\|^2 = 0, \ \ t \in [0, T]$$

Taking $u_\varepsilon(t, \omega) = \widehat{C}(t, \omega)\widetilde{\alpha}(p_\varepsilon(t, \omega))$, where $\widehat{C}$ is defined in (9) and $0 \le \widetilde{\alpha}(p) \le 1$ (see (15)), by a direct computation we obtain that $x_\varepsilon(t; x_0), t \in [0, T]$, fulfils (16).

The proof is complete.

As one might expect, if $\Lambda = L(g_0, g_1, \cdots, g_m)$ is finitely generated over $C^\infty(R^n)$ then the integral representation given in Lemma 2 has to be reconsidered. The main obstruction comes from the auxiliary system (see (4)) where the vector fields $q_j$ will not be globally defined any more and the Cauchy data $x(0) = x_0$ is entering the system (4).

For $p \in S_\rho(0) \subseteq R^M$ and $x_0 \in B \subseteq R^n$ (compact set) consider $q_j \in C^\infty(S_\rho(0) \times B; R^M)$ $j = 0, 1, \cdots, m$, and define stochastic differential equations

$$(21) \qquad dp = q_0(p; x_0)dt + \sum_{i=1}^{m} q_i(p; x_0) \circ dw^i(t), \ p(0) = 0, \ t \in [0, T].$$

$$(22) \qquad \begin{aligned} dp = \ & \widetilde{q}_0(p; x_0)dt \\ & + \sum_{i=1}^{m} \widetilde{q}_i(p; x_0) \circ dw^i(t) \qquad p(0) = 0, \end{aligned}$$

where $\widetilde{q}_j(p; x_0) = \widetilde{\alpha}(p)q_j(p; x_0)$ and $\widetilde{\alpha}$ isdefinedin(5)for $2N < \rho$.

**Lemma 4.**

*Suppose that the Lie algebra $\Lambda = L(g_0, g_1, \cdots, g_m)$ fulfils the assumption A.1 and there exists a system of generators $\{Y_1, \cdots, Y_M\} \subseteq \Lambda$ of complete vector fields. Then for a compact set $B \subseteq R^n$ there exist $S_\rho(0) \subseteq R^M$ and $q_j \in C^\infty(S_\rho(0) \times B; R^M)$ such that*

*c) $\widetilde{x}(t; x_0) \overset{\Delta}{=} G(\widetilde{p}(t; x_0); x_0)$ and the stopping time $\widetilde{\tau}$, $t \in [0, T], x_0 \in B$, is a local solution in (1), where $G$ is defined in (3), $\widetilde{p}(t; x_0), t \in [0, T]$, is the solution in (21) and*

$$\widetilde{\tau}(\omega; x_0) = \inf\{t \ge 0 : \|\widetilde{p}(t, \omega; x_0)\| \ge N\}$$

**Proof.**

For the given system of generators $\{Y_1, \cdots, Y_M\} \subseteq \Lambda$ and $x_0 \in B$ define $G(p; x_0)$ as in (3), i.e.,

$$(23) \qquad G(p; x_0) \overset{\Delta}{=} G_1(t_1) \circ \cdots \circ G_M(t_M)(x_0), p \in R^M.$$

The function in (23) is the solution in a gradient system

$$
\text{(24)} \quad
\begin{aligned}
\frac{\partial y}{\partial t_1} &= Y_1(y), \quad \frac{\partial y}{\partial t_2} = X_2(t_1; y), \cdots, \\
\frac{\partial y}{\partial t_M} &= X_M(t_1, \cdot, t_{M-1}; y), \qquad y(0) = x_0,
\end{aligned}
$$

and the algebraic representation for (24) is a nonsingular one only locally for $p \in S_\rho(0)$ uniformly with respect to $x_0 \in B$ (see Lemma 1.1.1). More precisely these holds

$$
\text{(25)} \quad
\begin{aligned}
&\{Y_1(y), \; X_2(t_1; y), \cdots, X_M(t_1, \cdots, t_{M-1}; y)\} \\
&\{Y_1(y), \cdots, Y_M(y)\} A(p; x_0), \text{ provided} \\
&y = G(p; x_0), \; p \in R^M, x_0 \in B, \text{ and } A(0; x_0) \\
&= I_M, \; A_{ij} \in C^\infty(R^M \times B)
\end{aligned}
$$

Define $S_\rho(0) \subseteq R^M$ such that $A(p; x_0)$ satisfies

$$
\text{(26)} \qquad \det A(p; x_0) \neq 0 \; (\forall) \; p \in S_\rho(0), \; x_0 \in B.
$$

and associate $q_j \in C^\infty(S_\rho(0) \times B; R^M)$ such that

$$
A(p; x_0) q_j(p; x_0) = a_j(p; x_0), \qquad j = 0, 1, \cdots, m,
$$

where $a_j(p; x_0) \in C^\infty(S_\rho(0) \times B; R^M)$ are found such that

$$
\text{(28)} \qquad g_j(G(p; x_0)) = \sum_{i=1}^{M} a_j^i(p; x_0) Y_i(G(p; x_0)), \; j = 0, 1, \cdots, m.
$$

Using (27) and (28) in (24) we obtain

$$
\text{(29)} \qquad \frac{\partial G}{\partial p}(p; x_0) q_j(p; x_0) = g_j(G(p; x_0)), \; j = 0, 1, \cdots, m,
$$

and by a direct computation we conclude that $\tilde{x}(t; x_0) \stackrel{\triangle}{=} G(\tilde{p}(t; x_0); x_0)$ and $\tilde{\tau}, \; t \in [0, T]$ is a local solution in (1).
The proof is complete.

## 4.3 Systems of Hyperbolic equations

For $g_i \in \mathbf{F}_n(\vec{g}_i \in \text{Der}(R^n))$ a linear system is defined

$$
\text{(1)} \qquad \left\langle \frac{\partial S}{\partial x}(x), g_i(x) \right\rangle = 0, i = 1, \cdots, m
$$

and a nontrivial solution can be found provided

(2)                          $\dim L(g_1, \cdot, g_m)(x_0) = k < n$

and $\Lambda = L(g_1, \cdot, g_m)$ is(f.g.o; $x_0$) for some $x_0 \in R^n$ fixed (see Proposition 3.3.2)

In addition, the domain of (1) is a $k$-dimensional integral manifold $M_{x_0}$ of $\Lambda$ and it is described analytically by

(3)          $M_{x_0} \overset{\triangle}{=} \{y \in R^n : y = y(\widehat{p}; x_0), \widehat{p} \in D_k \overset{\triangle}{=} \prod_1^k (-a_i, a_i)\}$

where

(4)      $y(\widehat{p}; x_0) = G_1(t_1) \circ \cdots \circ G_k(t_k)(x_0), \qquad \widehat{p} = (t_1, \cdots, t_k) \in D_k.$

Here $G_i(t)(x)$ is the local flow generated by an $Y_i$ of a $k$-minimal system of generators $\{Y_1, \cdots, Y_M\} \subseteq \Lambda$.

A Cauchy problem for (1) makes sense if the degenaracy in (2) and the $k$-minimal system of generators in (4) are preserved for any $x_0 \in \Omega \subseteq R$, where $\Omega$ is a bounded open conex set. More precisely, we assume that:

h) there exist a domain $\Omega \subseteq R^n$ and a system of generators $\{Y_1, \cdots, Y_M\} \subseteq \Lambda$ such that $\dim \Lambda(x_0) = k$ and $\{Y_1(x_0), \cdot, Y_k(x_0)\} \subseteq R^n$ are linearly independent for any $x_0 \in \Omega$, where $k < n$ is fixed.

Assuming (h), a Cauchy condition will be determined by a scalar function $\varphi \in C^\infty(R^d)$, where $d = n - k$, and any $x \in \Omega$ can be written as

(5)                      $x = \widehat{x} + w, \text{ where } \widehat{x} \in R^k, \text{ w} \in R^{n-k}$

For a fixed $x_0 \in \Omega$, there exists an open set $W(0) \subseteq R^{n-k}$ such that $x_0 + W(0) \subseteq \Omega$.

A Cauchy problem $C.P(x_0, W(0); \varphi)$ for (1) means to find a solution $S$ in (1) fulfilling $S(x_0 + w) = \varphi(w)$ for any $w \in W(0)$, where $x_0 \in \Omega, \varphi \in C^\infty(R^{n-k})$, and the open set $W(0) \subseteq R^{n-k}$ are given.

**Proposition 1.**
Let $g_i \in \mathbf{F}_n$ and $\Omega \subseteq R^n$ $i = 1, \cdots, m$ be given such that the condition (h) is fulfilled with $k = \dim L(g_1, \cdots, g_m)(x), x \in \Omega$. Then for $x_0 \in \Omega$ and $\varphi \in C^\infty(R^{n-k})$ there exists an open $W(0) \subseteq R^{n-k}$ such that the Cauchy problem $C.P(x_0, W(0); \varphi)$ has a unique solution for (1).

**Proof.**

By hypothesis the conditions in Proposition 3.3.2 are fulfilled for any $x_0 \in \Omega$ and in addition the domain $M_{x_0}$ of the system (1) is described explicitly by

$$(6) \qquad y(\widetilde{p}; x_0) = G_1(t_1) \circ \cdots \circ G_k(t_k)(x_0), \ \widehat{p} \in D_k = \prod_1^k (-a_i, a_i)$$

where $G_i(t)(x)$ is the flow generated by $Y_i$ given in (h). Let $x_0 \in \Omega$ be fixed and $\widehat{x}_0 \triangleq (x_{01}, \cdot, x_{0,k})$.

A nontrivial solution for (1) is found solving the equations.

$$(7) \qquad F(z) = y, \ \text{where } z = (\widehat{p}, x_{k+1}, \cdots, x_n), \ y \in V(x_0),$$

and

$$F(z) \triangleq y(\widehat{p}; \widehat{x}_0, x_{k+1}, \cdot, x_n)$$

From (7) we obtain a smooth mapping $z(y) = (\widehat{p}(y), x_{k+1}(y), \cdots, x_n(y)), \ y \in V(y_0)$ such that

$$(8) \qquad F(z(y)) = y \quad (\forall) \ y \in V(x_0).$$

$$(9) \qquad x_j(y(\widehat{p}; x_0 + w)) = x_{0j} + w_j, \ j = k+1, \cdot, n, \ (\forall) \ \widehat{p} \in D_k$$

where $w \triangleq (w_{k+1}, \cdots, w_n) \in W(0) \subseteq R^{n-k}$.

It shows that the system (1) has $(n-k)$ functionally independent solutions $x_j(x), \ x \in V(x_0), \ j = k+1, \cdots, n$ fulfilling the system (1) on the domain $M_{x_0+w}$ for each $w \in W(0)$. More precisely, the conditions in Theorem 3.2.3 (see Remark 3.2.3 also) are fulfilled and for each $w \in W(0)$ we obtain $q_i \in C^\infty(D_k; R^k), \ i = 1, \cdots, m$, such that

$$(10) \qquad \frac{\partial y}{\partial \widehat{p}}(\widehat{p}; x_0 + w)q_i(\widehat{p}) = g_i(y(\widehat{p}; x_0 + w)), \qquad i = 1, \cdots, m$$

Taking directional derivatives with respect to $q_i$ and applying (10) in (9) we obtain

$$(11) \qquad \left\langle \frac{\partial x_j}{\partial x}(y(\widehat{p}; x_0 + w)), g_i(y(\widehat{p}; x_0 + w)) \right\rangle = 0, \qquad i = 1, \cdots, m,$$

$$(\forall) \ \widehat{p} \in D_k \triangleq \prod_1^k (-a_i, a_i)$$

which means that $x_j(x)$, $j = k+1, \cdots, n$, fulfilling (9) are the solutions in
(1) on the domain

(12) $$M_{x_0+w} \overset{\triangle}{=} \{y \in R^n : y = y(\widehat{p}; x_0 + w), \widehat{p} \in D_k\}$$

for each $w \in W(0)$

Now, define $S(x) = \varphi(x_{k+1}(x) - x_{0k+1}, \cdots, x_n(x) - x_{0n})$, $x \in V(x_0)$ and
using (9) it follows that

(13) $$S(x_0 + w) = \varphi(w), \ w \in W(0) \subseteq R^{n-k}.$$

In addition $S(x), x \in V(x_0)$, fulfils the system (1) on the domain $M_{x_0+w}$,
i.e.,

(14) $$\left\langle \frac{\partial S}{\partial x}(x), g_i(x) \right\rangle = 0, \ \ i = 1, \cdots, m,$$

provided $x = y(\widehat{p}; x_0 + w)$, for $\widehat{p} \in D_k$.

The system (14) is obtained taking directional derivatives of $S(x)$ with
respect to $g_i(x)$, $i = 1, \cdots, m$, and using the system (11) for each $j = k+1, \cdots, n$.

The proof is complete.

**Remark.**

*The unique solution for the Cauchy problem $C.P(x_0, W(0); \varphi)$ can be
written $S(x) = \varphi(x_{k+1}(x) - x_{k+1}^0, \cdots, x_n(x) - x_n^0)), x \in V(x_0)$ where the
$(n-k)$ functionally independent solutions $x_j(x), j = k+1, \cdots, n$, of (1)
are defined in (8) and (9).  Let $\widehat{S}(x), x \in V(x_0)$, be an arbitrary solution of $C.P(x_0, W(0); \varphi)$.  Then $\widehat{S}(y(\widehat{p}; x_0 + w)) = \Psi(w)$, for some $\Psi \in
C^\infty(W(0))$, $(\forall) \widehat{p} \in D_k$, and in particular, for $\widehat{p} = 0$ we obtain*

$$\widehat{S}(x_0 + w) = \varphi(w) = \Psi(w), \ w \in W(0).$$

Therefore, $\widehat{S}(y(\widehat{p}; x_0 + w)) = \varphi(w), \ w \in W(0)$ and using the solution

$$(\widehat{p}(y), x_{k+1}(y), \cdots, x_n(y)), y \in V(x_0),$$

in (8) we obtain

$$\widehat{S}(y) = \varphi(x_{k+1}(y) - x_{k+1}^0, \cdots, x_n(y) - x_n^0), \ \ y \in V(x_0)$$

where $x_0 \overset{\triangle}{=} (x_1^0, \cdots, x_k^0, x_{k+1}^0, \cdots, x_n^0)$.

**Example 1.**

A linear system of hyperbolic equations is defined

(α)
$$-\frac{\partial \mathcal{U}}{\partial t}(t, x) + A_1 \frac{\partial \mathcal{U}}{\partial x_1}(t, x) + \cdots + A_n \frac{\partial \mathcal{U}}{\partial x_n}(t, x) = 0$$

where $\mathcal{U}(t, x) \in R^m$, $x \overset{\triangle}{=} (x_1, \cdots, x_n) \in R^n$, and $A_i$ is an $(m \times m)$ constant matrix, $i = 1, \cdots, n$.

We shall confine ourselves to consider solution $\mathcal{U}(t, x) = (\mathcal{U}_1(t, x), \cdots, \mathcal{U}_m(t, x))$ with the property $\mathcal{U}_i(t; x) = S(t, x) + k_i, k_i \in R$, for any $i = 1, \cdots, m$, and the system (α) can be rewriten as

(β)
$$-\frac{\partial S}{\partial t}(t, x) + \left\langle b_i, \frac{\partial S}{\partial x}(t, x) \right\rangle = 0, \quad i = 1, \cdots, m,$$

where the constant vector $b_i = (b_i^1, \cdots, b_i^n) \in R^n$ is defined by

$$b_i^j = \sum_{k=1}^{m} (A_j^T e_i)^k, \quad j = 1, \cdots, n,$$

and $e_1, \cdots, e_m \in R^m$ is the canonical basis.

Denote $g_i = \begin{pmatrix} -1 \\ b_i \end{pmatrix} \in R^{n+1}$, $y = (t, x) \in R^{n+1}$, and (β) fulfils the conditions in Proposition 1, i.e.,

(γ)
$$\left\langle \frac{\partial S}{\partial y}(y), g_i \right\rangle = 0, \quad i = 1, \cdots, m$$

where

$$\dim L(g_1, \cdots, g_m)(y) = \dim \text{ span } \{b_1, \cdots, b_m\} = k < n + 1, \quad y \in R^{n+1}$$

There is not much loose of generality if we assume $m \le n$, $\{b_1, \cdots, b_m\} \subseteq R^n$ are linearly independent and we consider the following situations:

I) $m = n$, $d = 1$, $\varphi \in C^\infty(R)$

II) $m < n$, $d = n + 1 - m$, $\varphi \in C^\infty(R^d)$

In the case $m = n$ there is a unique $g_0 = \begin{pmatrix} 1 \\ c \end{pmatrix} \in R^{n+1}$ such that

(1)
$$\langle c, b_i \rangle = 1, \quad i = 1, \cdot, n$$

For each $y_0 = (w_0, 0) \in R^{n+1}$ let $G(p; w_0) \stackrel{\triangle}{=} G_1(t_1) \circ \cdots G_n(t_n)(y_0)$, $p = (t_1, \cdot, t_n) \in R^n$, be the integral manifold associated with $\gamma$, i.e. $G(p; w_0) = y_0 + \sum_{i=1}^{n} t_i g_i$.

For arbitrary $y = (t, x) \in R^{n+1}$ we obtain the solution $(p(y), w(y))$ of $G(p(y); w(y)) = y$, i.e.,

$$w(y) = t + \sum_{1}^{n} t_i(x), \quad \sum_{1}^{n} t_i(x) b_i = x$$

and

(2)                                      $w(y) = t + \langle c, x \rangle, \quad x \in R^n, \ t \in R,$

where $c \in R^n$ fulfils (1). It is easily seen that $w(G(p; w_0)) = w_0$, $p \in R^n$, i.e., $w(y)$ is a solution in $(\gamma)$. For $y_0 = 0 \in R^{n+1}$ and given $\varphi \in C^\infty(R)$ we obtain

(3)                                   $S(t, x) = \varphi(t + \langle c, x \rangle), \quad (t, x) \in R^{n+1}$

as the unique solution for C.P. $(0, R; \varphi)$ of $(\beta)$.

In the general case $m < n$, we find $v_j = \begin{pmatrix} 1 \\ c_j \end{pmatrix} \in R^{n+1}$ $j = 0, 1, \cdots, d-1$,

such that $c_0 \in$ span $\{b_1, \cdots, b_m\} \stackrel{\triangle}{=} X, c_j \notin X$, $j = 1, \cdots, d-1$, and

(4)
$$\langle c_0, b_i \rangle = 1, \ i = 1, \cdots, m, \ \langle c_j, b_i \rangle = 0, \ j = 1, \cdots, d-1,$$
$$i = 1, \cdots, m, \ \langle c_{j_1}, c_{j_2} \rangle = 0 \text{ if } j_1 \neq j_2.$$

By definition $\{c_1, \cdots, c_{d-1}, b_1, \cdots, b_m\} \subseteq R^n$ is a base and for each $y_0 = \left( w_0, \sum_{j=1}^{d-1} w_j c_j \right) \in R^{n+1}$, let $G(p; w_0, w_1, \cdots, w_{d-1}) \stackrel{\triangle}{=} G_1(t_1) \circ \cdots \circ G_m(t_m)(y_0)$
be the $m$–dimensional integral manifold associated with $(\gamma)$, i.e.,

$$G(p; w_0, w_1, \cdots, w_{d-1}) = y_0 + \sum_{i=1}^{n} t_i g_i$$

For an arbitrary $y = (t, x) \in R^{n+1}$ we obtain the solution $(p(y), w(y))$ of $G(p(y); w(y)) = y$, i.e.,

(5)                          $\sum_{1}^{m} t_i(x) b_i = x_1, \quad w_0(y) = t + \sum_{1}^{m} t_i(x)$

$$w_j(y) = \langle c_j, x \rangle, j = 1, \cdots, d-1, \text{ where } x_1 \stackrel{\triangle}{=} \text{pr}_X x$$

Using (4) we rewrite

(6)    $w_0(y) = t + \langle c_0, x_1 \rangle$, where $x_1 \overset{\triangle}{=} \mathrm{pr}_X x$, $y = (t, x) \in R^{n+1}$

and it is easily seen that

(7)    $w_j(G(p; w_0, \cdots, w_{d-1})) = w_j$, $p \in R^m$, $j = 0, 1, \cdots, d-1$

Therefore $w_j(y)$, $j = 0, 1, \cdots, d-1$, in (5) are the functionally independent solutions in $(\gamma)$ and $(\beta)$.

Now, for $y_0 = 0 \in R^{n+1}$ and given $\varphi \in C^\infty(R^d)$ we obtain

(8)    $S(t, x) = \varphi(t + \langle c_0, x_1 \rangle, \langle c_1, x \rangle, \cdots, \langle c_{d-1}, x \rangle)$,

$x_1 = \mathrm{pr}_X x$, $(t, x) \in R^{n+1}$ as the unique solution for C.P. $(0, R^d; \varphi)$ of $(\gamma)$ and $(\beta)$.

## 4.4  Finite–Dimensional Nonlinear Filters

An optimal estimation of the state $x(t)$ of a nonlinear stochastic system given past observations $y^t \overset{\triangle}{=} \{y(s) : 0 \le s \le t\}$ is obtained as the conditional mean $\hat{x}(t) = E\{x(t)/y^t\}$. The dynamic of $\hat{x}(t), t \in [0, T]$, is usually determined by an Itô stochastic differential equation, where the drift and diffusion parts are non–anticipating process with respect to the given filtration $\{y^t, t \in [0, T]\}$, i.e. $\hat{x}(t)$, $t \in [0, T]$, is not determined by a recursive stochastic differential equation. More precisely, we consider a system of the form

(1)    a)   $dx = f(x)dt + \sum_1^m g_i(x)dw_i(t)$  $x(0) = x_0$, $x \in R^n$

    b)   $dy = h(x)dt + dv(t)$    $y(0) = 0$, $y \in R^d$

where $(w, v)$ is a standard $(m+d)$-dimensional Wiener process on the probability space $\{\Omega, \mathcal{F}, P\}$, and $f, g_i, h$ are vector valued functions.

The dynamic of $\hat{x}(t) = E\{x(t)/y^t\}$, $t \in [0, T]$, is based on (1) and usually has the form

(2)
$$d\hat{x}(t) = [\widehat{f}(x(t) - (\widehat{x(t)h^T}(x(t)) - \hat{x}(t)\widehat{h^T}(x(t)))\widehat{h}(x(t)]dt$$
$$- (\widehat{x(t)h^T}(x(t)) - \hat{x}(t)\widehat{h^T}(x(t)))dy(t), \ t \in [0, T],$$

where $\Lambda$ denotes conditional mean given past observations $y^t$.

Assuming that the conditional probability density $p(t, x)$ of $x(t)$ given $y^t$ exists, it fulfils the stochastic partial differential equation

(3)                 $dp = \mathcal{L}p dt + p(h(x) - \widehat{h}(x)(t))(dy(t) - \widehat{h}(x(t))dt)$

where

$$\mathcal{L}(\varphi) = -\sum_{i=1}^{m} \frac{\partial}{\partial x_i}(\varphi f_i) + \frac{1}{2}\sum_{i=1}^{n}\sum_{j=1}^{n} \frac{\partial^2(\varphi a_{ij})}{\partial x_i \partial x_j}$$

is the forward diffusion operator, $a(x) = G(x)G^T(x)$, $G(x) = (g_1(x), \cdots, g_m(x))$,
$f(x) = (f_1(x), \cdots, f_n(x))$.

Both the equation (2) and (3) are not recursive and appears to involve an infinite–dimensional computation in general. On the other hand, the Zakai equation for an unnormalized conditional density $\rho(t, x)$ corresponding to the scalar $y(t)$

(4)                 $d\rho(t, x) = \mathcal{L}\rho(t, x)dt + h(x)\rho(t, x)dy(t), \ t \in [0, T],$

is much simpler than (3) (see (4) as a bilinear differential equation in $\rho$ with $y$ considered as the input).

In addition, it shows that the Lie algebraic and differential geometric techniques developed for finite–dimensional systems of this type may be brought to bear here.

It might be so if the dynamic of $\widehat{x}(t), t \in [0, T]$ can be written in a recursive and depending pathwise on $y(t), t \in [0, T]$.

It is the purpose of the following to write the dynamic of $\widehat{x}(t), t \in [0, T]$, depending pathwise on $y(t), t \in [0, T]$, and to use differential geometric techniques for describing the input $(y(\cdot))$ output $(\widehat{x}(\cdot))$ mapping.

It will be proved that some statistic of the conditional mean of $x(t)$ given $y^t$ can be calculated with a finite–dimensional recursive estimator

(5)                 $d\widehat{x} = \widehat{f}(\widehat{x}, y(t))dt + \sum_{i=1}^{m} g_i(\widehat{x})dw_i(t), \ \widehat{x}(0) = x_0$

$E(c(x(t)/y^t)) = \gamma^t(x_0; y(\cdot))$ (see (19)) where $\gamma^t(x_0; y(\cdot))$ is a continuous functional on $R^n \times C([0, T]; R^d)$

Let

(6)     $Z(t) = \exp\left[\sum_{i=1}^{m} \int_0^t h_i(x(s))dy_i(s) - \frac{1}{2}\int_0^t |h(x(s))|^2 ds\right], \ t \in [0, T].$

where $x(t)$, $t \in [0, T]$, is the pathwise unique solution in (1. a) and $y(t)$, $t \in [0, T]$, is the given observation process fulfilling (1. b). Then $(w(t), y(t))$, $t \in [0, T]$, is a new standard $(m + d)$ dimensional Wiener process under a new probability measure

(7) $\qquad \overline{P}(A) = Z^{-1}(T)P(A)$, $A \in \mathcal{F}$, provided $h \in C_b(R^n)$.

It is a direct consequence of the Cameron-Martin-Giysanov formula if

$w(t)$ and $y(t) = v(t) + \displaystyle\int_0^t h(x(s))ds$ are viewed on a new probability space

$\{\Omega, \mathcal{F}, P_1\}$, where

(8) $\qquad \begin{aligned} P_1(A) &= \zeta(T)P(A), \\ \zeta(t) &= \exp - \left[\sum_{i=1}^d \int_0^t h_i(x(s))dv_i(s) + \tfrac{1}{2}\int_0^t |h(x(s))|^2 ds\right]. \end{aligned}$

It is easily seen that $Z^{-1}(t) = \zeta(t)$, $t \in [0, T]$, and therefore $\overline{P}(A) = P_1(A)$ for any $A \in \mathcal{F}$.

By definition, $Z(t)$, $t \in [0, T]$, is a continuous $\overline{P}$ martingale and

(9) $\qquad E\{c(x(t))/y^t\} = \overline{E}\{c(x(t))Z(t)/y^t\}$ $P$ a.s.

For $h \in C_b^2(R^n)$ let us rewrite (9) as a continuous functional on $y(s)$, $0 \le s \le t$, and integrating $\displaystyle\sum_{i=1}^d \int_0^t h_i(x(s))dy_i(s)$ by parts we obtain

(10)

$$\sum_1^d \int_0^t h_i(x(s))dy_i(s) = y(t)h(x(t))$$

$$- \int_0^t y(s)Lh(x(s))ds$$

$$- \sum_{i=1}^m \int_0^t < y(s) \bullet \nabla h(x(s)), g_i(x(s)) > dw_i(s),$$

where $y(s) \bullet \nabla h$ and $y(s) \bullet Lh$ are the gradient in $x$ of $y(s) \bullet h(x) \overset{\triangle}{=} \sum_{i=1}^d h_i(x)y_i(s)$ and respectively the parabolic operator $L$ in $x$ applied to $y(s) \bullet h$.

Here $L$ is associated to (1.a) and

$$(11) \qquad L \stackrel{\triangle}{=} \sum_1^n f_i(x) \frac{\partial}{\partial x_i} + \frac{1}{2} \sum_{i,j=1}^n a_{ij}(x) \frac{\partial^2}{\partial x_i \partial x_j},$$

with $G \stackrel{\triangle}{=} (g_1, \cdots, g_m)$, $a = GG^T$, $f = (f_1, \cdots, f_n)$. Let

$$(12), \quad e(s,x) = \frac{1}{2} < a(x)y(s) \cdot \nabla h(x), y(s) \cdot \nabla h(x) > -y(s) \cdot Lh(x) - \frac{1}{2}|h(x)|^2$$

where $\langle az, z \rangle = |Gz|^2$, for $z \in \mathrm{R}^n$. Using (10)–(12) in (6) we rewrite $Z(t)$ as

$$(13) \qquad Z(t) = \widehat{Z}(t) \exp y(t) \cdot h(x(t)) \exp \int_0^t e(s, x(s)) ds$$

and

$$(14) \qquad \widehat{Z}(t) = \exp - \left[ \sum_{i=1}^m \int_0^t \langle y(s) \cdot \nabla h(x(s)), g_i(x(s)) \rangle dw_i(s) \right.$$
$$\left. + \frac{1}{t} \int_0^t \langle a(x(s))y(s) \cdot \nabla h(x(s)), y(s) \cdot \nabla h(x(s)) \rangle ds \right]$$

**a) Distribution of $(w(\cdot), x(\cdot))$ conditioned on $y(\cdot)$**

Denote $\overline{\Omega} = C([0, T]; R^{m+n+d})$, $\overline{\mathcal{F}} =$ Borel-measurable sets in $\overline{\Omega}$, and probability measure $\overline{P}_{x_0}$ on $\overline{\Omega}$ generated by the solution $\{w(t), x(t; x_0), y(t); t \in [0, T]\}$ in (1). It is assumed that:

$i_1)f, g_i$ in (1.a) are globally Lipschitz continuous functions,

and therefore there is a pathwise unique solution $x(t; x_0)$, $t \in [0, T]$, for each $x_0 \in R^n$ fixed; $\overline{P}_{x_0}$ is the probability measure on $\overline{\Omega}$ provided $\{\Omega, \mathcal{F}, P_1\}$ is the basic probability space, where $P_1$ is defined in (8). Let $\overline{P}^y_{x_0}$ be the conditional distribution measure of $(w(\cdot), x(\cdot))$ given $y(\cdot)$. Then $\{w(t), y(t); t \in [0, T]\}$ is a standard vector valued Wiener process under the probability $\overline{P}_{x_0}$ and $\overline{P}^y_{x_0}$ lies in the space of probability measures on $\overline{\Omega}^1 = C([0, T]; R^{m+n})$. Moreover (9) can be rewriten

$$(15) \qquad \begin{aligned} E_{x_0}\{c(x(t))/yt\} &= \overline{E}_{x_0}\{c(x(t))Z(t)/yt\} \\ &= \overline{E}^y_{x_0}((x(t))Z(t)), \quad t \in [0, T], \quad \overline{P}_{x_0} a.s. \end{aligned}$$

and using (13) in (15) we obtain

(16)

$$E\{c(x(t))/y^t\} \quad = \overline{E}_{x_0}\{c(x(t))Z(t)/y^t\}$$

$$= \overline{E}^y_{x_0}(c(x(t))\widehat{Z}(t)\exp(y(t)\cdot h(x(t)))\exp\int_0^t e(s,x(s))ds,$$

$t \in [0,T]$, for all $y(\cdot) \in C([0,T];R^d)$ with $y(0) = 0$.

Assuming that $g_i$ in (1.a) fulfils

$i_2$) $g_i \in C_b(R^n)$, $i = 1, \cdots, m$,

then $\widehat{Z}(t), t \in [0,T]$, is a continuous martingale with respect to $\overline{E}_{x_0}$ allowing one to use $\widehat{P}_{x_0}(A) = \widehat{Z}(T)\overline{P}_{x_0}(A)$ as a new probability measure, or to modify the drift coefficient in (1.a) from $f$ to

$$\widehat{f}(t,x) = \quad f(x) - a(x)y(t)\cdot\nabla h(x)$$

(17)

$$= f(x) - \sum_{i=1}^m (\alpha_i(x)y(t))g_i(x), \ t \in [0,T]$$

where $\alpha_i(x) = (g_i^T(x)\nabla h_1(x),\cdots,g_i^T(x)\nabla h_d(x)), i = 1,\cdots,m$. Let us consider the solution $\widehat{x}(t)$, $t \in [0,T]$, of

(18) $$dx = \widehat{f}(t,x)dt + \sum_{i=1}^m g_i(x)dw_i(t), \ x(0) = x_0, \ t \in [0,T]$$

Then the conditional mean in (16) obeys to the following

(19)
$$E\{c(x(t))/y^t\} \quad = \overline{E}_{x_0}\{c(x(t))Z(t)/y^t\}$$
$$= \widehat{E}(c(\widehat{x}(t))\exp y(t)\cdot h(\widehat{x}(t))\exp\int_0^t e(s,\widehat{x}(s))ds,$$

where $\widehat{E}$ stands for the mean with respect to the probability measure

$$\widehat{P}(A) = \widehat{\zeta}(T)P(A)$$

Here the continuous martingale $\widehat{\zeta}(t), t \in [0,T]$, fulfils

(20) $$\widehat{\zeta}(t) = \exp - \left[ \sum_1^d \int_0^t h_i(\widehat{x}(s))dv_i(s) + \frac{1}{2}\int_0^t |h(\widehat{x}(s))|^2 ds \right]$$

and is obtained by replacing $x(s)$ with $\widehat{x}(s)$ in $\zeta(t)$ from (8).

Now it is the Lie algebraic structure of (18) which allow one to write the input $(y(\bullet), x_0)$ output (solution $\widehat{x}(\bullet)$)) mapping in a more explicit way.

**b) The Lie algebraic structure and integral representation of $\widehat{x}(t)$ in (18)**

Assuming that the vector fields $g_i$ in (18) are smooth $g_i \in C_b^\infty(R^n)$, and the Lie algebra $\Lambda = L(g_1, \cdots, g_m)$ is finite–dimensional we can represent any solution $y(t; x), t \in [0, T]$, of

$$(21) \qquad dy = \sum_1^m g_i(y) \circ dw_i(t), \quad y(0) = x.$$

Namely, using Lemma 4.2.2 and fixed basis $\{Y_1, \cdots, Y_N\} \subseteq \Lambda$, we define

$$(22) \qquad G(p; x) = G_1(t_1) \circ \cdots \circ G_N(t_N)(x), \ p = (t_1, \cdots, t_N) \in R^N,$$

where $G_i(t)(x)$ is the global flow generated by $Y_i$.

Then, for $p(t), \ t \in [0, T]$, the solution of

$$(23) \qquad dp = \sum_1^m q_i(p) \circ dw_i(t), \ p(0) = 0, \ p \in R^N$$

we obtain

$$(24) \qquad y(t; x) = G(p(t); x), \ t \in [0, T], \ x \in R^n,$$

where $0$ stands for Fisk–Stratonovich integral (see a.3) and $q_i(\bullet) \in C^\omega(R^N; R^N)$ is the unique solution of the equation

$$(25) \qquad \frac{\partial G}{\partial p}(p; x)q_i(p) = g_i(G(p; x)), \ p \in R^N, \ i = 1, \cdots, m.$$

By definition, $G(p(t); x)$ is a diffeomorphism with respect to $x \in R^n$ and the solution in (18) can be represented

$$(26) \qquad \widehat{x}(t; x_0) = G(p(t); \widehat{z}(t; x_0)), \ t \in [0, T],$$

provided $\widehat{z}(t; x_0), \ t \in [0, T]$, fulfils the ordinary differential equation

$$(27) \qquad \frac{dz}{dt} = \left[\frac{\partial G}{\partial x}(p(t); z)\right]^{-1} F(t, z), \ z(0) = x_0,$$

Here

(28) $$F(t, z) = \widehat{f}(t, G(p(t); z)) - \frac{1}{2} \sum_{i=1}^{m} \left( \frac{\partial g_i}{\partial x} g_i \right) (G(p(t); z)),$$

where $\widehat{f}(t, x)$ is defined in (17).

The continuous dependence of the solution $\widehat{x}(t; x_0)$ on $(x_0, y(\cdot))$ in (26) is determined by the usual continuous dependence theorem for an ordinary differential equation (see (27)) and in a pathwise way with respect to the continuous process $p(t)$, $t \in [0, T]$.

The above given computations can be summarized. Assume that vector functions $f, g_i, h$ are given such that:

i) $h \in C_b^2(R^n; R^d), g_i \in C_b^1(R^n; R^n)$ and $f \in C^1(R^n; R^n)$ is globally Lipschitz continuous, $i = 1, \cdots, m$.

**Proposition 1.**

*Assume that $f, g_i$, $h$, $i = 1, \cdots, m$ fulfil (i) and consider the solution $(x(t; x_0)$, $y(t))$, $t \in [0, T]$, of (1). Then the conditional mean $E\{c(x(t))/y^t\}$ can be expressed in a pathwise way with respect to $y(s), 0 \leq s \leq t$, and fulfils (19), where $c(\cdot) \in C_b(R^n)$ and $y^t$ is the $\sigma$-algebra generated by $\{y(s), 0 \leq s \leq t\}$. In addition, if $g_i \in C_b^\infty(R^n; R^n), i = 1, \cdots, m$, and the Lie algebra $L(g_1, \cdot, g_m)$ is finite dimensional then the recursive estimator $\widehat{x}(t; x_0), t \in [0, T]$, from (18) can be represented in a pathwise way as in (26).*

**Remark 1.**

*In the second part of Proposition 1 we have assumed implicitly that the solution $p(t), t \in [0, T]$, in (23) exists.*

*A more realistic situation is to consider a local solution $(p(t), \tau), t \in [0, T]$, for (23) (see Lemma 2.2) and as a consequence we obtain a local solution $(y(t; x), \tau), t \in [0, T]$, for (21) via $y(t; x) = G(p(t); x)$ in (24). Finally, we obtain a pathwise form of the solution $\widehat{x}(t; x_0)$ in (18) provided $0 \leq t \leq T \wedge \tau$. It shows the complexity of stochastic differential equations even if we suppose that $\Lambda = L(g_1, \cdots, g_m)$ is finite–dimensional. We only mention that a global solution $p(t), t \in [0, T]$, for (23) is meaningful if $\Lambda$ is a nilpotent algebra.*

**Remark 2.**

*It can be expected that for an infinite–dimensional Lie algebra $L(g_1, \cdot, g_m)$ the integral representation in (18) is not valid and the analysis is more technical. In a sense, it must be similar to what we have done for a finite–dimensional case but we need to deal with vector fields $q_i$ depending in a smooth manner on the initial condition $y(0) = x$ and it influences negatively the diffeomorphism property of the solution in (24).*

**c) Anticipating drift and integral representation of the solution for stochastic differential equations**

It was clearly seen that a finite–dimensional Lie algebra $L(g_1, \cdots, g_m)$ allow ones to reduce the pathwise smooth dependence of solutions in a non-linear filter problem in (1) to a problem for an ordinary differential equation defined in (27). As far as the final equation is an ordinary differential equation with parameters (see (27)) there is no obstruction in considering that the original drift $f(x)$ in (1.a) fulfils a more general assumption such as:

$i_1$) $f(t, x, \omega) : [0, T] \times R^n \times \Omega \to R^n$ is a measurable, Lipschitz continuous function in $x \in R^n$ for each $(t, \omega) \in [0, T] \times \Omega$ and

$$(29) \qquad \|f(t, x, \omega)\| \le c(1 + \|x\|) \ (\forall) \ (t, x, \omega) \in [0, T] \times R^n \times \Omega,$$

Then define the stochastic differential equation

$$(30) \qquad dx = f(t, x, \omega)dt + \sum_{i=1}^{m} g_i(x) \circ dw_i(t) \qquad x(0) = x_0 \in R^n$$

where $g_i \in C^\infty(R^n; R^n)$, $i = 1, \cdots, m$,

The main obstruction for a solution in (30) to be defined comes from the drift $f(t, x, \omega)$ fulfilling $(i_1)$ which can be avoided by separating the analysis of (30) into two subsystems starting with

$$(31) \qquad dy = \sum_{i=1}^{m} g_i(y) \circ dw_i(t), \quad y(0) = x \in R^n$$

A solution for (31) can be defined as a nonanticipating continuous process $y(t; x), t \in [0, T]$, for each $x \in R^n$ provided a stopping time $\tau$ is used. More precisely, assuming that the Lie algebra $\Lambda = L(g_1, \cdots, g_m)$ is finite–dimensional and $\{Y_1, \cdots, Y_N\} \subseteq \Lambda$ is a fixed base define the composition of the corresponding local flows

(32)
$$G(p; x) = G_1(\ t_1) \circ \cdots \circ G_N(t_N)(x),$$

$$p = (t_1, \cdots, t_N) \in D_N = \prod_1^N (-a_i, a_i), \ x \in V(x_0) \subseteq R^n,$$

where $G_i(t)(x)$ is the local flow generated by $Y_i$. A local solution $\{y(t;x), \tau\}$, $t \in [0,T]$, $x \in V(x_0)$, for (31) can be defined in general but the stopping time $\tau$ has to be independent of the initial condition $x \in V(x_0)$. It can be accomplished using the homomorphism correspondence between the original Lie algebra $L(g_1, \cdots, g_m)$ and $L(q_1, \cdots, q_m)$, where $q_i \in (D_N; R^N)$ is determined such that

$$(33) \qquad \frac{\partial G}{\partial p}(p;x)q_i(p) = g_i(G(p;x)), \quad i = 1, \cdots, m$$

and $G(p;x)$ is defined in (32).

The equations in (33) admit a unique solution if we notice that $G(p;x)$ in (32) is the solution in a gradient system

$$(34) \qquad \frac{\partial y}{\partial t_1} = Y_1(y), \frac{\partial y}{\partial t_2} = X_2(t_1; y), \cdots, \frac{\partial y}{\partial t_N} = X_N(t_1, \cdot, t_{N-1}; y)$$

and the conditions in Theorem 2.2.1 are fulfilled.

Then associate the following stochastic differential equation

$$(35) \qquad dp = \sum_{i=1}^{m} q_i(p) \circ dw_i(t), \quad p(0) = 0, \quad p \in D_N$$

and consider $\{p(t), \tilde{\tau}\}$, $t \in [0,T]$, a local solution of (35) where

$$(36) \qquad \tilde{\tau}(\omega) = \inf\{t \geq 0; |\tilde{p}(t, \omega)| \geq M\}, \text{ with } 2M < \rho = \min(a_1, \cdots, a_N)$$

Here $\tilde{p}(t)$, $t \in [0,T]$, is the continuous nonanticipating process resulting as a unique solution of the equation

$$(37) \qquad dp = \sum_{i=1}^{m} \tilde{q}_i(p) \circ dw_i(t), \quad p(0) = 0, \quad t \in [0,T],$$

where $\tilde{q}_i(p) = \tilde{\alpha}(p)q_i(p)$, and $\tilde{\alpha} \in C^\infty(R^N; R)$ is chosen such that $\tilde{\alpha}(p) = 1$ for $p \in S_M(0)$, $\tilde{\alpha}(p) = 0$ for $p \in R^N - S_{2M}^{(0)}$ and $0 \leq \tilde{\alpha}(p_{\leq}1)$ $(\forall)$ $p \in R^N$. It is easily seen that $\tilde{p}(t)$ fulfils

$$(38) \qquad \tilde{p}(t \wedge \tilde{\tau}) = \sum_{i=1}^{m} \int_0^{t \wedge \tilde{\tau}} q_i(\tilde{p}(s)) \circ dw_i(s), \quad t \in [0,T]$$

and therefore $\{p(t) \overset{\triangle}{=} \tilde{p}(t), \tilde{\tau}\}$, $t \in [0,T]$, is a local solution of (35). Now we are in position to define a local solution in (31).

**Proposition 2.**

*Assume that $g_i \in C^\infty(R^n; R^n)$, $i = 1, \cdots, m$, are given such that Lie algebra $L(g_1, \cdots, g_m) = \Lambda$ is finite–dimensional. For a fixed base $\{Y_1, \cdots, Y_N\} \subseteq \Lambda$ define a local diffeomorphism $G(p; x), x \in V(x_0)$ as in (32) and let $\{\widetilde{p}(t), \widetilde{\tau}\}$, $t \in [0, T]$, be a local solution in (35). Then $\{y(t; x) \stackrel{\triangle}{=} G(\widetilde{p}(t); x), \widetilde{\tau}\}$, $t \in [0, T]$, is a local solution in (31).*

**Proof.**

The local solution $\{p(t), \widetilde{\tau}\}$, $t \in [0, T]$, in (35) fulfils

$$(39) \qquad p(t \wedge \widetilde{\tau}) = \sum_1^m \int_0^{t \wedge \widetilde{\tau}} q_i(p(s)) \circ dw_i(s), \ t \in [0, T],$$

Define $y(t; x) = G(\widetilde{p}(t); x)$, where $G$ is given in (32), and applying the usual rule of Itô's stochastic differentiation we obtain that $y(t \wedge \widetilde{\tau}; x) \stackrel{\triangle}{=} G(\widetilde{p}(t \wedge \widetilde{\tau}); x)$ fulfils the following integral equation

$$(40) \qquad y(t \wedge \widetilde{\tau}; x) = x + \sum_1^m \int_0^{t \wedge \widetilde{\tau}} g_i(y(s; x)) \circ dw_i(s), \ t \in [0, T],$$

which means that $\{y(t, x), \widetilde{\tau}\}$, $t \in [0, T]$, is a local solution in (31). The proof is complete.

Now, a local solution for (30) can be defined provided the local solution $\{y(t; x), \widetilde{\tau}\}, x \in V(x_0)$, $t \in [0, T]$, in Proposition 2 is used. Namely, let $\{\widetilde{p}(t), \widetilde{\tau}\}$, $t \in [0, T]$, be the local solution for (35) defined in (36) and (37). By definition $\widetilde{p}(t) \in S_{2M}(0)$, $(\forall)\ t \in [0, T]$, and there exists $0 < T_1 \leq T$ such that the solution $\widetilde{z}(t, \omega), t \in [0, T]$, of the ordinary differential equation

$$(41) \qquad \frac{dz}{dt} = \left[\frac{\partial G}{\partial x}(\widetilde{p}(t); z)\right]^{-1} f(t, G(\widetilde{p}(t); z), \omega), \ z(0) = x_0$$

fulfils, $\widetilde{z}(t, \omega) \in V(x_0)$, for any $(t, \omega) \in [0, T_1] \times \Omega$. Denote

$$(42) \qquad \widetilde{x}(t) = G(\widetilde{p}(t); \widetilde{z}(t)), \ \widetilde{x}_\varepsilon(t) = G(\widetilde{p}_\varepsilon(t); \widetilde{z}_\varepsilon(t)), \ t \in [0, T_1],$$

where $G(p; x)$ is defined in Proposition 2 and $\widetilde{p}_\varepsilon(t)$, is the solution of an ordinary differential equation

$$(43) \qquad \frac{dp}{dt} = \sum_{i=1}^m \widetilde{q}_i(p) \frac{dv_i^\varepsilon(t)}{dt} = \sum_{i=1}^m \widetilde{\alpha}(p) q_i(p) \frac{dv_i^\varepsilon(t)}{dt}, \qquad p(0) = 0$$

Here $v^\varepsilon(t)$, $t \in [0, T]$, is the Langevin approximation of the Wiener process $w(t)$, $t \in [0, T]$, defined in (a.3) and $\tilde{z}_\varepsilon(t)$, $t \in [0, T]$, is the solution in (41) when $\tilde{p}(t)$ is replaced by $\tilde{p}_\varepsilon(t)$. It is proved in (a.3) that

(44)     $$\lim_{\varepsilon \to 0} E|\tilde{p}_\varepsilon(t \wedge \tilde{\tau}) - \tilde{p}(t \wedge \tilde{\tau})|^2 = 0 \text{ foreach } t \in [0, T_1]$$

and using the property $\tilde{z}(t), \tilde{z}_\varepsilon(t) \in V(x_0), \tilde{p}(t), \tilde{p}_\varepsilon(t) \in S_{2M}(0)$, for $t \in [0, T_1]$, we obtain

(45)
$$\lim_{\varepsilon \to 0} E|\tilde{z}_\varepsilon(t \wedge \tilde{\tau}) - \tilde{z}(t \wedge \tilde{\tau})|^2 = 0,$$
$$\lim_{\varepsilon \to 0} E|\tilde{x}_\varepsilon(t \wedge \tilde{\tau}) - \tilde{x}(t \wedge \tilde{\tau})|^2 = 0, \; (\forall) \, t \in [0, T_1].$$

The meaning of $\tilde{x}(t), t \in [0, T_1]$, is a solution of (30) is based on the approximating ordinary differential equation fulfilled by the sequence $\{\tilde{x}_\varepsilon(t), t \in [0, T_1]\}\varepsilon > 0$. By a straight computation we obtain that $(\tilde{p}_\varepsilon(t), \tilde{x}_\varepsilon(t))$ $t \in [0, T]$, is the solution of the following system

(46)
$$\frac{dp}{dt} = \sum_{i=1}^{m} \tilde{\alpha}(\tilde{p}_\varepsilon(t)) q_i(p) \frac{dv_i^\varepsilon(t)}{dt}, \quad p(0) = 0$$
$$\frac{dx}{dt} = \sum_{i=1}^{m} \tilde{\alpha}(\tilde{p}_\varepsilon(t)) g_i(x) \frac{dv_i^\varepsilon(t)}{dt} + f(t, x, \omega), \; x(0) = x_0.$$

where $\tilde{\alpha}(p_\varepsilon(t)) \triangleq \alpha_\varepsilon(t) \in [0, 1]$ and

(47)
$$\lim_{\varepsilon \to 0} E|\alpha_\varepsilon(t \wedge \tilde{\tau}) \quad -\tilde{\alpha}\tilde{p}(t \wedge \tilde{\tau}))|^2$$
$$= \lim_{\varepsilon \to 0} E|\alpha_\varepsilon(t \wedge \tilde{\tau}) - 1|^2 = 0$$

The last conclusion allows one to see that the approximating equation used to obtain $\tilde{x}_\varepsilon(t)$, $t \in [0, T_1]$, is very close to the usual one when a stochastic differential equation is replaced by an ordinary differential equation.

**Remark**

   *The above computations proved that the continuous process $\tilde{x}(t)$, $t \in [0, T_1]$, and a stopping time $\tilde{\tau}$ verifying (45) and respectively (36) could be used as an alternative solution for a stochastic differential equation with anticipating drift as in (30). The integral representation of the solution is based essentially on the diffeomorphism property of the local solution constructed in Proposition 2 and it is easily seen that it gives a local solution for (30) when an usual drift $f(t, x)$ is used.*

## 4.5    Affine Control Systems

In applications of Chapter 2 we have shown that a finite–dimensional Lie algebra $\Lambda = L(g_1, \cdots, g_m)$ is meaningful in construction of periodic feedback stabilizing control of affine systems without drift and for describing the manifold structure of nonholonomic constraints viewed as affine systems. An affine control system is defined on an open set $\mathcal{O} \subseteq R^n$

$$
(1) \qquad \frac{dx}{dt} = f(x) + \sum_{i=1}^{m} u_i g_i(x), \ x \in \mathcal{O}
$$

where $f, g_i \in C^\infty(\mathcal{O}; R^n)$, $i = 1, \cdots, m$, and $u = (u_1, \cdots, u_m) \in R^m$ is the control parameter.

   Differential geometric analysis of the associated Lie algebra $\Lambda = L(f, g_1, \cdots, g_m)$ can be relevant for various control problems. To begin with we shall confine ourselves mainly in describing the integral reprezentation of solutions in (1), integral manifolds associated with $\Lambda$ and decomposition of the control system (1) into a controllable and uncontrollable subsystems. These problems are significant for the control theory and have a direct connection with the gradient systems studied in the previous chapter provided the control Lie algebra $\Lambda$ is locally finitely generated on $\mathcal{O}$.

   We recall that a set $S \subseteq C^\infty(\mathcal{O}; R^n)$ of smooth vector fields is locally finitely generated if for every $x_0 \in \mathcal{O}$ there exists a neighbourhood $V \subseteq \mathcal{O}$ of $x_0$ and a finite set $\{Y_1, \cdots, Y_M\} \subseteq S$ with the property that every other vector field $Y \in S$ can be represented on $V$ in the form

$$
(2) \qquad Y(y) = \sum_{i=1}^{m} a_i(y) Y_i(y), \ y \in V
$$

where each $a_i$ is a real-valued smooth function defined on $V, a_i \in C^\infty(V; R)$. The global version of (2) is when $V = \mathcal{O}$.

   An integral manifold of the control Lie algebra $\Lambda$ passing through an arbitrary point $x_0 \in \mathcal{O}$ will be defined as a finite composition of local flows

$$
(3) \qquad
\begin{aligned}
G(p; x_0) &= G_1(t_1) \circ \cdots \circ G_k(t_k)(x_0), \qquad k \leq n, \\
p &= (t_1, \cdot, t_k) \in D_k = \prod_{1}^{k} (-a_i, a_i),
\end{aligned}
$$

where $G_i(t)(x)$, $t \in (-a_i, a_i)$, $x \in V(x_0)$ is the local flow generated by a vector field $Y_i \in \Lambda$.

   It allows one restrict to the minimum the concepts which are used in explaining the matters.

The above given definitions and the statements which are obtained in this section will have an obvious analog if the state space $\mathcal{O} \subseteq R^n$ is replaced by a smooth manifold (see a.4).

There is nothing new in the given definitions and as a matter of fact they are frequently used in control theory. There is one point to be stressed which emphasizes the utility of the gradient system associated to (3) and its nonsingular algebraic representation presented in the previous chapter.

It allows one to avoid local nonsingularity of $\Lambda$ and to construct integral manifolds as slices of a coordinate neighbourhood, i.e., embedded submanifolds.

## 4.6 Integral Representation of Solutions

The purpose of this section is to represent solutions of the affine control system as a composition of flows like in (3) where $p$ stands for a new dynamical control parameter. In the setting of the previous chapter the function $G(p; x_0)$ in (3) is an orbit passing through $x_0$, or of the origin; $x_0$. It is easily seen that a locally finitely generated control Lie algebra $\Lambda$ is (f.g.o; $x_0$) for every $x_0 \in \mathcal{O}$ and $V(x_0) \subseteq \mathcal{O}$ fixed provided an orbit of the origin $x_0$ is forced to remain in $V(x_0)$. The affine control system (1) under the locally finitely generated assumption on the corresponding control Lie algebra $\Lambda$ fulfils the hypotheses of Theorem 3.2.2. Let $x_0 \in \mathcal{O}$ and $V(x_0) \subseteq \mathcal{O}$ be fixed and choose a system $\{Y_1, \cdots, Y_M\} \subseteq \Lambda$ of generators for $\Lambda$ when restricted to the neighbourhood $V(x_0)$. Let $k = \dim \Lambda(x_0)$ and assume that $\{Y_1, \cdots, Y_M\}$ is $k$-minimal, i.e., $Y_1(x_0), \cdot, Y_k(x_0)$ are linearly independent in $R^n$ and $Y_j(x_0) = 0$ for $j = k+1, \cdots, M$. Now, consider the following orbit

$$
(4) \qquad \begin{aligned} G(p; x_0) &= G_1(t_1) \circ \cdots \circ G_k(t_k)(x_0), \\ p &= (t_1, \cdots, t_k) \in D_k = \textstyle\prod_1^k(-a_i, a_i) \end{aligned}
$$

where $G(p; x_0) \in V(x_0)$, $p \in D_k$, and $G_i(t)(x)$, $t \in (-a_i, a_i)$, $x \in V(x_0)$ is the local flow generated by the vector field $Y_i$ in the fixed system $B = \{Y_1, \cdots, Y_M\}$.

An immediate goal is to represent any solution $x_u(t, x_0)$ of affine control system (1) starting from $x_0$ at time $t = 0$ and corresponding to a piecewise continuous control function $u(t), t \in [0, a]$. It will be performed through the orbit in (4) as

$$
(5) \qquad x_u(t; x_0) = G(p_u(t; x_0)), \ t \in [0, a],
$$

where $p_u(t)$, $t \in [0, a]$, is the solution in a new affine control system defined on an open set $D_k \subseteq R^k$.

(6)
$$\frac{dp}{dt} = q_0(p) + \sum_1^m u_i(t)q_i(p)),$$
$$p(0) = 0, \ p \in D_k = \prod_1^k(-a_i, a_i)$$

Here the admisible control function $u(t)$, $t \in [0, a]$, is chosen piecewise continuous for the sake of simplicity and the vector fields $q_j \in C^\infty(D_k; R^k)$ are
found such that

(7)
$$\frac{\partial G}{\partial p}(p; x)q_i(p) = g_i(G(p; x_0)), \qquad i = 1, \cdots, m,$$
$$\frac{\partial G}{\partial p}(p; x_0)q_0(p) = f(G(p; x_0)), \qquad p \in D_k.$$

As expected (see Theorem 3.2.3) the new vector fields $q_j \in C^\infty(D_k, R^k)$
depend on the $x_0 \in Q$ fixed and for different initial states in (1) we obtain
different control systems in (6). In addition (see the Remark following Theo-
rem 3.2.3), the new control Lie algebra $L = L(q_0, q_1, \cdots, q_m) \subseteq C^\infty(D_k; R^d)$
fulfils the maximal rank condition

(8)                        $\dim L(q_0, q_1, \cdots, q_m)(p) = k$ forall $p \in D_k$.

The new Lie algebra $L$ is finitely generated, i.e.    there exist
$\{Q_1, \cdots, Q_M\} \subseteq L$ such that

(9)
$$\frac{\partial G}{\partial p}(p; x_0)Q_j(p) = Y_j(G(p; x_0)), \ j = 1, \cdots, M,$$

and any $Q \in L$ can be written

(10)
$$Q(p) = \sum_{i=1}^M \alpha_i(p)Q_i(p), \text{ forall } p \in D_k$$

where $\alpha_i \in C^\infty(D_k; R^k)i = 1, \cdots, M$, depend on $Q$.

On the other hand, the meaning of the homomorphism correspondence
$\lambda : \Lambda \to L$ between the two Lie algebras is fully determined by the expressions
in (7) and (9).

The analysis of the trajectories of the original affine control system with a
fixed initial state $x(0) = x_0 \in \mathcal{O}$ and corresponding to the admissible control
functions is entirely based on the above given homomorphism.

The role of the original Lie algebra $\Lambda = L(f, g_1, \cdots, g_m)$ in the study
of interactions between input (control function) and output (state) depends on the associated finitely generated and nonsingular Lie algebra
$L = L(q_0, q_1, \cdots, q_m)$.

As a consequence of this we see that, when studying the behaviour of a control system initialized at $x_0 \in \mathcal{O}$, we may and do regard as a natural state space the submanifold $M_{x_0} \subseteq \mathcal{O}$

$$(11) \qquad M_{x_0} = \{x = G(p; x_0) : p \in D_k\} \subseteq V(x_0)$$

instead of the whole $\mathcal{O}$.

**Proposition 1.**

Let $f, g_i \in C^\infty(\mathcal{O}; R^n)$, $i = 1, \cdots, m$, be given such that the Lie algebra $\Lambda = L(f, g_1, \cdots, g_m)$ is locally finitely generated. Let $x_0 \in \mathcal{O}$ be fixed and consider a $k$-minimal system $\{Y_1, \cdots, Y_M\} \subseteq \Lambda$ of generators, where $k = \dim \Lambda(x_0) \leq n$. Define the mapping $G(p; x_0), p \in D_k$ in (4) according to the fixed $k$-minimal system of generators. Then the set $M_{x_0}$ in (11) is an integral smooth $k$-dimensional manifold of $\Lambda$ passing through $x_0$.

**Proof.**

It is a direct consequence of Theorem 3.2.2 and the conclusions $(c_1)$ and $(c_2)$ allow one to see that $M_{x_0}$ is a locally $k$-dimensional Euclidian space and the tangent space at some $x = G(p; x_0)$ fulfils $T_x M_{x_0} = \Lambda(G(p; x_0))$.

**Remark 1.**

Note that $\Lambda = L(f, g_1, \cdots, g_m)$ is singular and $\dim \Lambda(x_0)$ is not constant when $x_0$ is changing. Thus it may happen that at two different initial states $x_1, x_2 \in \mathcal{O}$ one obtains two different associated control systems in (6) and the corresponding integral manifolds $M_{x_1}, M_{x_2} \subseteq \mathcal{O}$ with $\dim M_{x_i} = \dim \Lambda(x_i)$, $i = 1, 2$.

The statement that $M_{x_0}$ is a slice coordinate neighbourhood of $\mathcal{O} \subseteq R^n$, i.e., an embedded submanifold of $\mathcal{O}$, depends on the following considerations.

**Proposition 2.**

Let the hypothesis in Proposition 1 be fulfilled and $\dim \Lambda(x_0) = k$ for $x_0 \in \mathcal{O}$ fixed. Consider $M_{x_0} = \{x = G(p; x_0) : p \in D_k\} \subseteq V(x_0)$ where $V(x_0) \subseteq \mathcal{O}$ is fixed by hypotheses.

Then for each compact ball $S_\rho(x_0) \subseteq V(x_0)$ there exists a coordinate transformation $z = \varphi(x)$ defined in a neighbourhood $\mathcal{U}$ of $S_\rho(x_0) \cap M_{x_0}$, $\mathcal{U} \subseteq V(x_0)$, such that

$$M_{x_0} \cap S_\rho(x_0) = \{x \in S_\rho(x_0) : \varphi_j(x) = \varphi_j(x_0), \; j = k+1, \cdots, n\}.$$

**Proof.**

The arguments are contained in Theorem 3.2.2 and for $y(p) = G(p; x_0)$ we obtain that $\left\{ \frac{\partial y}{\partial t_1}(p), \cdots, \frac{\partial y}{\partial t_k}(p) \right\} \subseteq \Lambda$ are linearly independent and spans $\Lambda(y)(p)$ for any $p \in D_k$, where $G(p; x_0)$, $p \in D_k$, is defined in (4) according to a fixed $k$-minimal system of generators.

In addition, we may and do use the fact that $y(p)$ is the solution in a gradient system

$$
\begin{array}{cc}
\frac{\partial y}{\partial t_1} = Y_1(y), & \frac{\partial y}{\partial t_2} = X_2(t_1; y), \cdots, \\
\end{array}
$$

(12)

$$
\frac{\partial y}{\partial t_k} = X_k(t_1, \cdots, t_{k-1}; y), \qquad y(0) = x_0,
$$

$$
\text{for } p = (t_1, \cdot, t_k) \in D_k = \prod_1^k (-a_i, a_i)
$$

determined by smooth vector fields. For each $p_0 \in D_k$ fixed and $x \in V(y(p_0))$ we may find a unique solution $G(p; x_0)$, $p \in \mathcal{U}(p_0)$, of (12) fulfilling $G(p_0; x) = x$.

Recalling that $y(p)$, $p \in \mathcal{U}(p_0)$, is a solution of (12) we obtain $y(p) = G(p; y_0)$, $p \in \mathcal{U}(p_0)$, where $y_0 \overset{\Delta}{=} y(p_0)$. By definition $\frac{\partial G}{\partial x}(p; y_0) = I_n$ and denote $z = (p, \widehat{x})$, $\widehat{x} = (x_{k+1}, \cdots, x_n)$, $F(z) = G(p; y_1(p_0), \cdots, y_k(p_0), \widehat{x})$. There is no loss of generality assuming that

(13)          $\frac{\partial F}{\partial z}(z_0) \overset{\Delta}{=} \left( \frac{\partial G}{\partial p}, \frac{\partial G}{\partial \widehat{x}} \right) (p_0, y_0)$ is nonsingular.

Using (13) we can solve the algebraic equations

(14)                                    $F(z) = y$

in a neighbourhood of $(z_0, y_0)$, where $z_0 = (0, x_{k+1,0}, \cdots, x_{n_0})$, and we obtain a smooth unique solution $z = \varphi(y)$, for $y \in V_1(y_0) \subseteq V(y_0)$ such that

(15)                          $F(\varphi(y)) = y$ for all $y \in V_1(y_0)$

$$
\frac{\partial \varphi}{\partial y}(y), \; y \in V_1(y_0), \text{is nonsingular}
$$

In addition, the solution $\varphi(y) = (\varphi_1(y), \widehat{x}(y))$ in (15) fulfils $\varphi_1(y(p)) = p$, $\widehat{x}(y(p)) = \widehat{y}_0$ for any $p \in \mathcal{U}(p_0)$, where $\widehat{y}_0 = (y_{k+1,0}, \cdots, y_{n_0})$. Therefore, using the coordinate transformation $z = \varphi(y)$, we may describe the set $M_{x_0}$ arround to $y_0 = G(p_0; x_0)$ as the slice defined in (16). What is done for an arbitrary $p_0 \in D_k$ can be repeated for any $p \in D_k$ preserving the choice of $z = (p, \hat{x}), \hat{x} = (x_{k+1}, \cdots, x_n)$ because $\dfrac{\partial G}{\partial x}(p; x) = I_n$ (identity matrix) and $\dfrac{\partial G}{\partial t_j}(p; y(p)) = \dfrac{\partial y}{\partial t_j}(p)$, for $j = 1, \cdots, k$. The proof is complete.

## 4.7 Decomposition of affine control systems

The statements contained in the given above Propositions allow one to recognize that for a fixed initial state $x_0 \in \mathcal{O}$ the trajectories of the system (1) restricted to a neighbourhood $V(x_0) \subseteq \mathcal{O}$ evolve on the integral manifold $M_{x_0}$. The slice coordinate transformation structure given in Proposition 2 explains the meaning of the decomposition into controllable and uncontrollable subsystems for (1). More precisely, the following statements are in appropiate.

**Proposition 3.**
*Assume that $\Lambda = L(f, g_1, \cdots, g_m)$ is locally finitely generated and for $x_0 \in \mathcal{O}$, $V(x_0) \subseteq \mathcal{O}$, fixed defines the integral manifold $M_{x_0} \subseteq V(x_0)$ of $\Lambda$ as in Proposition 1. Then any trajectory $x_u(t; x_0)$, $t \in [0, a]$, of the affine control system (1) corresponding to an admisible control and remaining in $V(x_0)$ fulfils $x_u(t; x_0) \in M_{x_0}$ for any $t \in [0, a]$.*

**Proof**
By hypothesis we may and do consider a $k$-minimal system $\{Y_1, \cdots, Y_M\} \subseteq \Lambda$ of generators, where $\dim \Lambda(x_0) = k$.

Then associate the auxiliary control system in (6) and the orbit in (4) fulfilling (7). It is easily seen the one to one correspondance between solutions $x_u(t; x_0) \in V(x_0)$, $t \in [0, a]$, in (1) and $p_u(t)$, $t \in [0, a]$, verifying (6) with the same admisible control $u$. It can be restated as $x_u(t; x_0) = G(p_u(t); x_0)$, $t \in [0, a]$, which means $x_u(t; x_0) \in M_{x_0}$, $t \in [0, a]$, where $M_{x_0}$ is defined in Proposition 1.

The proof is complete.

The statement in Proposition 3 can be rephrased as follows. The control system in (1) with initial state $x_0 \in \mathcal{O}$ and restricted to $V(x_0) \subseteq \mathcal{O}$ evolves on the manifold $M_{x_0}$ and any point $G(p; x_0) \in M_{x_0}$ can be joined with $x_0$ by an admissible trajectory of (1) provided the corresponding parameter $p \in D_k$ can be joined with $0$ in $D_k$ using the same control function. This leads us directly to the following.

**Remark 3.**

*The auxiliary control system (6) fulfils the maximal rank condition*

$$\dim L(q_0, q_1, \cdots, q_m)(p) = k$$

*, for any $p \in D_k$, and it enables one to conclude that the corresponding reachable set*

$$R(0) = \{p \in D_k : p_u(a) = p \text{ for some } a > 0 \text{ and admissible control } u\} \subseteq R^k$$

*has the property $\int R(0) \neq \emptyset$. Let $R(x_0)$ be the corresponding reachable set of (1). A direct implication of $\int R(0) \neq \emptyset$ is that $\int R(x_0) \neq \emptyset$ where $R(x_0) \subseteq M_{x_0}$ is used.*

*Now a decomposition statement for the control system (1) initialized at $x_0 \in \mathcal{O}$ can be formulated.*

**Proposition 4.**

*Assume that $\Lambda = L(f, g_1, \cdots, g_m)$ is locally finitely generated and for $x_0 \in \mathcal{O}$, $V(x_0) \subseteq \mathcal{O}$, fixed such that $\dim \Lambda(x_0) = k \leq 0$, consider the integral manifold $M_{x_0} = \{x = G(p; x_0) : p \in D_k\}$ defined in Proposition 3. Then for any closed ball $S_\rho(x_0) \subseteq V(x_0)$ there exists a coordinate transformation $z = \varphi(x)$, $x \in S_\rho(x_0)$ such that the first $k$ components of $\varphi$ describe the dynamic of the control system (1) on $M_{x_0} \cap S_\rho(x_0)$ and the last $n - k$ components of $\varphi$, $\varphi_j(x), j = k+1, \cdots, n$, fulfils $\left\langle \dfrac{\partial \varphi_j}{\partial x}(x), h(x) \right\rangle = 0$ for any $h \in \Lambda$ and $x \in M_{x_0} \cap S_\rho(x_0)(\varphi_j(x) = \varphi_j(x_0)$, for any $x \in M_{x_0} \cap S_\rho(x_0)$, $j = k+1, \cdots, n)$.*

**Proof.**

By hypothesis we can use the statement in Proposition 3 and we obtain $M_{x_0}$ as the controllable part of the affine control system (1). In addition, the slice coordinate structure given in Proposition 2 for $M_{x_0}$ allows one to define $\varphi_j(x)$, $j = k+1, \cdots, n$ as the uncontrollable part, provided the state is restricted to $x \in S_\rho(x_0) \subseteq V(x_0)$ and $z = \varphi(x)$ is the coordinate transformation contained in Proposition 2.

**Remark 4.**

*The decomposition result stated in Proposition 4 is valid for any $x_0 \in \mathcal{O}$ provided $\Lambda = L(f, g_1, \cdots, g_m)$ is locally finitely generated and can be assimilated into a global decomposition statement for the affine control system in (1).*

**Remark 5.**

*A new situation may appear when dealing with approximations for the original control system (1) involving Lie brackets of the vector fields $f, g_i$. One result of this type is the classical Chow Theorem (1939) requiring the full rank condition of the original control Lie algebra $\Lambda = L(f, g_1, \cdots, g_m)$. In this respect the general sheme of the homomorphism correspondence $\lambda : \Lambda \to L$ when $\Lambda$ is locally finitely generated is useful because the new Lie algebra $L$ fulfils the required full rank condition and the conclusion of Chow's Theorem applied in (6) can be transfered to the original control system (1).*

**Bibliographical notes**

Integral representations for the solutions of stochastic differential equations have been considered by several authors and the analysis was based on the structure of the Lie algebra $\Lambda = L(f, g_1, \cdots, g_m)$. In Yamato (14) $\Lambda$ is a nilpotent algebra, Kunita (10) (11) worked with a solvable $\Lambda$ and Krener, Lobry (9) based their considerations on a special kind of solvable $\Lambda$.

The nonlinear filter problem and the pathwise representation with respect to given observation process $yt$ is found in Pardoux (12).

The way of stating the control problems considered here follows the description in Isidori (8) where the reader can find a more complete picture of this theory.

# Chapter 5

# Stabilization and Related Problems

In the last chapter of this book we discuss in more detail some relevant problems of control theory encompassing approximations, parametrization around a fixed flow and stabilization for affine control systems, controlled invariant Lie algebras' disturbance decoupling and singularly perturbed differential equations.

The chosen issues are strongly related to the theoretical tools developed in the previous chapters, and for the sake of simplicity we shall confine ourselves to considering them mainly as a part of the finite–dimensional Lie algebraic structure. The infinit–dimensional case is not excluded, provided the finite generation is assumed and there are situations where an informed reader may complete the analysis by himself.

As far as the stabilization problem for an affine control system with singularities is concerned, the analysis is performed on the associated controllable system and the conclusions receive an easy translation to the original one. The same association is meaningful for nonsingular approximations completed by an explicit formula related to the parametrization around a fixed flow, which are the main ingredients for a stabilization algorithm.

## 5.1  Equivalent Controllable Systems

Consider an affine control system defined on an open set $\mathcal{O} \subset R^n$. Let $\varphi^u(t; x_0)$, $0 \leq t \leq T(x_0, u) \leq T$ be the solution of the control system

$$(1) \qquad \frac{dx}{dt} = f_0(x) + \sum_{i=1}^{m} u_i(t) f_i(x), \ x(0) = x_0, \ x \in \mathcal{O},$$

117

where the admissible control $u(t) \overset{\Delta}{=} (u(t), \ldots, u_m(t)) \in R^m$, $t \in [0, T]$, is a bounded piecewise continuous function $(u(\cdot) \in \mathcal{A}_T)$, and $f_j$ are smooth vector fields on $\mathcal{O}$, i.e., $f_j \in C^\infty(\mathcal{O}; R^n)$. Assume that $f_0, \ldots, f_m$ are commutative vector fields. Then the solution of the equation (1) is represented as

$$(2) \qquad \varphi^u(t; x_0) = F_0(t) \circ F_1(v_1(t)) \circ \ldots \circ F_m(v_m(t))(x_0)$$

where $F_i(\tau)(x_0)$ is the local flow generated by the vector field $f_i$ and $v_i(t) = \int_0^t u_i(s)ds$, $i = 1, \ldots, m$.

To prove that $\varphi^u(t; x_0)$ defined in (2) is the solution in (1), with Cauchy data $x_0$ at $t = 0$ and corresponding to the fixed admissible control $u(\bullet) \in \mathcal{A}_T$, one relies on the commuting property of the flows $F_i(t_i)$, $F_j(t_j)$, provided the corresponding vector fields $f_i$ and $f_j$ commute. This is the case when diffusion equations behave like ordinary differential equations if $V(t)$ is a Wiener process. In the system theory or the control theory, $u(\bullet)$ in equation (1) is called the input and the solution $\varphi^u(\bullet; x_0)$ is called the output. The above example shows the way of representing the output by decomposition of the solution into local flows and it is an important problem in applications that we can compute the output from the input explicitly. We shall consider the problem in a general framework, and it gives the meaning of the integral representation associated with solutions for nonlinear control systems, which is the basis for defining equivalent controllable systems and invariant submanifolds.

We begin by discussing the integral representation of solutions for a nonlinear system of the form (1). Consider a point $x_0$ in the state space of (1) and we are looking for a finite composition of local flows

$$F(p; x_0) = F_1(t_1) \circ \ldots \circ F_k(t_k)(x_0), \ t_i \in (-a_i, a_i), \ i = 1, \ldots, k,$$

starting from the fixed $x_0 \in \mathcal{O}$ such that

$c_1)F_i(t)(x)$, $t \in (-a_i, a_i)$, $x \in V_i(x_0) \subseteq \mathcal{O}$, is the local flow generated by some $Y_i$ in the real Lie algebra $L(f_0, f_1, \ldots, f_m)$ determined by $\{f_0, \ldots, f_m\} \subseteq C^\infty(\mathcal{O}; R^n)$;

$c_2)$ Any solution $x^u(t; x_0)$, $t \in [0, T]$, of (1) associated with an admissible control $u(\cdot) \in \mathcal{A}_T$ can be represented uniquely $x^u(t; x_0) = F(p^u(t); x_0)$, where $p^u(t) \in R^k$, $t \in [0, \alpha], \alpha \le T_i$, is an absolutely continuous function.

Some considerations regarding invariant submanifolds and equivalent controllable systems associated with (1) are made in the previous chapter, but there are significant problems in control theory where the role played by the drift vector field $f_0$ is quite different compared with the control vector fields $f_1, \ldots, f_m$. An example is the stabilization problem associated with (1). It is one of the reasons that this time the study is focussed on the ideal $I_0(f_1, \cdot, f_m)$ generated by $f_0$ in the Lie algebra $L(f_0, \ldots, f_m)$, i.e. $I_0(f_1, \ldots, f_m)$ is the Lie algebra determined by a countable set of vector fields

$$C \stackrel{\triangle}{=} \{ad^k f_0(f_i) : i = 1, \ldots, m, \ k \geq 0\}$$

In addition, the computation of the control function is supposed to be as explicit as possible and the main hypothesis we shall use throughout this chapter is the following

i) $\wedge \stackrel{\triangle}{=} I_0(f_1, \ldots, f_m)$ is a finite dimensional Lie algebra and $f_0 \in I_0(f_1, \ldots, f_m)$. Let $\{Y_1, \ldots, Y_N\} \subseteq I_0(f_1, \ldots, f_m)$ be a fixed basis and for an $x_0$ fixed in the state space $\mathcal{O}$ of (1) denote $k = \dim \wedge(x_0)$. Without no loss of generality we may consider the fixed basis co be $k$–minimal with respect to $x_0 \in \mathcal{O}$ fixed, i.e.;

ii) $\{Y_1(x_0), \ldots, Y_k(x_0)\} \subseteq R^n$ are linearly independent and $Y_j(x_0) = 0$, $j = k + 1, \ldots, N$. According to the $k$–minimal basis $\{Y_1, \ldots, Y_N\} \subseteq \mathcal{O}$, we write $F_j(t)(x)$, $t \in (-a_j, a_j), x \in V_j(x_0) \subseteq \mathcal{O}$, for the local flow generated by the vector field $Y_j$. Starting with the fixed $x_0 \in \mathcal{O}$, it is meaningfull to consider the following restricted composition of flows

(3)
$$y(0) = F(p; x_0) = F_1(t_1) \circ F_2(t_2) \circ \ldots \circ F_k(t_k)(x_0)$$

where

$$k = \dim \wedge(x_0) \quad \text{and} \quad p \stackrel{\triangle}{=} (t_1, \ldots, t_k) \in D_k \stackrel{\triangle}{=} \prod_{i=1}^{k} (-a_i, a_i) \subseteq R^k$$

**Proposition 1.**

Let $f_i \in C^\infty(\mathcal{O}; R^n), i = 0, 1, \ldots, m$, be given such that the hypothesis *(i) is fulfilled and consider* $\{Y_1, \ldots, Y_N\} \subseteq I_0(f_1, \ldots, f_m) \stackrel{\triangle}{=} \wedge$ *a $k$–minimal basis where $x_0 \in \mathcal{O}$ is fixed and $k = \dim \wedge(x_0)$. Then any solution $x^u(t; x_0) \in V(x_0) \subseteq \mathcal{O}$, $t \in [0, T]$, of (1) corresponding to an admissible control $u(\cdot) \in \mathcal{A}_T$ can be represented uniquely by*

(4)
$$x^u(t; x_0) = y(p^u(t)), \ t \in [0, \alpha), \ 0 < \alpha \leq T,$$

*where $y(p), p \in D_k = \prod_{1}^{k} (-a_i, a_i)$, is defined in (3) and $p^u(t) \in D_k$, $t \in [0, \alpha)$, is an absolutely continuous function.*

**Proof.**

By hypothesis, $\wedge = I_0(f_1, \ldots , f_m)$ is a finite dimensional Lie algebra and the fixed $k$-minimal basis $\{Y_1, \ldots , Y_N\} \subseteq \wedge$ determine a global gradient system

(5) $\quad \dfrac{\partial y}{\partial t_1} = Y_1(y), \quad \dfrac{\partial y}{\partial t_2} = X_2(t_1; y), \quad \ldots , \quad \dfrac{\partial y}{\partial t_N} = X_N(t_1, \ldots , t_{N-1}; y),$

for all $y \in \mathcal{O}$ and $\widetilde{p} \overset{\triangle}{=} (t_1, \ldots , t_N) \in R^N$, admitting as a local solution the mapping $y(p), p \in D_k$ defined in (3). For, we notice that $\widetilde{y}(\widetilde{p}) \overset{\triangle}{=} F_1(t_1) \circ \ldots \circ F_k(t_k) \circ \ldots \circ F_N(t_N)(x_0)$ is a local solution of (5) for all $\widetilde{p} \overset{\triangle}{=} (t_1, \ldots , t_N) \in D_N \overset{\triangle}{=} \prod_1^N(-a_i, a_i)$ and as $x_0 \in \mathcal{O}$ is a stationary point for each $Y_j, j = k+1, \ldots , N$, we obtain $F_j(t_j)(x_0) = x_0$, $j = k+1, \ldots , N$, and consequently $y(p), p \in D_k$, defined in (3) coincides with $\widetilde{y}(\widetilde{p}), \widetilde{p} \in D_N$. On the other hand, using Theorem 2.3.1, the following vectors are linearly independent in $R^n$,

(6) $\quad \dfrac{\partial y}{\partial t_1} = Y_1(y(p)), \ldots , \dfrac{\partial y}{\partial t_k}(p) = X_k(t_1, \quad \ldots , \quad t_{k-1}; y(p))$

for any $p \in D_k$, and $q_i \in C^\infty(D_k; R^k)$, $i = 0, 1, \ldots , m$, will exist such that restricting ourselves to the mapping $y(p)$ in (3) we may recover the original vector fields $f_i$ as follows

(7) $\quad \dfrac{\partial y}{\partial p}(p)q_i(p) = f_i(y(p)), i = 0, 1, \ldots , m, \text{ forall } p \in D_k$

Now, consider the auxiliary control system in $R^k$

(8) $\qquad \dfrac{dp}{dt} = q_0(p) + \sum_{i=1}^m u_i(t)q_i(p), \quad p(0) = 0,$

restricted to the state space

$$p \in D_k \overset{\triangle}{=} \prod_1^k(-a_i, a_i) \subseteq R^k,$$

where $u(t) = (u_1(t), \ldots , u_m(t))$ is the admissible control defining the local solution $x^u(t; x_0) \in V(x_0) \subseteq \mathcal{O}, t \in [0, T]$, of the original system (1).

We obtain that $y(p^u(t))$, $t \in [0, \alpha)$, coincides with $x^u(t; x_0)$, $t \in [0, \alpha)$. If another absolutely continuous $p(t) \in D_k$, $t \in [0, \alpha)$, fulfils

$$x^u(t; x_0) = y(p(t)), \; t \in [0, \alpha), \; p(0) = 0,$$

then by derivation we obtain

$$\frac{dp}{dt} = q_0(p(t)) + \sum_{i=1}^{m} u_i(t)q_i(p(t)), \ t \in [0, \alpha)$$

except a finite set of points $t$. The uniqueness of solution in (8) allow one to obtain $p(t) = p^u(t)$, $t \in [0, \alpha)$, and the proof is complete.

**Remark 1.**
   With $x_0 \in \mathcal{O}$ fixed and $\{Y_1, \dots, Y_N\} \subseteq I_0(f_1, \dots, f_m)$ a $k$-minimal basis we may and do associate an invariant $k$-dimensional manifold $M_{x_0} \subseteq \mathcal{O}$ of (1) and using Theorem 2.3.1 it can be defined explicitly as follows

(9)                 $M_{x_0} = \{x \in \mathcal{O} : x = F(p; x_0), p \in D_k\}$

   In other words, following Proposition 1, the trajectories of the control system (1) which start from $M_{x_0}$ will remain in $M_{x_0}$ for all times in a neighbourhood of $t = 0$.

**Remark 2.**
   Let $q_i \in C^\infty(D_k; R^k)$, $i = 0, 1 \dots, m$, be defined such that (7) is fulfilled , where $y(p)$ is defined in (3). The auxiliary control system (see (8)),

(10)               $\frac{dp}{dt} = q_0(p) + \sum_{i=1}^{m} u_i(t)q_i(p). \ p(0) = 0, \ p \in D_k$

plays a key role in studying the original control system (1). It is Theorem 2.3.1 which allow one to conclude that the Lie algebra $L(q_0, \dots, q_m) \subseteq C^\infty(D_k; R^k)$ is finitely generated (of finite type) and according to the hypothesis (i) we deduce that the ideal $I_0(q_1, \dots, q_m)$ generated by $q_0$ into $L(q_0, \dots, q_m)$ fulfils

(11)               $\dim I_0(q_1, \dots, q_m)(p) = k$ for any $p \in D_k$

The abovementioned property (11) determines the controllability of the auxiliary system (10) and, relying on that, the control system (10) is called the equivalent controllable system.

**Proposition 2.**
   Let $f_i \in C^\infty(\mathcal{O}; R^n)$, $i = 0, 1, \dots, m$, be given such that the hypothesis (i) is fulfilled. For $x_0 \in \mathcal{O}$ fixed let $\{Y_1, \dots, Y_N\} \subseteq I_0(f_1, \dots, f_m)$ be a fixed basis as in Proposition 1. Then there is an one to one mapping between local

solutions $x^u(t; x_0)$ of the control system (1) and the local solutions $p^u(t)$ of the equivalent controllable system (10), for each admissible control $u(\cdot) \in \mathcal{A}_T$. In addition, the ideal $I_0(q_1, \ldots, q_m)$ is finitely generated and $q_0 \in I_0(q_1, \ldots, q_m)$

**Proof.**

As was mentioned in the proof of Proposition 1, the vector fields mapping defined in (7) is responsible for a direct computation of the one to one correspondence between two local solutions $x^u(t; x_0)$ and $p^u(t)$ corresponding to the same admissible control $u(\cdot)$. The second part in the statement of the Proposition can be deduced from the following sharper conclusion. The two Lie algebras $I_0(f_1, \ldots, f_m)$ and $I_0(q_1, \ldots, q_m)$ obey to the conclusions in Theorem 3.2.3 and $\{Q_1, \ldots, Q_N\} \subseteq I_0(q_1, \ldots, q_m)$ will exist such that

(12)     $$\text{span}\{Q_1(\text{p}), \ldots, Q_N(\text{p})\} = \text{R}^k \text{ forany p} \in \text{D}_k$$

and any $Q \in I_0(q_1, \ldots, q_m)$ can be written as

(13)     $$Q(p) = \sum_{i=1}^{N} a_i(p) Q_i(p), \ p \in D_k$$

with $a_i(\cdot) \in C^\infty(D_k; R)$, $i = 1, \ldots, N$, depending on $Q$. Let us mention that the conclusion (12) relies again on the mapping between vector fields in (7) and taking directional derivatives with respect to some vector fields in $(q_0, \ldots, q_m)$ we obtain $\{Q_1, \ldots, Q_m\} \subseteq I_0(q_1, \ldots, q_m)$ such that

(14)     $$\frac{\partial y}{\partial p}(p) Q_1(p) = Y_1(y(p)), \ldots, \frac{\partial y}{\partial p}(p) Q_N(p) = Y_N(y(p)),$$

where $y(p), p \in D_k$, is defined in (3).

Finally, by hypothesis $f_0 \in I_0(f_1, \ldots, f_m)$ and using (14) we obtain $q_0 \in C^\infty(D_k; R^k)$ such that

(15)     $$\frac{\partial y}{\partial p}(p) q_0(p) = f_0(y(p)) = \sum_{j=1}^{N} \alpha_j Y_j(y(p)),$$

with $\alpha_j \in R$. Therefore $q_0(p) \in I_0(q_1, \ldots, q_m)(p)$ for any $p \in D_k$ and the proof is complete.

**Remark 3.**

The statements in Propositions 1 and 2 are based on the local solution $y(p), p \in D_k$, defined in (3), and the resulting controllable system in (10) appeared defined locally for $p \in D_k$. It was necessary for the equivalence *mapping between solutions and to point out the manifold structure supporting solutions of the control system (1). It is the purpose of the next Proposition to associate a new full rank Lie algebra independently of the state $x_0 \in \mathcal{O}$ and defined globally on the space of parameters $p \in R^M$ which is relevant for stabilization problem that will be considered later on.*

**Proposition 3.**
   *Assume the hypothesis (i) fulfilled for $I_0(f_1, \ldots f_m)$ and let $\{Y_1, \ldots, Y_M\} \subseteq I_0(f_1, \ldots, f_m)$ be a fixed basis. Then there exist $q_i \in C^\omega(R^M; R^M)$ (analytical vector fields) $i = 0, 1, \ldots, m$, and a homomorphism $\lambda : I_0(f_1, \ldots, f_m) \to I_0(q_1, \ldots, q_m)$ such that*

$c_1$) $\dim I_0(q_1, \ldots, q_m)(p) = M$ *for all* $p \in R^M$

$c_2$) $q_0 \in I_0(q_1, \ldots, q_m)$

**Proof.**
   The statement is more or less an application of the Theorem 2.1.2 and a recalling of the main arguments would be useful. The existence of the homomorphism $\lambda$ is based on the gradient system associated to the fixed basis $\{Y_1, \ldots, Y_M\} \subseteq I_0(f_1, \ldots, f_m)$.
   In addition, the global existence and nonsingular algebraic representation of the gradient system allow one to obtain new vector fields $q_j \in C^\omega(R^M; R^M)$ such that $\lambda(q_i) = f_i, i = 0, 1, \ldots, m$.
   Then, in a natural manner, the mapping $\lambda$ is extended to the Lie algebras $I_0(f_1, \ldots, f_m)$ and $I_0(q_1, \ldots, q_m)$. More precisely, consider the gradient system

(16) $$\frac{\partial y}{\partial t_1} = Y_1(y), \frac{\partial y}{\partial t_2} = X_2(t_1; y), \ldots, \frac{\partial y}{\partial t_M} = X_M(t_1, \ldots, t_{M-1}; y),$$

where the vector fields $X_j(p_j) \in C^\infty(\mathcal{O}; R^n), p_j \overset{\triangle}{=} (t_1, \ldots, t_{j-1})$ are defined as in Theorem 2.1.1 by convergent exponential series

$$X_1 = Y_1,$$

(17) $$X_2(t_1) = (\exp t_1 adY_1)(Y_2), \ldots, X_M(t_1, \ldots, t_{M-1})$$

$$= (\exp t_1 adY_1) \ldots (\exp t_{M-1} adY_{M-1})(Y_M)$$

The hypotheses in Theorem 2.1.2 are fulfilled and the following global algebraic representation is valid

(18) $\{Y_1, X_2(t_1), \ldots, X_M(t_1, \ldots, t_{M-1})\} = \{Y_1, \ldots, Y_M\}A(p), \ p \in R^M$

where the analytical $(M \times M)$ matrix $A(p)$, fulfils $A(0) = I_M$ (identity) and is nonsingular for any $p \in R^M$ as in Theorem 2.1.2. On the other hand, there is no real obstruction in applying Theorem 2.2.1 for our finite–dimensional Lie algebra $I_0(f_1, \ldots, f_m) \subseteq C^\infty(\mathcal{O}; R^n)$ because the open set $\mathcal{O} \subseteq R^n$, replacing the full space $R^n$, is not essential in the algebraic representation in (18). Starting with (18) and noticing that a basis $\{Y_1, \ldots, Y_M\} \subseteq I_0(f_1, \ldots, f_m)$ is an $M$–minimal system of generators we may and do rewrite $I_0(f_1, \ldots, f_m) \overset{\triangle}{=} \Lambda$, as a Lie algebra determined by the fixed basis. Now, applying Theorem 2.2.1 we obtain a homomorphism $\lambda : \Lambda \to L(Q_1, \ldots, Q_M)$, where $Q_i \in C^\omega(R^M; R^M)$ are such that

(19) $$\dim L(Q_1, \ldots, Q_M)(p) = M \text{ for any } p \in R^M.$$

In particular, using $f_j \in \Lambda$, let $q_j \in C^\omega(R^M; R^M)$ be such that

(20) $$\lambda(f_j) = q_j, \ j = 0, 1, \ldots, m.$$

The conclusion $(c_2)$ in Theorem 2.2.1 and the equations (20) allow one to see that any $f \in I_0(f_1, \ldots, f_m)$ can be associated with a $q \in I_0(q_1, \ldots, q_m)$ fulfilling

(21) $$\lambda(f) = q$$

As far as $L(Y_1, \ldots, Y_M) = I_0(f_1, \ldots, f_m)$ and $Q_i \in I_0(q_1, \ldots, q_m)$, from (21) we obtain

(21) $$L(Q_1, \ldots, Q_M) = I_0(q_1, \ldots, q_m)$$

and the proof is complete.

**Remark 4.**

*Under the hypothesis (i) in Proposition 3 we may establish a mapping acting only in one direction, from the solutions of an auxiliary controllable system to the solutions of the original control system (1). In other words, fixing a basis $\{Y_1, \ldots, Y_M\} \subseteq I_0(f_1, \ldots, f_m)$ we associate a controllable system in the whole space $R^M$*

(a) $$\frac{dp}{dt} = q_0(p) + \sum_{i=1}^{m} u_i(t) q_i(p), \ p(0) = 0, \ p \in R^M,$$

*such that a local solution $p^u(t) \in D_M, t \in [0, \alpha], \alpha \leq T$, corresponding to an admissible control $u(\cdot) \in A_T$ defines a solution in (1) as $x^u(t; x_0) =$*

$F(p^u(t); x_0)$, $t \in [0, \alpha]$, *where* $F(p; x_0)$ *is the composition of local flows,*

$$F(p; x_0) = F_1(t_1) \circ \ldots, \circ F_M(t_M)(x_0),$$

(b)

$$p \triangleq (t_1, \ldots, t_M) \in D_M \triangleq \prod_{i=1}^{M}(-a_i, a_i),$$

*and* $F_i(t_i; x)$, $t_i \in (-a_i, a_i)$, $x \in V_i(x_0)$, *is the local flow generated by* $Y_i$ *in the fixed basis.*

In the case that each $Y_i$ is a complete vector field there is no restriction for the domain mapping in (b), i.e. $D_M = R^M$. The controllable system in (a) is defined independently of any initial condition $x_0 \in \mathcal{O}$ and admissible control used for solutions in (1). It will be relevant for avoiding the special analysis involved for describing singularities of the Lie algebra $I_0(f_1, \ldots, f_m)$.

The above given remarks can be summarized in the following:

**Proposition 4.**

*Assume* $f_i \in C^\infty(\mathcal{O}; R^n)$, $i = 0, 1, \ldots, m$, *be given such that the Lie algebra* $I_0(f_1, \ldots, f_m)$, *meet the hypothesis (i). Let* $\{Y_1, \ldots, Y_M\} \subseteq I_0(f_1, \ldots, f_m)$ *be a fixed basis and associate* $q_i \in C^\omega(R^M; R^M)$ *and* $\lambda : I_0(f_1, \ldots, f_m) \to I_0(q_1, \ldots, q_m)$ *as in Proposition 3. Then, for each local solution* $p^u(t) \in D_M, t \in [0, T]$, *of the auxiliary controllable system (a), we obtain a local solution* $x^u(t; x_0) = F(p^u(t); x_0), 0 \leq t \leq T$, *of (1), where* $F(p; x_0), p \in D_M$, *is defined in (b). In addition, using the same mapping* $F(p; x_0)$ *in (b) and for* $Q_j \in L(q_1, \ldots, q_m), j = 1, \ldots, N$, *fixed, a local solution of the extended auxiliary system.*

(c)
$$\frac{dp}{dt} = q_0(p) + \sum_{j=1}^{N} v_j(t)Q_j(p), \ p(0) = 0, \ t \in [0, T], \ p \in D_M$$

*gives rise to a local solution* $x^v(t; x_0) \in \mathcal{O}, t \in [0, T]$ *of the extended original system*

(d)
$$\frac{dx}{dt} = f_0(x) + \sum_{j=1}^{N} v_j(t)Y_j(x), \ x(0) = x_0, x \in \mathcal{O}$$

*where* $Y_j \in L(f_1, \ldots, f_m)$ *are fixed such that* $\lambda(Y_j) = Q_j$.

**Proof.**

The first part in the statement follows by a direct computation of the derivative of the function $F(p^u(t); x_0)$ with respect to $t \in [0, T]$, provided we notice that the homomorphism $\lambda$ between the two Lie algebras fulfils

(22)
$$\frac{\partial F}{\partial p}(p; x_0)q_i(p) = f_i(F(p; x_0)), \quad i = 0, 1, \ldots, m$$
$$\frac{\partial F}{\partial p}(p; x_0)\lambda(f)(p) = f(F(p; x_0)), \quad \text{for any } f \in I_0(f_1, \ldots, f_m)$$

In particular, the mapping $F(p; x_0), p \in D_M$, defined in (b) is a local solution of the gradient system

(23)
$$\frac{\partial y}{\partial t_1} = Y_1(y), \quad \frac{\partial y}{\partial t_2} = X_2(t_1; y), \quad \ldots,$$
$$= \frac{\partial y}{\partial t_M} = X_M(t_1, \ldots, t_{M-1}; y), \quad p \in R^M, \ y \in \mathcal{O}$$

and along with each local solution of (23) we obtain (22) fulfilled. Actually, for $p_* \in R^M$ and $x_* \in \mathcal{O}$ fixed arbitrarily there exist an open set $\mathcal{U}(p_*) \subseteq R^M$, containing $p_*$, and a local solution $\hat{y}(p), p \in \mathcal{U}(p_*)$, of (23) with Cauchy data $\hat{y}(p_*) = x_*$.

As in Proposition 3, the algebraic representation of the gradient system (23) is used and we obtain

(24)
$$\frac{\partial \hat{y}}{\partial p}(p) = \{Y_1(\hat{y}(p)), \ldots, Y_M(\hat{y}(p))\}A(p), \quad \text{for all } p \in \mathcal{U}(p_*)$$

The nonsingular $(M \times M)$ matrix $A(p)$ in (24) is defined for any $p \in R^M$ and let $q_j(p), p \in R^M, j = 0, 1, \ldots, m$ be the solutions of the algebraic equations

(25)
$$A(p)q_j(p) = a_j$$

where $a_j \in R^M$ are the coordinates of $f_j$ with respect to the fixed basis $\{Y_1, \ldots, Y_M\} \subseteq I_0(f_1, \ldots, f_m)$.

Then multiplying with $q_j(p)$ in (24) allows one to see

(26)
$$\frac{\partial \hat{y}}{\partial p}(p)q_j(p) = f_j(\hat{y}(p)), \quad \text{for all} \quad p \in \mathcal{U}(p_*), j = 0, 1, \ldots, m.$$

From (26), by a standard procedure we pass from the individuals $q_j, f_j$, to the corresponding Lie product such that

(27)
$$\frac{\partial \hat{y}}{\partial p}(p)q(p) = f(\hat{y}(p)), \quad \text{for any } f \in I_0(f_1, \ldots, f_m)$$

and $q = \lambda(f) \in I_0(q_1, \ldots, q_m)$. In particular, for $Y \in L(f_1, \ldots, f_m)$ we obtain $Q = \lambda(Y) \in L(q_1, \ldots, q_m) \subseteq I_0(q_1, \ldots, q_m)$ such that

$$\text{(28)} \qquad \frac{\partial \widehat{y}}{\partial p}(p)Q(p) = Y(\widehat{y}(p)), \text{for all } p \in \mathcal{U}(p_*)$$

and using $F(p; x_0), p \in D_M$, defined in (b) of Remark 3 as a local solution $\widehat{y}(p)$ for which (28) holds we obtain

$$\text{(29)} \qquad \begin{aligned} \frac{\partial F}{\partial p}(p; x_0)q_i(p) &= f_i(F(p; x_0)), \quad p \in D_M, i = 0, 1, \ldots, m, \\ \frac{\partial F}{\partial p}(p; x_0)Q_j(p) &= Y_j(F(p; x_0)), \quad p \in D_M, j = 1, \ldots, N. \end{aligned}$$

Here $Q_j \in L(q_1, \ldots, q_m)$, and $Y_j \in L(f_1, \ldots, f_m)$ are fixed in the statement.

Denote $x^u(t; x_0) = F(p^u(t); x_0)$ and $x^v(t; x_0) = F(p^v(t); x_0)$ for $t \in [0, T]$ where $p^u(t), p^v(t) \in D_M$ are the solutions of the auxiliary controllable system (a) and respectively the extended auxiliary system (c).

Finally, it is a straightderivation with respect to the time variable $t \in [0, T]$ which allow one to conclude that $x^u(t; x_0)$ is a solution for the original system and $x^v(t; x_0)$ agrees with the extended original system (d).

The proof is complete.

*Comment on auxiliary control systems.*

The clasical Chow Theorem (1939) tell us that if $q_0 = 0$ (see $f_0 = 0$) then any point $p \in D_M$ can be joined with the origin $0 \in D_M$ using an admissible solution of the system (a); by reversing time the trajectory is traversed in both directions. It is based on the fact that solutions in (c) fulfil this property (see $\dim L(q_1, \ldots, q_m)(p) = M$ for any $p \in D_M$) and in addition each trajectory of (c) can be approximated in a good way with solutions of the auxiliary controllable system.

$$\frac{dp}{dt} = \sum_{i=1}^{m} u_i(t)q_i(p), \ p \in D_M.$$

When $f_0 \neq 0$, and consequently $q_0 \neq 0$, then approximations of the type already mentioned are relevant around the fixed trajectory $F_0(t)(x_0), t \in [0, T]$, where the local flow $F_0(t)(x), x \in V(x_0) \subseteq \mathcal{O}$, is generated by the vector field $f_0$. Good estimates are expected if the finite–dimensional Lie algebra $\Lambda = I_0(f_1, \ldots, f_m)$ fulfils the full rank condition, i.e., $\dim \Lambda(x) = n$ for all $x \in \mathcal{O}$, which is an embarrassing hypothesis. It is the full rank condition of the auxiliary controllable system (a) (see $\dim I_0(q_1, \ldots, q_m)(p) = M$ for all $p \in R^M$) which allow one to obtain good estimates around the fixed trajectory $Q_0(t)(0)$, $t \in [0, T]$, generated by the vector field $q_0$ and it will give rise to good estimates for the original control system (1). This is why we shall focuss the next analysis on approximations for auxiliary control system in (a).

## 5.2   Approximations, Small Controls

A simple and useful parametrization around a fixed trajectory is obtainable provided a special controllability condition is imposed, and from this point of view the auxiliary controllable system (a) is more appropriate than the original one in (1).

On the other hand, the approximations and small controls which are introduced here require no controllability conditions, and they can be performed on the general affine control systems without any need of the hypothesis (i) or the correspondence $\lambda : I_0(f_1, \dots, f_m) \to (q_1, \dots, q_m)$. Finally, the conclusions are based on the parametrization around a fixed trajectory for which small controls and approximations are necessary ingredients. This is why the analysis is performed on affine control systems resembling the auxiliary controllable system of the form

$$(*) \qquad \frac{dp}{dt} = q_0(p) + \sum_{i=1}^{m} u_i(t) q_i(p), \ p(0) = 0, \ p \in D_M = \prod_{i=1}^{M} (-a_i, a_i)$$

where $q_i \in C^\infty(R^M, ; R^M)$, $i = 0, 1, \dots, m$, and assume that the solution $p^0(t) \in D_M$, $t \in [0, T]$, of

$$(**) \qquad \qquad \frac{dp}{dt} = q_0(p), \ p(0) = 0$$

exists. The reachable set $R(T) \subseteq D_M$ at the instant $t = T$ associated with $(*)$ is given by $R(T) = \{p \in D_M : p = p^u(T)$, for some $p^u(\cdot)$ solution of $(*), u(\cdot) \in \mathcal{A}_T\}$ where $\mathcal{A}_T$ is the admissible class of bounded piecewise continuous controls $u(\cdot), \dots, u_m(\cdot)$ defined on $[0, T]$. Let us mention that a simple and explicit parametrization of the reachable set $R(T)$ around $p^0(T)$ induces a similar parametrization of the reachable set $R_{x_0}(T)$ of the original system (1) around the fixed point $x^0(T)$ provided we use the mapping $F(p; x_0), p \in D_M$, defined in (b) under the hypothesis (i) and there holds

$$R_{x_0}(T) = F(R(T); x_0).$$

In doing this we need to produce trajectories $p^u(t), t \in [0, T]$, restricted to the open set $D_M$ which can be determined by using small controls in

$L_\infty([0,T]; R^m)$. On the other hand, we are supposed to produce approximating solutions for extended controllable system defined in (c). This will bring another type of "small controls" which have been used for producing a Lie bracket starting with the vector fields composing a Lie bracket.

### a) Small controls

As far as the reference trajectory fulfils $p^0(t) \in D_M$, $t \in [0,T]$, it is obvious that small controls $u(\cdot) \in \mathcal{A}_T$ in the space $L_\infty([0,T]; R^m)$, i.e., $\|u\|_\infty \stackrel{\triangle}{=} \|u(t)\| \le \eta$ for some sufficiently small $\eta > 0$, will allow one to obtain admissible solutions $p^u(\cdot)$ in (∗) with respect to the fixed open set $D_M = \prod_{i=1}^{M}(-a_i, a_i)$. In other words, $p^u(t)$ belongs to $D_M$ for all $t \in [0,T]$ provided $u(\cdot) \in \mathcal{U}_\infty \subseteq \mathcal{A}_T$ where

(1) $$\mathcal{U}_\infty \stackrel{\triangle}{=} \{u(\cdot) \in \mathcal{A}_T : \|u\|_\infty \le \eta\} \text{ for some } \eta > 0.$$

In the set $\mathcal{U}_\infty$ are included admissible controls which are pointwise restricted to a small ball around the origin in $R^m$ and they are not appropriate for approximating a trajectory driven by a Lie bracket. For this type of approximation, which encompasses the trajectories in the extended auxiliary system (c), we need small controls measuring the smallness by the integral of $u(\cdot)$, i.e., $\int_0^T u(t)dt$ must be small in $R^m$. In the differential geometric setting a Lie bracket is produced using two piecewise constant scalar functions $u_1(\cdot), u_2(\cdot)$ with the range in the set $\{-1, 0, 1\}$, but it can be accomplished also by using piecewise polynomial scalar functions which belong to $C^1([0,T]; R)$. We adopt the smooth type of controls and, to begin with, let $S^0$ be the space consisting of scalar polynomial functions defined on the interval $[0,1]$ and satisfying

($\alpha$) $$\int_0^1 s(t)dt = 0, \qquad s(0) = s(1) = 0, \qquad \frac{ds}{dt}(0) = \frac{ds}{dt}(1) = 0$$

Let $s_1(\cdot), s_2(\cdot) \in S^0$ be such that

($\beta$) $$\int_0^1 s_1(t)\hat{s_2}(dt) = 1 \quad \left(\text{see} \int_0^1 s_2(t)\hat{s_1}(t)dt = -1\right)$$

where $\widehat{s}(t) \stackrel{\triangle}{=} \int\limits_0^t s(\tau)d\tau.$ These functions will be fixed in the sequel and they could be taken as polynomials of sixth and fifth degree respectively. From the fixed functions $s_1(\cdot), s_2(\cdot)$ we obtain two admissible controls $u_1(t, h), u_2(t, h), t \in [0, T]$, depending on the parameter $h = T/N$, by shifting $s_1(\cdot)$ and $s_2(\cdot)$ from $[0, 1]$ on each interval $[kh, (k + 1)h], k = 0, 1, \ldots, N - 1,$ where $N$ is a natural number. More precisely, write

$$s_i^k(t, h) = s_i((t - kh)/h), \ t \in [kh, (k + 1)h] \stackrel{\triangle}{=} I_k, \ i = 1, 2,$$

and

$$(\gamma) \qquad u_i(t, h) = \frac{1}{\sqrt{h}} s_i^k(t, h), \ t \in I_h, k = 0, 1, \ldots, N - 1, \ i = 1, 2.$$

The new functions $u_1(\cdot, h), u_2(\cdot, h)$ agree with the properties $(\alpha)$ and $(\beta)$, i.e. $u_i(kh, h) = \dfrac{du_i}{dt}(kh, h) = 0, \ k = 0, 1, \ldots, N,$ and $u_i(\cdot) \in C^1([0, T]; R), \ i = 1, 2.$

In addition, the following integral conditions hold

$$\int\limits_{I_k} u_i(t, h)dt = 0, \ k = 0, 1, \cdot, N - 1, \ \text{for i} = 1, 2,$$

and

$$(2) \qquad \int\limits_{I_k} u_1(t, h)\widetilde{u}_2(t, h)dt = h \ (\text{or} \int\limits_{I_k} u_2(t, h)\widetilde{u}_1(t, h)dt = -h)$$

for any $k = 0, 1, \cdot, N - 1,$ where

$$\widetilde{u}(t, h) = \int\limits_0^t u(\tau, h)d\tau, t \in [0, T].$$

Each of the two admissible controls is small in the integral form provided $h \stackrel{\triangle}{=} T/N$ is small, i.e.,

$$(3) \qquad \max_{t \in [0, T]} \left| \int\limits_0^t u_i(\tau) \right| \leq C_0\sqrt{h}, \ \text{for some constant } C_0 > 0,$$

and it can be proved by noticing that

$$\int_0^t u(\tau, h)d\tau = \sum_{k=0}^{j-1} \int_{I_k} u(t, h)dt + \int_{jh}^t u(\tau, h)d\tau = \int_{jh}^t u(\tau, h)d\tau$$

for $t \in [jh, (j+1)h]$, and

$$\max_{t \in [jh,(j+1)h]} \left| \int_0^t u_i(\tau, h)d\tau \right| \leq \int_{jh}^{(j+1)h} |u_i(t, h)|dt = \sqrt{h} \int_0^1 |s_i(t)|dt$$
$$\leq C_0\sqrt{h}$$

Denote by $W_h$ the set of functions in $C^1([0, T]; R)$ fulfilling (3) for some constant $C_0 > 0$. The effect of the controls $u_i(\cdot, h) \in W_h$ in a dynamical system can be measured by a new vector field defined as a Lie bracket $[q_1, q_2]$ of the two vector fields multiplying the given control functions. Namely, consider the equation:

(4) $$\frac{dp}{dt} = q_0(p) + u_1(t, h)q_1(p) + u_2(t, h)q_2(p), p(0) = 0$$

where $q_j \in C^\infty(R^M; R^M)$, and let $\widetilde{p}(t), t \in [0, T]$, be the solution of

(5) $$\frac{dp}{dt} = q_0(p) + [q_1, q_2](p), \ p(0) = 0$$

fulfilling $\widetilde{p}(t) \in D_M = \prod_{i=1}^M (-a_i, a_i)$, for all $t \in [0, T]$.

**Remark 1.**
    *There is no loss of generality considering that the vector fields $q_j$ in (4) and (5) are bounded on $R^M$, otherwise we take a closed sufficiently large ball $S_\rho(0) \subseteq R^M$ such that the existing solution $\widetilde{p}(\cdot)$ fulfils $\widetilde{p}(t) \in$ int $S_\rho(0)$ for all $t \in [0, T]$. Define new bounded vector fields $\widetilde{q}_j(p) = \widetilde{\alpha}(p)q_j(p)$, where $\widetilde{\alpha}(\cdot) \in C^\infty(R^M; R)$ is chosen such that $\widetilde{\alpha}(p) = 1$ for $p \in S_\rho(0), \widetilde{\alpha}(p) = 0$ for all $p \in R^M \backslash S_{2\rho}(0)$, and $0 \leq \widetilde{\alpha}(p) \leq 1$ in the rest. The result is that the new dynamic containing $\widetilde{q}_j$ in the place of $q_j$ will give rise to a solution which remains in $S_\rho(0)$ for sufficiently small $h$, and it is the reason why we may consider the vector fields are bounded. Let $p_h(t), t \in [0, T]$, be the solution of (4) and it is worth to point out the behaviour of $p_h(\cdot)$ with respect to $h$.*

On the first interval $I_0 = [0, h]$, and using the integral form of the solution we obtain

(6)
$$
\begin{aligned}
p_h(h) &= \int_{I_0} q_0(p_h(t))dt + \int_{I_0} u_1(t,h)q_1(p_h(t))dt + \int_{I_0} u_2(t,h)q_2(ph(t))dt \\
&= \int_{I_0} q_0(ph(t))dt + \left( \int_{I_0} u_1(t,h)dt \right) q_1(0) + \left( \int_{I_0} u_2(t,h)dt \right) q_2(0) \\
&\quad + \int_{I_0} u_1(t,h) \left( \int_0^t \frac{\partial q_1}{\partial p}(ph(\sigma))\frac{dph(\sigma)}{d\sigma}d\sigma \right) dt \\
&\quad + \int_{I_0} u_2(t,h) \left( \int_0^t \frac{\partial q_2}{\partial p}(ph(\sigma))\frac{dph}{d\sigma}(\sigma)d\sigma \right) dt
\end{aligned}
$$

The conditions (2) and (3) used in (6) allow one to see that:

(7)  $\quad p_h(h) = \int_0^h q_0(p_h(t))dt + h\left[ \frac{\partial q_1}{\partial p}(0)q_2(0) - \frac{\partial q_2}{\partial p}(0)q_1(0) \right] + hO(\sqrt{h}),$

where $0(\eta)$ denotes a function fulfilling $\|(0\eta)\| \leq C\eta$ with a fixed constant $C > 0$ not depending on $t \in [0, T]$ and $h$. The last expression can be rewriten as

(8)  $\quad p_h(h) = \int_0^h q_0(p_h(t))dt + \int_0^h [q_1, q_2](p_h(t))dt + hO\left( \sqrt{h} \right),$

where

$$
[q_1, q_2](p) \triangleq \frac{\partial q_1}{\partial p}(p)q_2(p) - \frac{\partial q_2}{\partial p}(p)q_1(p)
$$

is the Lie bracket associated with the vector fields $q_i \in C^\infty(R^M; R^M)$. Repeating the above given computations on each interval $I_k, k = 0, 1, \ldots, N-1$, we obtain

(9)  $\quad p_h(jh) = \int_0^{jh} q_0(ph(t))dt + \int_0^{jh} [q_1, q_2](p_h(t))dt + jhO\left( \sqrt{h} \right)$

for any $j \in \{0, 1, \ldots, N\}$. Using

$$
\left| \int_{kh}^t u_i(\sigma, h)d\sigma \right| \leq C_0\sqrt{h} \qquad \text{for any } t \in I_k
$$

(see (3)), the equation (9) can be extended for all $t \in [0, T]$ and we obtain:
(10)

$$p_h(t) = \int_0^t q_0(p_h(\sigma))d\sigma + \int_0^t [q_1, q_2](p_h(\sigma))d\sigma + O\left(\sqrt{h}\right), \quad \text{for all } t \in [0, T]$$

Now, let $\widetilde{p}(t)$, $t \in [0, T]$, be the solution of

(11) $$\frac{dp}{dt} = q_0(p) + [q_1, q_2](p), \ p(0) = 0$$

fulfilling $\widetilde{p}(t) \in D_M$ for all $t$.

A standard argument shows that $\varphi(t) \stackrel{\triangle}{=} \|p_h(t) - \widetilde{p}(t)\|$ satisfies

(12) $$\varphi(t) \le C_1 \int_0^t \varphi(\sigma)d\sigma + \left\| O\left(\sqrt{h}\right) \right\|, \quad t \in [0, T].$$

Gronwall's Lemma applied to (12) implies

(13) $$\|p_h(t) - \widetilde{p}(t)\| \le \|O(\sqrt{h})\| \exp C_1 \quad \text{for all } t \in [0, T],$$

and therefore we may conclude that

(14) $$\max_{t \in [0,T]} \|p_h(t) - \widetilde{p}(t)\| \le C_2\sqrt{h}, \quad \text{for some constant } C_2 > 0$$

The conclusion (14) can be interpreted as an approximation in the space $C([0, T]; R^M)$ of $\widetilde{p}(\cdot)$ by solutions of the control system (4) using small controls $u_i(\cdot, h) \in W_h$ in a sense specified by the norm in (3).

**b) Approximations using small controls in $\mathcal{U}_\infty \cup W_h$**
The above given computations can be summarized as follows:

**Lemma 1**

Let $q_j \in C^\infty(R^M; R^M)$ and $D_M = \prod_{i=1}^M (-a_i, a_i), j = 0, 1, 2$, be given and denote $\widetilde{p}(t, v), t \in [0, T]$, the solution of

(*) $$\frac{dp}{dt} = q_0(p) + v(t)[q_1, q_2](p), \ p(0) = 0,$$

fulfilling $\widetilde{p}(t) \in D_M$ for all $t \in [0, T]$, where $v(\cdot) \in C^1([0, T]; R)$. Let $u_i(t, h)$ be defined as in $(\gamma), i = 1, 2$, and consider $p_h(t, v), t \in [0, T]$, the solution of

(**) $$\frac{dp}{dt} = q_0(p) + v(t)u_1(t, h)q_1(p) + u_2(t, h)q_2(p), \quad p(0) = 0.$$

*Then $p_h(t,v) \in D_M$ for all $t \in [0,T]$, and*

*(c)*        $\max\limits_{t \in [0,T]} \|p_h(t,v) - \widetilde{p}(t,v)\| \leq C_2 \sqrt{h} for some constant C_2 > 0$

*provided $h \overset{\Delta}{=} T/N$ is sufficiently small and*

$$\|v\|_1 \overset{\Delta}{=} \max\limits_{t \in [0,T]} \left( \|v(t)\| + \left\| \frac{dv}{dt}(t) \right\| \right) \leq C_1$$

*for some constant $C_1 > 0$.*

### Sketch of the proof

The only difference between the systems in Lemma 1 and those used in the previous computations (see (4) and (5)) is the appearance of a function $v(\cdot) \in C^1([0,T]; R)$, i.e., $\|v\|_1 \leq C_1$. The systems $(*)$ and $(**)$ in Lemma 1 can be written as in (4) and respectively (5) by adding a new state variable and denote

$$r = (t,p) \in R^{M+1},$$

$$\widetilde{q}_0(r) = \begin{pmatrix} 1 \\ q_0(p) \end{pmatrix}, \quad \widetilde{q}_1(r) = \begin{pmatrix} 0 \\ v(t)q_1(p) \end{pmatrix}, \quad \widetilde{q}_2(r) = \begin{pmatrix} 0 \\ q_2(p) \end{pmatrix}.$$

With these notations, we rewrite $(*)$ as

$(\sharp)$        $$\frac{dr}{d\tau} = \widetilde{q}_0(r) + [\widetilde{q}_1, \widetilde{q}_2](r), \ r(0) = 0$$

and $(**)$ receives the following form

$(\sharp\sharp)$.        $$\frac{dr}{d\tau} = \widetilde{q}_0(r) + u_1(\tau, h)\widetilde{q}_1(r) + u_2(\tau, h)\widetilde{q}_2(r), \qquad r(0) = 0$$

Noticing that $\widetilde{q}_j$ and $\dfrac{\partial \widetilde{q}_j}{\partial r}, j = 0, 1, 2$, are bounded provided $q_j$, $\dfrac{\partial q_j}{\partial p}$ fulfils a boundedness condition and $\|v\|_1 \leq C_1$, then Remark 1 holds for the enlarged systems describing the dynamics of the new variable $r = (t,p)$. As far as $[\widetilde{q}_1, \widetilde{q}_2](r) = \begin{pmatrix} 0 \\ v(t)[q_1, q_2](p) \end{pmatrix}$, we obtain that the estimate in (14) is fulfilled for the corresponding solutions in $(\sharp)$ and $(\sharp\sharp)$ which represents the conclusion in our statement. The proof is complete.

In Lemma 1 only a one scalar function $v(\cdot) \in C^1)[0,T]; R)$ is considered. An induction argument completing the computations in Lemma 1 will allow one to handle with a finite set of Lie products with different lengths multiplied by some scalar functions in $C^1([0,T];R)$. The following result regarding approximations can be found as a particular case of the analysis performed in [17].

**Lemma 2.**
   *Let* $q_i \in C^\infty(R^M;R^M), i = 0,1,\ldots,m$, *be given and consider a set of functions* $S_1 = \{v \in C^1([0,T];R) : \|v\|_1 \le c_1\}$. *Let* $Q_1,\ldots,Q_N \in L(q_1,\ldots,q_m)$ *be fixed and for each* $v_1,(\cdot),\ldots,v_N(\cdot) \in S_1$ *consider* $p_\varepsilon(t,v), t \in [0,T]$, *the solution of*

$$(*) \qquad \frac{dp}{dt} = v_0 q_0(p) + \varepsilon \sum_{j=1}^N v_j(t) Q_j(p), p(0) = 0,$$

*fulfilling* $p_\varepsilon(t,v) \in D_M \overset{\triangle}{=} \prod_{i=1}^M (-a_i, a_i), t \in [0,T]$, *for* $\varepsilon \in [0,\varepsilon_0], v_0 \in [0,\varepsilon_1]$.
*Then there exist* $u_{i_j}(\cdot,h), u_i(\cdot,h) \in W_h \subseteq C^1([0,T];R), i = 1,\ldots,m, j = 1,\ldots,N$, *such that the solution* $p_\varepsilon(t,v,h), t \in [0,T]$, *of*

$$(**) \qquad \frac{dp}{dt} = v_0 q_0(p) + \varepsilon \sum_{j=1}^N v_j(t) u_{ij}(t,h)\widetilde{q}_j(p) + \sum_{i=1}^m u_i(t,h)q_i(p), \; p(0) = 0$$

*fulfils* $p_\varepsilon(t,v,h) \in D_M$ *for all* $t \in [0,T]$, *and*

$$(c) \qquad \max_{t \in [0,T]} \|p_\varepsilon(t,v,h) - p_\varepsilon(t,v)\| \le C_2\sqrt{h}$$

*for some constant* $C_2 > 0$ *uniformly with respect to* $v_i(\cdot) \in S_1, \varepsilon \in [0,\varepsilon_0], v_0 \in [0,\varepsilon_1]$, *where* $\widetilde{q}_j$ *denotes the left first vector field in the Lie product* $Q_j = [\widetilde{q}_j,\ldots,]$.

**Remark 2.**
   *It is easily seen that the conclusion in Lemma 2 allow one to approximate solutions of the extended system* $(*)$ *by using small controls* $u(\cdot,h) \in W_h$ *and the given vector fields* $q_j, j = 0,1,\ldots,m$. *In addition, the explicit estimate in (c) shows in particular that for* $t = T$ *there holds*

$$p_\varepsilon(T,v,h) = p_\varepsilon(T,v) + o(\delta,v), \text{ where } \lim_{\delta \downarrow 0} \frac{o(\delta,v)}{\delta} = 0$$

*uniformly in* $\varepsilon \in [0,\varepsilon_0], v_0 \in [0,\varepsilon_1], v = (v_1,\ldots,v_N)$ *with* $v_i(\cdot) \in S_1$, *provided* $\delta = h^r$ *and* $r \in (0,1/2)$.

**Remark 3.**

*There is no loss of generality considering the controls $u_i(\cdot, h) \in W_h, i = 1, \ldots, m$, in Lemma 2 with the following additional property satisfied*

$$(***) \qquad u_i(T - t, h) = -u_i(t, h), \ t \in [0, T], i = 1, \ldots, m.$$

*As Lemma 1 shows, the controls $u_i(\cdot, h), i = 1, \ldots, m$, may appear only by decomposition of a Lie product which contains at least two vector fields. The property $(***)$ is consistent with the situation when some $Q_j \in L(q_1, \ldots, q_m)$ of nonzero length appears in the system $(*)$ and in this case the approximation result in Lemma 2 can be stated on the interval $[0, T/2]$ and with $h = T/2N$. By definition, $u_{i_j}(T/2, h) = 0$ and $u_i(T/2, h) = 0$ for $j = 1, \ldots, N$, $i = 1, \ldots, m$. Taking $u_{i_j}(t, h) = 0$ for $t \in [T/2, T], j = 1, \ldots, N$, and redefining $u_i(t, h) = -u_i(T - t; h)$ for $t \in [t/2, T], i = 1, \ldots, m$, we obtain periodic controls $u_i(\cdot)h)$ in the class $W_h$ and fulfilling $(***)$ for each $i = 1, \ldots, m$. With this modification, the solution of $(**)$ defined on $[0, T]$ will approximate the corresponding solution in $(*)$ provided $v_j(t) = 0$ for all $t \in [T/2, T]$ and $j = 1, \ldots, N$. There is one obvious advantage in considering smooth controls $u_i(\cdot, h)$ which obey to the condition $(***)$. They produce smooth periodic solutions $\varphi(t, p), \ t \in [0, T]$, for the control system*

$$(15) \qquad \frac{dp}{dt} = \sum_{i=1}^{m} u_i(t, h) q_i(p), \ t \in [0, T],$$

*i.e. $\varphi(0, p) = \varphi(T, p) = p$ provided $h$ is sufficiently small and $p$ belongs to a ball $B(0) \subseteq D_M$ centered at origin.*

*In addition, the solution $P(t, p) = (P_1(t, p), \ldots, P_M(t, p))$, of the linear hyperbolic equation*

$$(16) \qquad \frac{\partial P}{\partial t}(t, p) + \sum_{i=1}^{m} u_i(t, h) \frac{\partial P}{\partial p}(t, p) q_i(p) = 0, \ t \in [0, T],$$

*exists and is a periodic one, i. e., $P(0, p) = P(T, p) = p$ for any $p \in B(0) \subseteq D_M$, provided $h$ is sufficiently small. Actually $P(t, p)$ can be defined as $P(t, p) = \varphi_{t,p}(T)$, where $\varphi_{t,p}(s), s \in [t, T]$, is the solution in (15) with Cauchy data $\varphi(t) = p$.*

**Lemma 3**

*Assume the conditions in Lemma 2 fulfilled and let $\overline{p}_\varepsilon(t, v, h)$ be the solution of $(**)$ where $u_i(t, h), i = 1, \ldots, m$, meet $(***)$ in Remark 3. Let $P(t, p), t \in [0, T], p \in B(0) \subseteq D_M$, be the periodic solution of (16). Then the* *following representation holds*

$$(c_1) \qquad p_\varepsilon(T, v, h) = \int_0^T \frac{\partial P}{\partial p}(t, p_\varepsilon(t, v, h))[v_0 q_0(p_\varepsilon(t, v, h))$$

$$+\varepsilon \sum_{j=1}^N v_j(t) u_{ij}(t, h) \tilde{q}_j(p_\varepsilon(t, v, h))]dt$$

*and*

$$(c_2) \qquad \lim_{\delta \downarrow 0} \frac{p\delta(T, v, h) - p\delta(T, v)}{\delta} = 0 \qquad uniformlyin \ v_0 \in [0, \varepsilon_1],$$

$v = (v_1, \ldots, v_N)$ *with* $\|v_j\|_1 \le C_1$, *provided* $\varepsilon = \delta = h^{1/3}$ *and* $p_\delta(t, v), t \in [0, T]$, *is the solution of* $(*)$ *in Lemma 2.*

### Proof

By definition, the smooth periodic solution $P(t, p)$, $t \in [0, T]$, obeys to (16) for any $p \in B(0) \supseteq D_M$ and, in particular, for $p = p_\varepsilon(t, v, h)$ we obtain that $P(t, p_\varepsilon(t, v, h))$, $t \in [0, T]$, meet

$$P(t, p_\varepsilon(T, v, h) = p_\varepsilon(T, v, h), P(0, v, h)) = p_\varepsilon(0, v, h) = 0$$

Computing the derivative of $P(t, p_\varepsilon(t, v, h))$ with respect to $t \in [0, T]$, and using (16) we obtain $p_\varepsilon(T, v, h)$ represented as in conclusion $(c_1)$.

The conclusion $(c_2)$ is a rewritting of (c) in Lemma 2 provided $\varepsilon = \delta = h^{1/3}$ and $p\delta(t, v)$ is the corresponding solution in the control system $(*)$. The proof is complete.

### Remark 4

*The conclusions in Lemma 3 allow one to see a simple parametrization of the reachable set* $R(T)$ *associated with auxiliary control system*

$$(17) \qquad \frac{dp}{dt} = v_0 q_0(p) + \sum_{i=1}^m u_i(t) q_i(p),$$

$$p(0) = 0, t \in [0, T], v_0 \in [0, \varepsilon_1]$$

where the admissible small controls $u_i(\cdot), i = 1, \ldots, m$, are taken in the set $\mathcal{U}_\infty \bigcup W_h$. A more explicit form will be obtained provided the solution $p\delta(\cdot, v)$ in conclusion $(c_2)$ is linearized around the solution $p^0(t)$, $t \in [0, T]$, of

$$(18) \qquad \frac{dp}{dt} = v_0 q_0(p), p(0) = 0, v_0 \in [0, \varepsilon_1],$$

for each $v_0 > 0$ fixed.

It is a direct computation which yields

(19) $$p\delta(t, v) = p^0(t) + \delta p(t, v) + o(\delta), t \in [0, T],$$

where $\lim\limits_{\delta \downarrow 0} \dfrac{o(\delta)}{\delta} = 0$ uniformly in $(t, v)$, provided that $p(t, v)$ is the solution of the following linear equation

(20) $$\frac{dp}{dt} = v_0 \frac{\partial q_0}{\partial p}(p^0(t))p + \sum_{j=1}^{N} v_j(t)Q_j(p^0(t)), \quad p(0) = 0,$$

or in integral form,

(21) $$p(t, v) = X(t) \int\limits_0^t \sum_{j=1}^{N} v_j(s)Y(s)Q_j(p^0(s))ds, \qquad t \in [0, T],$$

where $X(t)$ and $Y(t) = X^{-1}(t)$ are the $(M \times M)$ matrix solutions of the following linear equations

$$\frac{dX}{dt} = v_0 \frac{\partial q_0}{\partial p}(p^0(t))X, \qquad X(0) = I_M$$

$$\frac{dY}{dt} = -Y \frac{\partial q_0}{\partial p}(p^0(t))v_0, \qquad Y(0) = I_M$$

Therefore the conclusion $(c_2)$ in Lemma 3, for each $v_0 \in [0, \varepsilon_1]$ fixed, can be written as

(22) $$\lim\limits_{\delta \downarrow 0} \frac{p\delta(T, v, h) - p^0(T)}{\delta} = p(T, v),$$

where $p(\cdot, v)$ is represented in (21). The controls $v_j(\cdot)$ in representing $p(T, v)$ can be taken in a more concrete form in dependence on $Q_j \in L(q_1, \dots, q_m)$ as follows

$$v_j(t, v) = \langle X(T)Y(t)Q_j(p^0(t)), v \rangle \, t \in [0, T], j = 1, \dots, N,$$

where $v \in R^M$ designates the new parameter. Finally, $p(t, v)$ in (22) obtain the expression
(23)

$$p(T, v) = \left[ \int\limits_0^T X(T) \left( \sum_{j=1}^{N} Y(t)Q_j(p^0)(t))Q_j^*(p^0(t))Y^*(t))Y^*(t) \right) X^*(T)dt \right] v$$
$$= G(T)v, \quad v \in R^M,$$

where the $(M \times M)$ matrix $G(T)$ is the extended Grammian

(24) $$G(T) = \sum_{j=1}^{N} \int_0^T X(T)Y(t)Q_j(p^0(t))Q_j^*(p^0(t))Y^*(t)X^*(T)dt$$

associated with (17).

c) **Some conclusions**

For the given smooth vector fields $q_j \in C^\infty(R^M; R^M)$, $j = 0, 1, \ldots, m$, and the corresponding affine control system

($\alpha$) $$\frac{dp}{dt} = v_0 q_0(p) + \sum_{i=1}^{m} u_i(t)q_i(p), p(0) = 0, t \in [0, T],$$

evolving in the open set

$$p \in D_M = \prod_{i=1}^{M} (-a_i, a_i) \subseteq R^M$$

we shall confine ourselves to consider only the case when the extended Grammian matrix in (22) is strictly positive definite on $R^M$. To this end, assume that there exist $Q_1, \ldots, Q_N \in L(q_1, \ldots, q_m)$, $N \geq M$, such that

($h$) $$\dim \text{span } \{ad^k q_0(Q_j)(0), j = 1, \ldots, N, k \geq 0\} = M.$$

It is easily seen that the Grammian in (24) is strictly positive definite on $R^M$ provided (h) is fulfilled.

The first conclusion tells us that, for each $V_0 \in [0, \varepsilon_1]$ fixed, the reachable set $R(T, v_0)$ at the instant $t = T$ the control system ($\alpha$) has a nonempty interior with respect to the topology of $R^M$ and

($c_1$) $$p^0(T) \in \text{int } R(T, v_0)$$

This conclusion is obtained in a standard way from (22) by applying a fixed point theorem when the parametrization of $p(T, v)$ is that given by (23).

As far as the parameter $v_0 = 1$ in ($\alpha$) is concerned we may handle this situation by a time transformation which reduces the system ($\alpha$) to

($\beta$) $$\frac{dp}{dt} = q_0(p) + \sum_{i=1}^{m} u_i(t)q_i(p),$$
$$p(0) = 0, t \in [0, v_0 T]$$

As a consequence, the conclusion $(c_1)$ written for $(\beta)$ tell us that:

$\widetilde{c}_1)$ $\bar{p}^0(T_1) \in$ int $R(T_1)$ provided $T_1$ sufficiently small, where $\bar{p}^0(t)$, $t \in [0, T]$, is the solution of $\dfrac{dp}{dt} = q_0(p), p(0) = 0$, and $R(T_1)$ is the reachable set at the time $t = T_1$ associated with the control system $(\beta)$.

The next conclusion expresses the condensed form we can obtain for $p_\delta(T, v, h)$ in (22) with respect to $v \in R^M$. More precisely, the parametrization of $p(T, v)$ in (23) may influence essentially the writting in (22) provided the hypothesis (h) is assumed. Let $v_0 \in [0, \varepsilon_1]$ be fixed. Then there exists admissible controls $u_i(\cdot, v) \in \mathcal{U}_\infty \cup W_h, i = 1, \ldots, m$, depending on $v \in R^M$ such that the corresponding solution $p_\delta(t, v, h), t \in [0, T]$, of $(\alpha)$ obeys to

$c_2)$ $p_\delta(T, v, h) = p^0(T) + \delta[G(T) + S(\delta, v)]v, v \in R^M, \|v\| \leq C_1$ where the $(M \times M)$ matrix $S$ fulfils $\lim\limits_{\delta \downarrow 0} S_(\delta, v) = \odot$ (zero matrix) uniformly with respect to $\|v\| \leq C_1$ for some $C_1 > 0$

Similarly, when the parameter $v_0 = 1$ is involved then the conclusion $(c_2)$ changes into:

$\widetilde{c}_2)$ $p_\delta(T_1, v, h) = \bar{p}^0(T_1) + \delta[\widetilde{G}(T_1) + \widetilde{S}(\delta, v)]v, \|v\| \leq C$, provided $T_1 > 0$ is sufficiently small, where $\lim\limits_{\delta \downarrow 0} \widetilde{S}(\delta, v) = \odot$ uniformly with respect to $\|v\| \leq C$.

Here $\bar{p}^0(t), t \in [0, T]$, denotes the solution of

$$\frac{dp}{dt} = q_0(p), \qquad p(0) = 0,$$

and $\widetilde{G}(T_1)$ is defined as in (24) replacing $p^0(t)$ by $\bar{p}^0(t)$, $t \in [0, T_1]$.

**Proof of conclusion $(c_2)$**

The conclusion $(c_2)$ is more or less a direct consequence of the conclusions in Lemma 3 which in the particular case of $q_0 = 0$ take the following simpler form, $p^0(0) = 0$,

(25)     $\lim\limits_{\delta \downarrow 0} \dfrac{p_\delta(T, v, h) - p_\delta(T, v)}{\delta} = 0$, uniformly with respect to $\|v\| \leq C$

where $p_\delta(t, v), t \in [0, T]$, is the solution of the system

(26)     $$\frac{dp}{dt} = \delta \left( \sum_{j=1}^{M} Q_j(0) Q_j^*(0) \right) v, p(0) = 0$$

(see $N = M$ required by the hypothesis (h)), where $\{Q_1, \ldots, Q_M\} \subseteq L(q_1, \ldots,$
$q_m)$, meet

(27)          $\{Q_1(0), \ldots, Q_M(0)\} \subseteq R^M$ are linearly independent,

In addition, the controls $v_j(t), j = 1, \ldots, M$, in Lemma 3 are taken according to $p^0(\cdot) = 0$ as

(28) $$v_j(t) = Q_j^*(0)v, \ t \in [0, T], \text{ for } \|v\| \le C$$

and $p_\delta(T, v, h)$ received the expression
(29)

$$p_\delta(T, v, h) = \delta \left( \sum_{j=1}^{M} \int_0^T \frac{\partial P}{\partial p}(t, p_\delta(t, v, h))\widetilde{q}_j(p_\delta(t, v, h)) \, Q_j^*(0)u_{ij}(t, h)dt)v \right)$$
$$= \delta R(\delta, v)v,$$

for some $(M \times M)$ matrix $R(\delta, v)$

Now, $p_\delta(T, v)$ corresponding to $p^0(\cdot) = 0$ takes the form

(30) $$p\delta(t, v) = \delta(G(T) + S_0(\delta, v))v,$$

where $G(T) \overset{\triangle}{=} T(\sum_{j=1}^{M} Q_j(0)Q_j^*(0))$ and

(31) $$\lim_{\delta \downarrow 0} S_0(\delta, v) = \odot \text{ uniformly with respect to } \|v\| \le C$$

Using (30), from (25) we obtained that the matrix $R(\delta, v)$ in (29) can be written more explicitly and there holds

(32) $$p\delta(T, v, h) = \delta(G(T) + S(\delta, v))v, \ \|v\| \le C,$$

where $G(T) = T(\sum_{k=1}^{M} Q_j(0)Q_j^*(0))$ is strictly positive definite and

(33) $$\lim_{\delta \downarrow 0} S(\delta, v) = \odot \text{ uniformlywithrespectto } \|v\| \le C$$

The expression in (32) shows that the conclusion $(c_2)$ is fulfilled provided $q_0 = 0$. In the general case, $q_0 \ne 0$ and the reference trajectory $p^0(t), t \in [0, T]$, is nonzero for each $v_0 \in (0, \varepsilon]$ fixed.

It is easily seen that

(34) $$\lim_{\delta \downarrow 0} \frac{p\delta(T, v) - p\delta(T, o, h)}{\delta}$$
$$\lim_{\delta \downarrow 0} \frac{p_\delta(T, v) - p_\delta(T, 0) + p_\delta(T, 0) - p_\delta(T, 0, h)}{\delta}$$

and using Lemma 2 we obtain

(35) $$\lim_{\delta \downarrow 0} \frac{p\delta(T, o, h) - p\delta(T, 0)}{\delta} = \lim_{\delta \downarrow 0} \eta(\delta) = 0,$$

provided $h^{1/3} = \delta$. Therefore, the limit in the left hand side of (34) exists and it is equal to the following

$$(36) \quad \lim_{\delta \downarrow 0} \frac{p_\delta(T, v) - p_\delta(T, 0, h)}{\delta} = \lim_{\delta \downarrow 0} \frac{p_\delta(T, v) - p_\delta(T, 0)}{\delta} = G(T)v$$

uniformly with respect to $\|v\| \le C$.

On other hand, there holds

$$(37) \quad p_\delta(T, v, h) - p_\delta(T, 0, h) = \delta \left[ \int_0^1 \frac{\partial p\delta}{\partial v}(T, \vartheta v, h) d\vartheta \right] v$$

and using (36) we obtain that the $(M \times M)$ matrix

$$R(\delta, v) \triangleq \int_0^1 \frac{\partial p\delta}{\partial v}(T, \vartheta v, h) d\vartheta$$

has the following structure

$$(38) \quad R(\delta, v) = G(T) + \widetilde{S}(\delta, v)$$

where $\lim_{\delta \downarrow 0} \widetilde{S}(\delta, v) = \odot$ uniformly with respect to $\|v\| \le C$. Inserting (38) into (37) we obtain

$$(39) \quad p_\delta(T, v, h) = p_\delta(T, 0, h) + \delta(G(T) + \widetilde{S}(\delta, v))v$$

where the behaviour of $p_\delta(T, 0, h)$ around the fixed final value $p^0(T) \triangleq p_\delta(T, 0)$ is measured in (35). Therefore (35) and (39) allow one to see

$$(40) \quad p_\delta(T, v, h) = p^0(T) + \delta[G(T)v + \widetilde{S}(\delta, v)v + \eta(\delta)]$$

and applying a fixed point theorem we obtain $\|v_0(\delta)\| \le \frac{1}{2}C$ as the solution for the following algebraic equation

$$(41) \quad G(T)v_0(\delta) + \widetilde{S}(\delta, v_0(\delta))v_0(\delta) + \eta(\delta) = 0, \ \delta \in (0, \delta_0)$$

Solving (41) we find $v_0(\delta), \|v_0(\delta)\| \le \frac{1}{2}C$ such that

$$(42) \quad p_\delta(T, v_0(\delta), h) = p_0(T), \text{ for each } \delta \in (0, \delta_0)$$

Define $v(\delta) = v_0(\delta) + v$, with $\|v\| \le \frac{1}{2}C = C_1$, and compute the value of the mapping $p_{\delta(T,v(\delta),h)}$ in (40). Using (42), by standard manipulations we obtain

$$
(43) \qquad
\begin{aligned}
p_\delta(T, v(\delta), h) &= p^0(T) + \delta[G(T) + \widetilde{S}(\delta, v(\delta))]v \\
&\quad + \delta[\widetilde{S}(\delta, v(\delta)) - \widetilde{S}(\delta, v_0(\delta))]v_0(\delta)
\end{aligned}
$$

and as far as $G(T)$ in (38) does not depend on $v$ we may write

$$
(44) \qquad [\widetilde{S}(\delta, v(\delta)) - \widetilde{S}(\delta, v_0(\delta))]v_0(\delta) = [R(\delta, v_0(\delta))]v_0(\delta) = K(\delta, v)v
$$

where $\lim\limits_{\delta \to 0} K(\delta, v) = \odot$ uniformly with respect to $\|v\| \le C_1$.

The last two expressions allow one to rewrite (43) as

$$
(45) \qquad p_\delta(T, v(\delta), h) = p^0(T) + \delta[G(T) + S(\delta, v)]v
$$

where $\lim\limits_{\delta \downarrow 0} S(\delta, v) = \odot$ uniformly with respect to $\|v\| \le C_1$, and for $v(\delta) = v_0(\delta) + v$ taken such that $v_0(\delta)$ solves the algebraic equation (41).

Denote $p_\delta(T, v, h) = p_\delta(T, v(\delta), h)$ for $v \in R^M, \|v\| \le C_1$ and the conclusion $(c_2)$ is obtained. The proof is complete.

The last conclusion is related to the Chow's Theorem mentioned several times in this text but never proved. It is worth mentioning that conclusion $(c_2)$, in the particular case $q_0 = 0$, is a basic fact for obtaining

$c_3)$ (Chow's Theorem) Assume $\dim L(q_1, \dots, q_m)(p) = M$ for any $p \in D_M = \prod\limits_{i=1}^{M}(-a_i, a_i)$. Then any two points $p_1, p_2 \in D_M$ can be joined in a time $T > 0$ by an admissible trajectory of the control system $\dfrac{dp}{dt} = \sum\limits_{i=1}^{m} u_i(t)q_i(p)$

**Sketch of the proof**

What is done at the time $t = 0$ and stipulated in $(c_2)$ can be repeated for any $p \in D_M$ provided $\dim L(q_1, \dots, q_m)(p) = M$ is fulfilled. Taking two arbitrary points $p_1, p_2 \in D_M$, denote as $[p_1, p_2]$ the closed interval joining the two points. For each $p \in [p_1, p_2]$ we find an open ball $S_\rho(p)$ that is joined with $p$ by a trajectory of the control system which can be traversed in both directious by solutions of the same control system. From the open covering $S_\rho(p)$, $p \in [p_1, p_2]$, of $[p_1, p_2]$, we substract a finite one which allows one to obtain $p_1$ and $p_2$ joined by a finite sequence of curves determined by the centers of the covering spheres and some common points of every two aside balls. This polygonal curve is the admissible trajectory for the control system joining $p_1$ and $p_2$.

The proof is complete.

## 5.3   Nonlinear Control Systems

A nonlinear control system evolving on an open set $\mathcal{O} \subseteq R^n$ is defined by

(1)
$$\frac{dx}{dt} = f(x, u), \quad x \in \mathcal{O}, \quad u \in \mathcal{U} \subseteq R^n$$

where the control parameter $u$ is restricted to a control set $\mathcal{U}$ which is assumed to meet $S_\rho(0) \subseteq \mathcal{U}$ for an open ball centered at the origin $S_\rho(0) \subseteq R^m$.

There is no loss of generality by fixing the origin in an open set contained in the control set $\mathcal{U}$ provided there exists at least one such point. Moreover, for the time being we restrict ourselves to consider only smooth admissible controls, i.e., $u(\cdot) \in C^1([0, T]; U)$, which allow one to convert the original nonlinear system into an affine control system. Later on, from the integral representation of solutions it will be a routine matter to extend a control $u :$ $[0, T] \to \mathcal{U}$ with continuous derivative $\dfrac{du}{dt}(t)$, $\quad t \in [0, T]$, to the one which is only a piecewise continuous function $u : [0, T] \to \mathcal{U}$. Since a smooth control $u(\cdot) \in C^1([0, T]; \mathcal{U})$ can be differentiated we denote as $y = (x, u) in R^{n+m}$ the new state variable and the associated affine control system evolving on the open set $G(\mathcal{O} \times S_\rho(0))$ is defined as

(2)
$$\frac{dy}{dt} = f_0(y) + \sum_{i=1}^{m} v_i(t) f_i(y), t \in [0, T], y(0) = y(0) \triangleq (x, 0) \in G,$$

where the new vector fields $f_j \in C^\infty(G; R^{n+m})$, meet

(3)
$$f_0(y) = \begin{pmatrix} f(x, u) \\ 0 \end{pmatrix}, f_i(y) = \begin{pmatrix} 0 \\ e_i \end{pmatrix}, \quad i = 1, \dots, m$$

and $e_1, \dots, e_m \in R^m$ is the canonical basis.

Here it is implicitly accepted that the original function $f$ is smooth, i.e., $f \in C^\infty(G; R^n)$. On the new control function $v(\cdot) = (v_1(\cdot), \dots, v_m(\cdot))$ there is no pointwise restriction and it will be taken as a continuous function fulfilling an integral restriction

(4)
$$u(t) = \int_0^t v(s) ds \in S_\rho(0), \quad t \in [0, T]$$

which is contained in the evolution of the new state variable $y \in G$.

Any solution $y(t) = (x(t), u(t)) \in G, t \in [0, T]$, of (2), corresponding to a control $v(\cdot) \in C([0, T]; R^m)$ determines an admissible control $u(\cdot) \in$

$C^1([0,T]; U)$ and the corresponding controlled trajectory $x(t) \in \mathcal{O}, t \in [0,T]$, fulfilling (1). Conversely, the correspondence from the solutions of (1) to solutions of (2) holds provided $u(\cdot) \in C^1([0,T]; S_\rho(0))$.

Now, all that is relevant for the associated affine system (2) has to be translated for the original system (1). Apart from the admissible class $v(\cdot) \in W_h$ of controls which allow one to produce small trajectories $u(t) \in S_\rho(0)$, it will be necessary to make explicit the Lie algebra corresponding to the ideal $I_0(f_1, \ldots, f_m)$ of the affine control system (2). By definition, $I_0(f_1, \ldots f_m)$ is the Lie algebra determined a countable set of the vector fields

$$C = \{\mathrm{ad}^k f_0(f_i); \; i = 1, \ldots, m, k \geq 0\}$$

A direct computation allow one to see

$$\mathrm{ad}\, f_0(f_i)(y) = (\; g_i(y) \;) \; \text{ with } g_i(y) \in \mathbb{R}^n$$

defined as

$$(5) \qquad\qquad g_i(x, u) = \frac{\partial f(x, u)}{\partial u_i}, \quad i = 1, \ldots, m$$

Denote $D_u(f)$ the set of smooth functions $G(x, u) \in \mathbb{R}^n$ which are produced using a multiindex derivative

$$(6) \qquad \begin{aligned} D_u^\alpha f(x, u) &\overset{\triangle}{=} \partial^{|\alpha|} f / \partial_{u_1}^{\alpha_1} \ldots \partial_{u_m}^{\alpha_m} (x, u) \\ |\alpha| &\overset{\triangle}{=} \alpha_1 + \ldots + \alpha_m, \alpha_i \geq 0 \end{aligned}$$

In other words, $D_u(f) \subseteq C^\infty(G; \mathbb{R}^n)$ is defined as

(7)
$$D_u(f) = \{g \in C^\infty(G; \mathbb{R}^n) : g = D_u^\alpha f, \text{ for some } \alpha = (\alpha_1, \ldots, \alpha_m), \alpha_i \geq 0\}$$

Considering functions $g \in C^\infty(G; \mathbb{R}^n)$, it makes sense to define a Lie bracket corresponding only to the state variable $x \in \mathcal{O}$ and denote

$$(8) \qquad \begin{aligned} \mathrm{ad}_x f(g)(\cdot, u) &\overset{\triangle}{=} \frac{\partial f}{\partial x}(\cdot, u) g(\cdot, u) - \frac{\partial g}{\partial x}(\cdot, u) f(\cdot, u), \; u \in S_\rho(0) \\ \mathrm{ad}_x^{k+1} f(g)(\cdot, u) &= \mathrm{ad}_x f(\mathrm{ad}_x^k f(g)(\cdot, u)), \qquad k \geq 0 \end{aligned}$$

With the notations in (5) and (6) we see easily that the ideal $I_0(f_1, \ldots, f_m)$ of affine system (2) can be described, using the following set of vector fields in $C^\infty(\mathcal{O}; \mathbb{R}^n)$

$$(9) \qquad\qquad \widehat{C} = \{\mathrm{ad}_x^k f(g)(\cdot, u), \; g \in D_u(f), \; k \geq 0, u \in S_\rho(0)\}.$$

More precisely, if $\Lambda$ is the Lie algebra in $C^\infty(\mathcal{O}; R^n)$ determined by the set $\widehat{C}$ then $I_0(f_1, \ldots, f_m)$ appears as

$$(10) \quad I_0(f_1, \ldots, f_m) = \left\{ \begin{pmatrix} v \\ 0 \end{pmatrix} : v \in \Lambda \right\} \cup \left\{ \begin{pmatrix} 0 \\ e_1 \end{pmatrix}, \ldots, \ldots, \begin{pmatrix} 0 \\ e_m \end{pmatrix} \right\}$$

The assumption that $I_0(f_1, \ldots, f_m)$ is finite–dimensional restricted to our case has the following interpretation

11) The Lie algebra $\Lambda = L(\widehat{C})$ determined by the vector fields of $\widehat{C} \subseteq C^\infty(\mathcal{O}; R^n)$ is finite–dimensional. Actually, the set $\widehat{C}$, determinin $\Lambda$, consists of vector fields in $C^\infty(\mathcal{O}; R^n)$ which are obtained by taking multi–index derivatives with respect to the control variable $u$ applyed to $f$ and in addition applying operation $ad_x^k f, k \geq 0$, to such functions for each $u \in S_\rho(0)$ fixed. Then let $u \in S_\rho(0)$ be varying and we obtain the whole set $\widehat{C} \subseteq C^\infty(\mathcal{O}; R^n)$. In the case that the original system (1) is driven by an affine system, i.e.,

$$(12) \qquad f(x, u) = g_0(x) + \sum_{i=1}^{m} u_i g_i(x)$$

then the corresponding set $\widehat{C}$ is, as expected, the following

$$(13) \qquad \widehat{C} = \{ad^k g_0(g_i), \ i = 1, \ldots, m \ k \geq 0\}$$

and the Lie algebra $\Lambda = L(\widehat{C})$ coincides with the ideal generated by $g_0$ into $L(g_0, g_1, \ldots, g_m)$, i.e.

$$(14) \qquad \Lambda = I_0(g_1, \ldots, g_m).$$

Even in this particular case the associated affine control system in (2) is necessarily provided the original control is restricted by $u \in S_\rho(0) \subseteq \mathcal{U}$.

**Remark.**

*In the general case it is difficult to handle an infinite set of vector fields depending on $u \in S_\rho(0)$ and let us mention that if the original $f(x, u)$ in (1) is a polynomial with respect to the control $u$ then it holds*

$$(15) \qquad f(x, u) = g_0(x) + \sum_{j=1}^{\mathcal{I}} w_j g_j(x),$$

$\mathcal{I} \geq m$ for some polynomial scalar functions $w_j(u)$, where $g_j \in C^\infty(\mathcal{O}), j = 0, 1, \ldots, \mathcal{I}$ Again, the Lie algebra $\Lambda = L(\widehat{C})$ has a simpler structure as in (12). Namely, assume (13) fulfilled. Then

$$(16) \qquad \Lambda = L(\widehat{C}) = I_0(g_1, \ldots, g_{\mathcal{I}})$$

where $I_0(g_1, \ldots, g_{\mathcal{I}})$ is the ideal generated by $g_0$ into the Lie algebra $L(g_0, g_1, \ldots, g_{\mathcal{I}})$. Noticing that this case resembles the previous affine case in (12), it is worth proceeding further by taking $w = (w_1, \ldots, w_{\mathcal{I}})$ as a new control in the original system (1) and to associating the corresponding affine system (2) according to this interpretation.

## 5.4 Stabilization of Affine Control Systems

Here we are concerned with the asymptotic stabilization of the complete affine control system.

$$(1) \qquad \frac{dx}{dt} = g_0(x) + \sum_{i=1}^{m} u_i g_i(x), \qquad x \in R^n, \ x(0) = x_0, \ t \geq 0$$

where $g_i$ are smooth vector fields on $R^n$, $g_i \in$ Vect $(R^n)$, and $u = (u_1, \ldots, u_m)$ $\in R^m$ is the control parameter without fulfilling any pointwise constraint of the type $u \in S_\rho(0) \subseteq R^m$. As expected, the analysis is based on the control Lie algebra $L(g_0, g_1, \ldots, g_m) \subseteq$ Vect $(R^n)$ and its algebraic properties. The analysis is carried out by focussing on the finding of stabilizing controls, as much as possible, in explicit form, and the singularities of the Lie algebra $\Lambda = L(g_0, g_1, \ldots, g_m)$ are manipulated by the following hypothesis:

a) A special case $g_0 = 0$ was considered as an application in a previous chapter and the degeneracy of the corresponding Lie algebra $\Lambda = L(g_1, \ldots, g_m)$ had to be measured by a simple target set $V$, where the Euclidian norm $\|x\|^2$ was used.

When a nonvanishing drift $g_0 \neq 0$ is involved we are forced to perform computations around the local flow $G_0(t; x_0)$ determined by the vector field $g_0$ and, in addition, a good behaviour of the flow $G_0(t; x_0)$ is necessary. It is accomplished by partially allowing a Lyapunov type of function $L(x)$ measuring the good behaviour of $G_0(t; x_0)$ and a basis of $\Lambda = L(g_0, \ldots, g_m)$ taken in the ideal $I_0(g_1, \ldots, g_m)$ determined by $g_0$ into $\Lambda$.

The meaning of stabilization depends on the degeneracy of the Lie algebra $\Lambda$ and the origin in $R^n$ has to be replaced by the target set

$$(2) \qquad V = \{x \in R^n, \left\langle \frac{\partial L}{\partial x}(x), Y_j(x) \right\rangle = 0, \ j = 1, \ldots, N\},$$

where $\{Y_1, \ldots, Y_N\} \subseteq I_0(g_1, \ldots, g_m)$ is a fixed basis. Here $L : R^n \to R$, is a

$C^2$ function fulfilling

(3)     $\left\langle \dfrac{\partial L}{\partial x}(x), g_0(x) \right\rangle \leq 0$, for any $x \in B_0 \overset{\triangle}{=} \{x \in R^n, L(x) \leq C_0, C_0 > 0\}.$

In the case that the ideal $I_0(g_1, \ldots, g_m)$ is rich enough, i.e., if for each $x \neq 0$ there exists a $g \in I_0(g_1, \ldots, g_m)$ will exist such that $\left\langle \dfrac{\partial L}{\partial x}(x), g(x) \right\rangle \neq 0$ then $V = \{0\}$ and the usual asymptotic stability follows.

Another change involves the control function which is used to stabilize the system (1). Actually, for each initial state $x_0 \in R^n$ fixed, the solutions in (1) are supported by a $C^\infty$ embedded manifold $M(x_0) \subseteq R^n$ with $\dim M(x_0) = \dim \Lambda(x_0)$. Smooth stabilizing controls are appropriate on each leaf but discontinuities may appear moving from one leaf to another.

To avoid such difficulties we are relying on controls defined as periodic functions $u(t, x_0)$, of $t \geq 0$, and $C^\infty$ with respect to initial condition $x_0 \in R^n$. The time dependence of the stabilizing controls appears as necessary provided the fixed basis $\{Y_1, \ldots, Y_N\} \subseteq I_0(g_1, \ldots, g_m)$ contains some vector fields of the form $Y_j = ad^{k_j} g_0(Z_j)$, where $Z_j \in L(g_1, \ldots, g_m), Z_j \notin \{g_1, \ldots, g_m\}$. When the fixed basis $\{Y_1, \ldots, Y_N\} \subseteq I_0(g_1, \ldots, g_m)$ has the form $Y_j = ad^{k_j} g_0(g_{ij})$, $j = 1, \ldots, N$, and $i_j \in \{1, \ldots, m\}$, then there is a real hope of obtaining stabilizing controls depending only on the flow $G_0(t, x_0)$, $t \in [0, T]$.

### a) The algorithm describing stabilizing controls.

It is assumed that the ideal $I_0(g_1, \ldots, g_m)$ determined by the vector field $g_0 \in C^\infty(R^n; R^n)$ into the Lie algebra $L(g_0, g_1, \ldots, g_m) \subseteq C^\infty(R^n; R^n)$ meets the following:

i) $I_0(g_1, \ldots, g_m)$ is a finite dimensional space and $g_0 \in I_0(g_1, \ldots, g_m)$;

ii) There exists a basis $\{Y_1, \ldots, Y_N\} \subseteq I_0(g_1, \ldots, g_m)$ such that the first $M \leq N$ vector fields $Y_j$ belong to $L(g_1, \ldots, g_m)$ and the rest of them have the form $Y_{M+j} = ad^{k_j} g_0(Y_{ij})$ for $j = 1, \ldots, N - M, i_j \in \{1, \ldots, M\}, k_j \geq 0.$

The hypothesis (ii) connects $g_0$ to the Lie algebra $L(g_1, \ldots, g_m)$ in a very specific manner, and when taking $g_0 = 0$ it will not be necessary.

Let $G_0(t; x_0), t \in [0, T], x_0 \in V(0) \subseteq R^n$, be the local flow generated by the vector field $g_0$ and assume that there exists a $C^2$ scalar function $L : R^n \to R$ such that

iii) $B_0 = \{x \in R^n : L(x) \leq C_0\} \supseteq V(0)$ is bounded for some $C_0 > 0$ and

$$\left\langle \frac{\partial L}{\partial x}(x), g_0(x) \right\rangle \leq 0 \text{ for any } x \in B_0.$$

The hypothesis (iii) incorporates a dissipation property of the fixed flow $G_0(t; x_0)$ with respect to the chosen function $L$, and if a trajectory starts in $B_0$ (see $x_0 \in B_0$) then $G_0(t; x_0) \in B_0$ for any $t \in [0, T]$. On the other hand, the hypotheses (i) and (ii) allow one to obtain a homomorphism correspondence $\lambda : L(g_0, g_1, \ldots, g_m) \to L(q_0, q_1, \ldots, q_m)$, where $q_i \in C^\infty(R^N; R^N)$, $i = 0, 1, \ldots, m$, are found as follows. Let $\{Y_1, \ldots, Y_N\} \subseteq \Lambda \overset{\Delta}{=} I_0(g_n, \ldots, g_m)$ be the fixed basis fulfilling (ii) and consider the mapping

$$(3) \qquad \begin{aligned} y(p) &\overset{\Delta}{=} G(p; x_0) = G_1(t_1) \circ \ldots \circ G_N(t_N)(x_0), \\ p &= (t_1, \ldots, t_N) \in \mathcal{U}(0) \subseteq R^N, \qquad x_0 \in V(0) \subseteq R^n, \end{aligned}$$

where $G_i(t)(x_0)$ is the local flow generated by the vector field $Y_i, i = 1, \ldots, N$.

The mapping defined in (3) is the local solution of the gradient system

$$(4) \qquad \frac{\partial y}{\partial t_1} = Y_1(y), \qquad \frac{\partial y}{\partial t_2} = X_2(t_1; y), \ldots, \frac{\partial y}{\partial t_N} = X_N(t_1, \ldots, t_{N-1}; y)$$

with Cauchy condition $y(0) = x_0$, and following the global algebraic representation included in the first part we rewrite the vector fields in (4) as

$$(5) \quad \{Y_1(y), X_2(t_1; y), \ldots, X_N(t_1, \ldots, t_{N-1}; y)\} = \{Y_1(y), \ldots, Y_N(y)\}A(p)$$

for all $y \in R^n, p \in R^N$, where the $(N \times N)$ analytical matrix $A(p)$ meets

$$(6) \qquad A(0) = I_N \text{ (identity)}, \qquad \det A(p) \neq 0 \text{ for any } p \in R^N.$$

Let the constant vectors $a_i \in R^N, i = 0, 1, \ldots, m$, be fixed such that

$$(7) \qquad \sum_{j=1}^{N} a_i^j Y^j = g_i, \text{ for } i = 0, 1, \ldots, m$$

where $g_i \in C^\infty(R^n; R^n)$ define the control system (1) and $\{Y_1, \ldots, Y_N\} \subseteq I_0(g_1, \ldots, g_m)$ is the fixed basis.

Now, let $q_i(p) \in R^N$ be the unique solution of the algebraic equations

$$(8) \qquad A(p)q_i(p) = a_i, \qquad i = 0, 1, \ldots, m,$$

where $a_i \in R^N$ meet (7).

By definition, there holds

$$(9) \qquad \{Y_1(y), \ldots, Y_N(y)\}a_i = g_i(y), \qquad i = 0, 1, \ldots, m$$

and using (8) in (5) we obtain

(10)
$$\frac{\partial y}{\partial p}(p)q_i(p) = q_i(y(p)),$$

for any $p \in \mathcal{U}(0) \subseteq R^N$ and $i = 0, 1, \ldots, m$, where $y(p)$ is the local solution defined in (3). Based on what has been done in the first part of this chapter, it is a routine matter to extend the algebraic equations in (10) for any $Y_j$ in the fixed basis and we obtain $h_j \in L(q_0, \ldots, q_m)$ such that

(11)
$$\frac{\partial y}{\partial p}(p)h_j(p) = Y_j(y(p)), j = 1, \ldots, N, \ p \in \mathcal{U} \subseteq R^N$$

Now, using the hypothesis (ii) we may and do rewrite $h_j$ in a specific manner as

(12)
$$h_j \in L(q_1, \ldots, q_m), \qquad \text{for } j = 1, \ldots, M \le N,$$

and

$$h_{M+j} = ad^{k_j}q_0(h_{i_j}) \qquad \text{for } j = 1, \ldots, N - M, i_j \in \{1, \ldots, M\}$$

where the integers $k_j \ge 0$ are the same with those in (ii). In addition, there holds

(13)
$$h_j(0) = l_j, \qquad j = 1, \ldots, N$$

where $\{l_1, \ldots, l_N\} \subseteq R^N$ is the canonical basis, and

(14)    $\dim L(q_0, q_1, \ldots, q_m)(p) = N, \qquad \text{span } \{h_1(p), \ldots, h_N(p)\} = R^N,$

for all $p \in R^N$. A solution $p^u(t), t \in [0, T]$, of the auxiliary control system

(15)
$$\frac{dp}{dt} = q_0(p) + \sum_{i=1}^{m} u_i(t)q_i(p), \qquad p(0) = 0,$$

determines a corresponding solution $x^u(t), t \in [0, T]$, of (1) via the mapping $G(p; x_0)$ in (3), and there holds

(16)    $x^u(t; x_0) = G(p^u(t); x_0) = G(p^u(t); x_0), \ x^u(0; x_0) = x_0,$

using the same control function $u(\cdot)$ as in (15). Any approximation involving Lie brackets of the vector fields $q_j$ in (15) induces a corresponding approximation of the solution in the original system (1) and it involves the same algebraic computation of the Lie brackets among the original vector fields. $g_j$.

In particular, the computations around the local flow $Q_0(t;p), t \in [0,T]$, $p \in \mathcal{U}(0) \subseteq R^N$, generated by the vector field $q_0$ in (15) are more relevant because the rank property in (14) allow one to obtain a nonsingular parametrization of the reachable set $R(T)$ of (15) around the point $Q_0(T;0)$ which does not depend on the Cauchy data $x_0 \in R^n$ of the original system (1). As the nonsingular parametrization of $R(T)$ is concerned we are using an extended auxiliary control system

$$(17) \quad \frac{dp}{dt} = q_0(p) + \varepsilon \sum_{j=1}^{M} h_j(p)w_j(t,v), \qquad p(0) = 0, \ t \in [0,T], \varepsilon \in [0,\varepsilon_0],$$

where the scalar functions $w_j(t,v)$ are smooth and defined explicitly by

$$w_j(t,v) = \left\langle \left[\frac{\partial Q_0}{\partial p}(t;0)\right]^{-1} h_j(Q_0(t;0)), v_0(\delta) + v \right\rangle$$
$$\triangleq \widetilde{h}_j^*(t,0)(v_0(\delta) + v)$$

and $\widetilde{h}_j^* \in R^N$ is a row vector obtained by transposition of

$$\widetilde{h}_j(t;p) = \left[\frac{\partial Q_0}{\partial p}(t,p)\right]^{-1} h_j(Q_0(t;p)), j = 1,\ldots,M.$$

Here the vector fields $h_j \in L(q_1,\ldots,q_m)$ are those appearing in the extended control system (17) and they are defined in (11) with the properties listed in (12) and (13).

Each solution in (17) can be approximated by solutions of the control system (15) uniformly with respect to $\varepsilon \in [0,\varepsilon_0]$, and $v \in R^N, \|v\| \leq C_0$, provided $T > 0$ is sufficiently small. Moreover, there is a useful representation of the reachable set $R(T)$ around the point $Q_0(T;0)$ given in the conclusion $(c_2)$ of the previous section (see conclusions of §2) which allow one to see that an open sphere centred at $Q_0(T;0)$ can be included in a reachable set $R(T)$. More precisely, denote $B_\delta \subseteq R(T)$ the set consisting of:

$$(18) \ B_\delta = \{p \in R^M; p = Q_0(T;0) + \delta[W(T) + S(\delta,v)]v, v \in R^M, \|v\| \leq C_1\}$$

where the $(M \times M)$ matrix $S(\delta,v)$ satisfies

$$(19) \qquad\qquad \lim_{\delta \downarrow 0} S(\delta,v) = \odot \ (\text{zeromatrix}),$$

uniformly with respect to $\|v\| \leq C_1$ forsomeconstant $C_1 > 0$, and $W(T)$ is the extended Grammian matrix

$$(20) \qquad W(T) = \sum_{j=1}^{N} \int_0^T X(T)Y(t)h_j(p_0(t))h_j^*(p_0(t))Y^*(t)X^*(T)dt$$

with $p_0(t) \triangleq Q_0(t;0)$, and $Y(t) = X^{-1}(t)$, $t \in [0,T]$, satisfying

$$\frac{dX}{dt} = \frac{\partial q_0}{\partial p}(p_0(T))X, \qquad X(0) = I_M \text{ (identity)}$$

The nonsingularity of the parametrization related to the reachable set $R(T)$ is expressed by the strict positivity of the matrix $W(T)$ in (22), i.e.,

$$(21) \qquad \langle W(T)v, v\rangle \geq \gamma\|v\|^2 \qquad \text{for all } v \in \mathrm{R}^M, \text{ where } \gamma > 0.$$

In addition, any point $p \in B_\delta$ in (18) can be realized as a final value at $t = T$, of a set of admissible trajectories of the auxiliary control system in (15) and it holds,

$$(22) \qquad \begin{aligned} p &= p_\delta(T, v) \text{ foreach } p \in B_\delta, \\ p &= p_0(T) + \delta[W(T) + S(\delta, v)]v, \qquad \|v\| \leq C \end{aligned}$$

where $p_\delta(t, v)$, $t \in [0,T]$, satisfies

$$(23) \qquad \begin{aligned} \frac{dp}{dt} &= q_0(p) + \delta \sum_{j=1}^{M} \tilde{u}_{i_j}(t, \delta)w_j(t, v)q_{i_j}(p) \\ &+ \sum_{i=1}^{m} u_i(t, \delta)q_i(p), \ p(0) = 0, \end{aligned}$$

with $i_j \in \{1, \ldots, m\}$, $w_j(t, v)$ defined in (17) and $\tilde{u}_{i_j}(t, \delta), u_i(t, \delta), t \geq 0$ some periodical functions in $t \geq 0$ with the common period $T > 0$. Now the solutions $p_\delta(t, v), t \in [0,T]$, of (23) are translated into solutions $x_\delta(t, v; x_0), t \in [0,T]$, of (1) via the mapping in (16) for any $x_0$ in the bounded set $B_0 \subseteq R^n$ defined in (iii). In particular, we obtain

$$(24) \qquad x_\delta(T, v; x_0) = G(p_\delta(T, v), x_0), \qquad \delta \in (0, \delta_0), \|v\| \leq C_1, x_0 \in B_0$$

and using the Leibnit–Newton formula

$$(25) \qquad \begin{aligned} x_\delta(T, v; x_0) &= G(p_0(T); x_0) + \delta\langle\frac{\partial G}{\partial p}(p_0(T); x_0), W(T)v\rangle \\ &+ \delta\langle\frac{\partial G}{\partial p}(p_0(T); x_0), S(\delta, v)v\rangle + 0(\delta)\|v\|^2 \end{aligned}$$

where $\lim\limits_{\delta\downarrow 0}\dfrac{0(\delta)}{\delta} = 0$ uniformly with respect to $\|v\| \le C_1$. A similar compu-
tation is performed when the Lyapunov type of function $L$, associated with
the drift vector field $g_0$, is applied to the values of admissible solutions in (1).
By a straight computation we see easily that:
(26)

$$
\begin{aligned}
L(x_\delta(T,v;x_0)) &= L(G_0(T;x_0)) + \delta\langle\frac{\partial L}{\partial x}(G_0(T;x_0)), \frac{\partial G}{\partial p}(p_0(T),x_0)S(\delta,v)v\rangle \\
&\quad +\delta\langle\frac{\partial L}{\partial x}(G_0(T;x_0)), \frac{\partial G}{\partial p}(p_0(T);x_0)W(T)v\rangle + 0(\delta)\|v\|^2
\end{aligned}
$$

where the scalar function $0(\delta)$ fulfils $\lim\limits_{\vartheta\downarrow 0}\dfrac{0(\delta)}{\delta} = 0$, uniformly with respect to
$\|v\| \le C_1$.

Now, using the algebraic representation in (5) and the local solution
$G(p;x_0)$ of the gradient system in (4) we obtain

(27)    $$\frac{\partial G}{\partial p}(p_0(T);x_0) = \{Y_1,\dots,Y_N\}(G_0(T;x_0))A(p_0(T))$$

where $\det A(p_0(T)) \ne 0$.

The vector $\nu \in R^N, \|\nu\| \le C_1$, can be chosen such that the value in (26)
decreases compared with $L(G_0(T;x_0))$ uniformly with respect to $x_0$ in the
bounded set $B_0$. Let us define, for any $x_0 \in B_0$, the mapping

(28)

$$
\begin{aligned}
v(x_0) &= -\left(\frac{\partial G}{\partial p}(p_0(T);x_0)\right)^* \frac{\partial L}{\partial x}(G_0(T;x_0)) \\
&= -A^*(p_0(T))\begin{pmatrix} \langle Y_1(G_0(T;x_0)), \frac{\partial L}{\partial x}(G_0(T;x_0))\rangle \\ \vdots \\ \langle Y_N(G_0(T;x_0)), \frac{\partial L}{\partial x}(G_0(T;x_0))\rangle \end{pmatrix}
\end{aligned}
$$

where "$*$" means transposition.

Inserting (28) into (26) and using that the matrix $(N\times N)W(T)$ is strictly
positive definite we obtain

(29)        $$L(x_{\tilde\delta}(T,v(x_0);x_0)) \le L(G_0(T;x_0)) - \frac{\tilde\delta}{2}\gamma\|v(x_0)\|^2$$

for $\tilde\delta > 0$ sufficiently small and $\gamma > 0$ a constant, uniformly with respect to $x_0$
in the bounded set $B_0$. Using hypothesis (iii) it holds $L(G_0(T;x_0)) \le L(x_0)$

and (29) becomes

(30)  $$L(x_{\widetilde{\delta}}(T, v(x_0); x_0)) \leq L(x_0) - \widetilde{\delta}\frac{\gamma}{2}\|v(x_0)\|^2,$$

for $\widetilde{\delta} > 0$ sufficiently small and uniformly with respect to $x_0 \in B_0$, provided $v(x_0)$ is defined as in (28).

The meaning of the stabilization we are looking for is based on the last estimate written in (30) and we need to extend the definition of a solution $x_{\widetilde{\delta}}(t, v(x_0); x_0)$ from $t \in [0, T]$, to the whole $R_+ = [0, \infty)$, preserving on each $[kT, (k+1)T]$ what has been done for $t \in [0, T]$.

### b) Main results and the proofs

The solutions $x_{\delta}(t, v; x_0), t \in [0, T]$, $x_0 \in B_0$, of (1) are defined for each $v \in R^N, \|v\| \leq C_1$, according to

(31)    $x_{\delta}(t, v; x_0) = G(p_{\delta}(t, v); x_0), \delta \in (0, \delta_0), \|v\| \leq C_1, x_0 \in B_0.$

where $G(p; x_0)$ is the mapping in (16) and $p_{\delta}(t, v), t \in [0, T]$, obeys to the auxiliarly control system in (23). By a direct computation we obtain the differential system fulfilled by $x_{\delta}(\cdot, v; x_0)$ as follows

(32)
$$\frac{dx}{dt} = g_0(x) + \delta \sum_{j=1}^{M} \widetilde{u}_{i_j}(t, \delta) w_j(t, v) g_{i_j}(x)$$
$$\sum_{i=1}^{m} u_i(t, \delta) g_i(x), t \in [0, T], x(0) = x_0,$$

where $i_j \in [1, \ldots, m]$ and periodically defined control functions $\widetilde{u}_{i_j}, u_i$ have the common period $T$ sufficiently small. Choosing $v = v(x_0)$ defined for all $x_0 \in B_0$ as in (28) the solution $x_{\delta}(t, v; x_0), t \in [0, T]$ in (32) exists for each $x_0 \in B_0$ and the estimate in (30) shows that we may conclude

(33)        $x_{\widetilde{\delta}}(T, v(x_0); x_0) \overset{\triangle}{=} x_1 \in B_0,$      for any $x_0 \in B_0.$

Write $\widetilde{x}(t, v; x_0) = x_{\widetilde{\delta}}(t, v; x_0),$   $t \in [0, T]$, and extend the definition of the solution on $t \in [T, 2T]$ starting with $x_1(x_0)$ in (33) as the new initial conditions.

In other words, $\widetilde{x}(t, v; x_0)$, for $t \in [T, 2T]$, is the solution of (32) with Cauchy condition $x(T) = x_1(x_0)$ where the parameter $v$ is replaced by

(34)                $v = v(x_1),$      $x_1 \in B_0$

and $v(\cdot)$ is the fixed mapping defined in (28). After a time $T$, i.e. on $[2T, 3T]$, we extend the definition of $\widetilde{x}(t, v; x_0)$ by changing the parameter $v$ into $v(x_2)$, where $x_2 = \widetilde{x}(T, v(x_1); x_1)$ and it allows one to associate, for $x_0 \in B_0$, a periodically defined solution $\widetilde{x}(t; x_0)$, of (32) for any $t \geq 0$, which is bounded. It makes sense to consider the $\omega$–limit set

$$(35) \quad \widehat{\Omega}_+(x_0) = \{\widehat{x} \in B_0 : \widehat{x} = \lim_{n \to \infty} \widetilde{x}(t_n; x_0), t_n = k_n T \text{ for some } \{k_n\} \uparrow \infty\}$$

for each $x_0 \in B_0$.

Along with $\Omega_+(x_0)$ consider the target set

$$(36) \qquad V = \{x \in R^n : \langle \frac{\partial L}{\partial x}(x), Y_j(x)\rangle = 0, \ j = 1, \dots, N\}$$

where $\{Y_1, \dots, Y_N\} \subseteq I_0(g_1, \dots, g_m)$ is the fixed basis fulfilling the hypotheses (i)–(iii).

Then the main result regarding the algorithm of stabilizing controls is the following

**Theorem 1.** *Let the vector fields $g_i \in C^\infty(R^n; R^n), i = 0, 1, \dots, m$, be given such that the hypotheses (i)–(iii) are fulfilled. Let $C_0 > 0$ be a fixed constant and define the corresponding bounded set $B_0$ as in (iii). Then there exist periodic controls $\widetilde{u}_i(t, x_0)$, with respect to $t \geq 0$ with common period $T > 0, C^\infty$ with respect to $x_0 \in B_0$, $i = 1, \dots, m$, such that the periodically defined solution $\widetilde{x}(t; x_0)$, $t \geq 0$ of the control system*

$$(*) \qquad \frac{dx}{dt} = g_0(x) + \sum_{i=1}^{m} \widetilde{u}_i(t, x_0) g_i(x), \quad x(0) = x_0$$

*fulfils $G_0(T, \widehat{\Omega}_+(x_0)) \subseteq V \cap B_0$ for any $x_0 \in B_0$. Here the $\omega$–limit set $\widehat{\Omega}_+(x_0)$ is associated with $\widetilde{x}(t; x_0), t \geq 0$, as in (35), the target set $V$ is defined in (36) and $G_0(t; y), t \in [0, T], y \in V(0) \subseteq R^n$, is the local flow generated by the drift vector field $g_0$.*

**Proof.**

For the mapping $v(x_0)$, $x_0 \in B_0$, defined in (28) and the periodic controls $\widetilde{u}_{i_j}(t, \widetilde{\delta}), u_i(t, \widetilde{\delta})$ in (32) we may redefine the control system in(32) as in the statement of the Theorem provided the scalar functions $w_j(t; v(x_0)) \overset{\triangle}{=} \widetilde{w}_j(t, x_0), j = 1, \dots, M$, defined on $t \in [0, T]$, smoothly are shifted on each interval $(kT, (k + 1)T]$ in a natural manner. In this way the resulting controls $\widetilde{u}_i(t, x_0)$ are no longer $C^1$ with respect to $t \geq 0$, but they are piecewise continuous and periodic with respect to the variable $t$.

Let $\widetilde{x}(t; x_0)$, $t \geq 0$ be the periodically defined solution of $(*)$ for each $x_0 \in B_0$. Then the computation performed for $t \in [0, T]$ led us to the estimate in (30). The estimate in (30) is repeated for $\widetilde{x}(t; x_0)$ on each interval $[kT, (k+1)T]$ and we obtain

(37)  $L(\widetilde{x}((k+1)T; x_0)) \leq L(x_0) - \widetilde{\gamma}(\|v(x_0)\|^2 + \|v(x_1)\|^2 + \ldots + \|v(x_k)\|^2),$

where $x_j \overset{\triangle}{=} \widetilde{x}(jT; x_0)$, $j = 0, 1, \ldots, k$, and $\widetilde{\gamma} = \widetilde{\delta}/2 > 0$. It allows one to conclude that the series in the right hand side is convergent, and therefore, there holds

(38)  $\qquad \lim_{k \to \infty} v(x_k) = 0, \qquad$ where $x_k \overset{\triangle}{=} \widetilde{x}(kT; x_0) \in B_0$

In addition, the limit points of the bounded sequence $\{x_k\}_{k \geq 0} \subseteq B_0$ are in the same closed set $B_0$. It gives the first conclusion with respect to the sequence of values $\widetilde{x}(kT; x_0)$, $\quad k \geq 0$, associated with the periodically defined solution of $(*)$ i.e.

(39)  $\qquad \{\widehat{x} \in B_0 : \widehat{x} = \text{limit point of } \{\widetilde{x}(kT; x_0), k \geq 0\}\} \subseteq G_0(-T; V)$

where the target set $V$ is defined in (36), or equivalently

(40)  $\qquad\qquad\qquad G_0(T; \widehat{x}) \in V \cap B_0$

for any $\widehat{x}$ in the set of limit points associated with the bounded sequence $\{\widetilde{x}(kT; x_0) : k \geq 0\}$.

The proof is complete.

**Remark 1.**

*The statement in Theorem 1 is a partial answer to the stabilization problem we are looking for. It is restricted by using only a part $\widehat{\Omega}_+(x_0) \subseteq \Omega_+(x_0)$ of the usually accepted as the $\omega$-limit set associated with a bounded solution. It is the purpose of the next theorem to include the whole set $\Omega_+(x_0)$ into the conclusions. Define*

(41)  $\qquad \Omega_+(x_0) = \{x \in B_0 : x = \lim_{k \to \infty} \widetilde{x}(t_k; x_0), \text{ for some } \{t_k\} \uparrow \infty\}.$

*By the previous hypotheses (see (ii) and (iii) ), it is imposed that the set $B_0$ defined in (iii) is invariant for the flow $G_0(t; x_0), t \in [0, T], x_0 \in B_0$, generated by the drift vector field $g_0$. From the hypothesis (i) we deduce that there holds*

(42)  $\qquad \left\langle \dfrac{\partial L}{\partial x}(x), g_0(x) \right\rangle = 0, \text{ for any x in the target set V (see(36))}$

It does not mean that the flow $G_0(t; x_0)$ leaves invariant the target set $V$ invariant, and to obtain a complete stabilizability conclusion we need the following:

(iv) The scalar function $L$ and the vector field $g_0$ obey the following : $L(G_0(t; \hat{x})) = $ constfor $t \in [0, T] \Rightarrow g_0(\hat{x}) = 0$ provided $\hat{x} \in B_0$ (La Salle's principle).

Now the final result can be stated using a slight modification of the controls used in Theorem 1. With the mapping $v(x_0)$, $x_0 \in B_0$, defined in (28) associate the scalar function

$$(43) \qquad\qquad a(x_0) = \|v(x_0)\|^2, x_0 \in B_0$$

and multiplying each control $\tilde{u}_j(t, x_0)$ in the Theorem 1 by $a(x_0)$ we obtain the new system

$$(**) \qquad \frac{dx}{dt} = g_0(x) + \sum_{i=1}^{m} a(x_0)\tilde{u}_i(t, x_0)g_i(x), \qquad x(0) = x_0$$

controlled by the new controls $\hat{u}_i(t, x_0) \overset{\triangle}{=} a(x_0)\tilde{u}_i(t, x_0)$, $i = 1, \ldots, m$, denoted $S_0 = \{x \in R^n : g_0(x) = 0\}$.

**Theorem 2.**

*Let the vector fields $g_i \in C^\infty(R^n; R^n)$, $i = 0, 1, \ldots, m$, be given such that hypotheses (i)–(iv) are fulfilled. Let $c_0 > 0$ be fixed and consider the corresponding bounded set $B_0$ defined in (iii). Then there exist periodic controls $\hat{u}_i(t, x_0)$ of a common period $T > 0$ with respect to $t \geq 0$, and $C^\infty$ for $x_0 \in B_0$ fulfilling $\hat{u}_i(t, x_0) = 0$, for $x_0 \in V$, $i = 1, \ldots, m$, such that the periodically defined solution $\tilde{x}(t; x_0), t \geq 0$ of the system $(**)$ fulfils $\Omega_+(x_0) \subseteq S_0 \cap V \cap B_0$ for any $x_0 \in B_0$. Here the $\omega$–limit set $\Omega_+(x_0)$ is associated with $\tilde{x}(t; x_0)$ as in (41) and the target set $V$ is given in (36).*

**Proof.**

The same simple arguments used for Theorem 1 are suitable for conclusions and applying Theorem 1 we obtain $L(G_0(t; \hat{x})) = $ const., $t \in [0, T]$, for each $\hat{x} \in \hat{\Omega}_+(x_0)$. Now, it is the hypothesis (iv) which allows one to conclude that $g_0(\hat{x}) \equiv 0$, for $\hat{x} \in \hat{\Omega}_+(x_0)$ and $\hat{\Omega}_+(x_0) \subseteq S_0 \cap V \cap B_0$ for any $x_0 \in B_0$. In addition, we notice that the scalar function $a(x_0) \geq 0$ multiplying $\tilde{u}_i(t, x_0)$ does not involve essential modifications in the nonsingular parametrization and the used extended Grammian $W(T)$ remains strictly positive definite. Now for an arbitrary sequence $\{t_k\} \uparrow \infty$ we rewrite $t_k = n_k T + \tilde{t}_k$ as $\{n_k\} \uparrow \infty$

and $\tilde{t}_k \in [0, T]$, and as $\lim_{t_k \uparrow \infty} \tilde{x}(t_k; x_0) = \hat{x} \in B_0 \cap S_0$ we obtain $\hat{x} \in V$ also and the proof is complete.

## 5.5   Controlled Invariant Lie Algebras

To begin with, we shall recall the deterministic disturbance decoupling. Consider two affine control systems evolving on the same open set $\mathcal{O} \subset R^n$. Let $\varphi(t; x_0)$ and $\Psi(t; x_0)$, $0 \leq t \leq T$ be the corresponding solutions of the control systems

$$(1) \qquad \frac{dx}{dt} = f_0(x) + \sum_{i=1}^{m} u_i f_i(x), \qquad x(0) = x_0, \; u = (u_1, \ldots, u_m) \subset R^m$$

$$(2) \qquad \frac{dy}{dt} = \sum_{i=1}^{d} p_j(t) g_j(y), \qquad y(0) = x_0, \qquad y \in \mathcal{O}$$

where the admissible control $p(t) = (p_1(t), \ldots, p_d(t)) \in R^d \; t \in [0, T]$, is a bounded piecewise continuous function and $u = (u_1, \ldots, u_m) \in R^m$ is a constant vector.

We consider only smooth vector fields $f_i, g_j \in C^\infty(\mathcal{O}; R^n)$ and one may wander about the affine control system governing the composition of the two solutions

$$(3) \qquad z(t; x_0) = \varphi(t; \Psi(t; x_0)) \quad \text{for } 0 \leq t \leq T$$

where $T$ is fixed such that $\varphi(t; y)$ exists for any $y = \Psi(t; x_0)$ and $0 \leq t \leq T$.
By a direct computation we obtain:

**Proposition 1.**
 $z(t; x_0)$, $0 \leq t \leq T$, *defined in (3) is the solution of the following affine control system*

$$\frac{dz}{dt} = f_0(z) + \sum_{i=1}^{m} u_i f_i(z) + \left( \frac{\partial \varphi(-t; z)}{\partial z} \right)^{-1} \left[ \sum_{j=1}^{d} p_j(t) g_j(\varphi(-t; z)) \right],$$

*where* $y = \varphi(-t; z)$ *is the diffeomorphism unique solution of* $\varphi(t; y) = z$, *for* $z$ *in a neighbourhood of* $z_0(t) \overset{\triangle}{=} \varphi(t; x_0)$, $0 \leq t \leq T$.

**Remark 1.**

One may recognize the expression $\left(\dfrac{\partial\varphi}{\partial z}(-t;z)\right)Y(\varphi(-t;z))$ as defining the differential map $\varphi(t)_*$ of $\varphi(t;y), y \in \mathcal{O}$, acting on the vector field $Y$. Namely

$$\varphi(t; h(t; z)) = z$$

implies

$$\left(\frac{\partial h}{\partial z}(t; z)\right)^{-1} = \frac{\partial \varphi}{\partial y}(t, h(t; z))$$

and

$$h(t; z_0(t)) = \Psi(t; x_0).$$

Therefore, the equation for $z$ can be written

(4)
$$\frac{dz}{dt} = f_0(z) + \sum_{i=}^{m} u_i f_i(z) + \sum_{i=1}^{m}(\varphi(t)_* g_j(z) p_j(t),$$

and if $g_k$ commute with all $f_j$ then $\varphi(t)_* g_k(z) = g_k(z)$.

**Corollary**

Assume that each $g_k$ commutes with all $f_j$, $j = 0, 1, \ldots, m$. Then the composite solution $z(t; x_0)$ in (3) fulfils the following affine control system

(5)
$$\frac{dz}{dt} = f_0(z) + \sum_{i=1}^{m} u_i f_i(z) + \sum_{j=1}^{d} p_j(t) g_j(z), \qquad z(0) = x_0.$$

In the general case the computation of $\varphi(t)_* g$ as a function of $t$ involves the whole Lie algebra $L(g, f_0, \ldots, f_m)$, as will be emphasized later on.

**Remark 2.**

The differential map $\varphi(t)_*$ can be used in (4) provided the admissible control $u \in R^m$ is an arbitrarily fixed constant vector. In the case where that we accept all bounded piecewise continuous functions $(u(\cdot) \in \mathcal{A}_T)$ as the admissible class of controls for (1) then the picture in (4) has to be changed to one involving an integral representation of solutions in (1) using a local solution for a gradient system, and the differential $\varphi(t)_*$ will be replaced accordingly.

### 5.1 Deterministic disturbance decoupling

One may assimilate $\Psi(t; x_0)$ as an input perturbation for the composite solution $z(t; x_0)$ in (3) which negatively affects the behaviour of the system (1). It is of particular interest for rendering some outputs $h \in C^1(\mathcal{O}; R^l)$ of the system (1), independent of the unkown perturbation $p_t(y) = \sum\limits_{j=1}^{d} p_j(t)g_j(y)$ and according to the canonical system (4) it can be specified as

$$(6) \qquad \sum_{j=1}^{d} \frac{\partial h}{\partial z}(z)(\varphi(t)_* g_j(z)p_j(t)) = 0, \text{ for all } \quad z \in \mathcal{O}.$$

We see from this that large oscillations of the disturbance $p(t) = (p_1(t), \dots, p_d(t))$ has no effect on the output and the behaviour of $h(z(t; x_0)), x_0 \in \mathcal{O}$, can be measured just by replacing $z(t; x_0), x_0 \in \mathcal{O}$, with $\varphi(t; x_0), x_0 \in \mathcal{O}$.

Usually, the algebraic relations in (6) will stand for $s_0$ called disturbance decoupling and they obtain a simpler formulation if the perturbing vector fields $g_j$ fulfil the following hypotheses:

i) $\langle \frac{\partial h_i}{\partial z}(z), g_j(z) \rangle = 0, z \in \mathcal{O}$, for all $i \in \{1, \dots, l\}, j \in \{1, \dots, d\}$
i.e., $g_j \in \ker(dh)$ for any $j = 1, \dots, d$.
ii) the Lie algebra $\Lambda = L(g_1, \dots, g_d)$ is controlled invariant, i.e., the Lie bracket $[f_i, g] \in \Lambda$ for any $g \in \Lambda$ and $i = 0, 1, \dots, m$.

Actually, the controlled invariance of the whole Lie algebra $\Lambda$ expressed in (ii) is obtained from the invariance of the basic vector fields $\{g_1, \dots, g_d\}$, and Jacobi's identity allows one to extend it for all $g \in \Lambda$. On the other hand, the hypothesis (i) stated only for the basic vector fields $\{g_1, \dots, g_l\}$ holds for any $g \in \Lambda$ because $\frac{\partial h_i}{\partial z}(z)$ is the usual gradient vector field associated with a scalar function. Therefore, both hypotheses (i) and (ii) can be stated only for the basic vector fields $\{g_1, \dots, g_d\}$ as:

$$(7) \qquad\qquad g_j \in \ker d\, h \text{ and} [f_i, g_j] \in \Lambda \overset{\triangle}{=} L(g_1, \dots, g_d)$$

for any $j = 1, \dots, d$, and $i = 0, 1, \dots, m$.

We recall that a Lie algebra $\Lambda \subseteq C^\infty(\mathcal{O}; R^n)$ is of finite type if there exists a system of generators $\{Y_1, \dots, Y_M\} \subseteq \Lambda$ such that any $g \in \Lambda$ can be written

$$g(x) = \sum_{i=1}^{M} a_i(x)Y_i(x), \qquad x \in \mathcal{O}, \text{ forsome } a_i(\cdot) \in C^\infty(\mathcal{O}; R), \qquad i = 1, \dots, M.$$

**Proposition 2.**

*Assume the conditions (7) are fulfilled. Then the disturbance decoupling conditions in (6) are satisfied provided the Lie algebra $\Lambda(g_1, \ldots, g_d)$ is of finite type.*

**Proof.**

The fulfilment of (6) requires no to explicit the mapping $(\varphi(t)_* g_j)(z), z \in \mathcal{O}$, for each $j = 1, \ldots, d$. By definition,

$$(8) \qquad (\varphi(t)_* g_j(x_0)) = \left( \frac{\partial \varphi}{\partial z}(-t; x_0) \right)^{-1} g_j(\varphi(-t; x_0))$$

for any $t \in I_a = -[a, a]$, and $x_0 \in \mathcal{O}$ fixed, where $\varphi(t; z), z \in V(x_0) \subseteq \mathcal{O}, t \in I_a$, is the local flow generated by the vector field $f(x) \triangleq f_0(x) + \sum\limits_{i=1}^{m} u_i f_i(x)$ with $(u_1, \ldots, u_m) \in R^m$ fixed. As a function of time $t \in I_a$, the expression in (8) can be associated with the Baker–Campbell–Hausdorff formula. If the corresponding series is convergent to an element of $\Lambda(x_0)$ for each $t \in I_a$ then the conclusion (6) will follow, provided the right hand side in (8) is expressed in an explicit manner with respect to the assumed generator system $\{Y_1, \ldots, Y_M\} \subseteq \Lambda$. The computation involves arguments similar to those used on other occasions, but we prefer now to do it directly. Denote $H(f) = \left[ \frac{\partial \varphi}{\partial z}(-t; x_0) \right]^{-1}, t \in I_a$, and it fulfils the following linear matrix equation

$$(9) \qquad \frac{dH}{dt} = H \frac{\partial f}{\partial x}(\varphi(-t; x_0)), \qquad H(0) = I_n \text{ (identity)}.$$

Let $\{Y_1, \ldots, Y_M\} \subseteq \Lambda$ be a fixed system of generators and for $g_j, [f, g_j] \in \Lambda$ we may and do write

$$(10) \qquad \begin{aligned} g_j(\varphi(-t; x_0)) &= \sum_{k=1}^{M} b_j^k(t, x_0) Y_k(\varphi(-t; x_0)) \\ &= \{Y_1, \ldots, Y_M\}(\varphi(-t; x_0)) b_j(t; x_0) \end{aligned}$$

where $b_j(\cdot, x_0) \in C^\infty(I_a; R^M)$, and

$$(11) \qquad [f, g_j](\varphi(-t; x_0)) = \sum_{k=1}^{M} a^k(t; x_0) Y_k(\varphi(-t; x_0)), \ t \in I_a$$

for some $a(\cdot; x_0) \in C^\infty(I_a; R^n)$

Then the expression in the right hand side of (8) is represented as

$$(12) \qquad \begin{aligned} H(t) g_j(\varphi(-t; x_0)) &= H(t)\{Y_1, \ldots, Y_M\}(\varphi(-t; x_0)) b_j(t; x_0) \\ &\triangleq N(t) b_j(t; x_0) \end{aligned}$$

and we shall try to obtain a useful representation for the matrix

$$(13) \qquad N(t) \overset{\triangle}{=} H(t)\{Y_1, \ldots, Y_M\}(\varphi(-t; x_0)), t \in I_a.$$

A Leibnitz–Newton formula applied to $N(t)$ requires the algebraic representation of the mapping $adf(Y_j), j = 1, \ldots, M$, and we obtain

$$(14)$$
$$\mathrm{ad}f\{Y_1, \ldots, Y_M\}(\varphi(-t; x_0)) = \{Y_1, \ldots, Y_M\}(\varphi(-t; x_0))A(t; x_0), \qquad t \in I_a,$$

where the $(M \times M)C^\infty$ matrix $A(t; x_0)$ with entries $a_{ij}(\cdot) \in C^\infty(I_a; R)$ is found using $\mathrm{ad}f(Y_j) \in \Lambda$ provided $\mathrm{ad}f(g_k) \in \Lambda$ for all $k = 1, \ldots, d$. For each $Y_j$ we fix $a_j(\cdot) \in C^\infty(I_a; R^M)$ such that

$$(15) \qquad \mathrm{ad}f(Y_j)(\varphi(-t; x_0)) = \{Y_1, \ldots, Y_M\}(\varphi(-t; x_0))a_j(t; x_0),$$

and the matrix $A(t; x_0)$ in (14) is composed by the column vector functions $a_1(t; x_0), \ldots, a_M(t, x_0)$. Now the rectangular matrix $N(t)$ is rewriten as

$$(16) \qquad N(t) = \{Y_1, \ldots, Y_M\}(x_0)I_n + \int_0^t N(s)A(s; x_0)ds,$$

where

$$\frac{dN}{ds}(s) = N(s)A(s),$$

is used. Let $\mathcal{Z}(t, x_0)t \in I_a$, be the $(M \times M)$ matrix unique solution of the linear equations

$$(17) \qquad \frac{dZ}{dt} = ZA(t; x_0), \qquad t \in I_a, \;\; Z(0) = I_M \text{ (identity)}$$

where the matrix $A(t; x_0)$ defines the integral equation (16). Then, necessarily, the solution in (16) is expressed as

$$(18) \qquad N(t) = \{Y_1, \ldots, Y_M\}(x_0)\mathcal{Z}(t; x_0), t \in I_a,$$

where $Z(t; x_0)$ is the solution of (17).

The equation (18) is proved first for the corresponding approximating sequences $\{N_k(\cdot)\}$ and $\{Z_k(t)\}$ of (16) and respectively (17).

Taking the limit we obtain (18) fulfilled for the unique solutions $N(\cdot)$ and $\mathcal{Z}(\cdot)$.

Using (18) and (12) we rewrite the mapping in (8) as

$$(19) \qquad \begin{aligned} (\varphi(t)_* g_j)(x_0) = \{Y_1(x_0), \ldots, Y_M(x_0)\}\mathcal{Z}(t; x_0)b_j(t; x_0) \\ t \in I_a, \quad j = 1, \ldots, d., \end{aligned}$$

where $b_j(\cdot) \in C^\infty(I_a; R^M)$ is defined in (10).

Using the hypothesis (7) we obtain

(20) $$\langle \frac{\partial h_i}{\partial z}(z), g(z) \rangle = 0, i = 1, \dots, l, z \in \mathcal{O},$$

for all $g \in \Lambda \triangleq L(g_1, \dots, g_d)$ and in particular

(21) $$\left\langle \frac{\partial h_i}{\partial z}(z), Y_k(z) \right\rangle = 0, i = 1, \dots, l, z \in \mathcal{O}, \text{ for any } k = 1, \dots, M.$$

Therefore, using (19) we may conclude

(22) $$\langle \frac{\partial h_i}{\partial z}(x_0), (\varphi(t)_* g_j)(x_0) \rangle = 0 \text{ for each } x_0 \in \mathcal{O}, t \in I_a, \text{ and for all} \\ i \in \{1, \dots, l\}, j \in \{1, \dots, d\}$$

This shows that the disturbance decoupling conditions in (16) are fulfilled and the proof is complete.

**Remark 3.**

*In Proposition 2 an arbitrarily fixed constant control* $u = (u_1, \dots, u_m) \in R^m$ *is used and the controlled vector field* $f(x) = f_0(x) + \sum_{i=1}^m u_i f_i(x), x \in \mathcal{O}$, *determines the behaviour of the corresponding local flow* $\varphi(t; x_0), t \in I_a, x_0 \in V \subseteq \mathcal{O}$. *The disturbance decoupling problem is solved provided the algebraic representation in (19) is written for each* $x_0$ *in a neighbourhood* $V \subseteq \mathcal{O}$ *and the disturbing term in the equation (4) for the composite solution* $z(t; x_0) = \varphi(t; \Psi(t; x_0))$ *takes the form*

(23) $$\sum_{j=1}^d (\varphi(t)_* g_j)(z) p_j(t) = \{Y_1(z), \dots, Y_M(z)\} Z(t; z) \left( \sum_{j=1}^d b_j(t; z) p_j(t) \right)$$

*where* $\{Y_1, \dots, Y_M\} \subseteq L(g_1, \dots, g_d) = \Lambda$ *is a fixed system of generators for a finite type of Lie algebra* $\Lambda$.

*Here, the new entries as the* $(M \times M)$ *matrix* $Z$ *and the vector functions* $b_j$ *are smooth on* $(t, z) \in I_a \times V$. *The general structure in (23) for the perturbing term in (4) suggests that the dynamic for the composite solution* $z(t; x_0)$ *in (4) can be rewriten*

(24) $$\frac{dz}{dt} = f_0(z) + \sum_{i=1}^m u_i f_i(z) + \sum_{j=1}^d p_j(t) w_j(t, z), \qquad z(0) = x_0, t \in I_a,$$

*for some smooth vector functions* $w_j \in C^\infty(I_a \times V; R^n)$ *obeying to* $w_j(t, \cdot) \in L(g_1, \dots, g_d)$ *for each* $t \in I_a, j = 1, \dots, d$. *This is why we can start considering the equation in (24) as the original perturbed equation with vector fields*

$w_j(t, \cdot)$ in a Lie algebra $L(g_1, \ldots, g_d)$ of the finite type which is invariant with respect to the controlled vector fields $f_j, j = 0, 1, \ldots, m$.

The disturbance decoupling problem becomes more relevant for applications if a Lipshitz continuous control on $\mathcal{O}$

$$(25) \qquad\qquad u_i = \tilde{u}_i(x), i = 1, \ldots, m$$

is used such that the solution $\tilde{\varphi}(t, x_0)$ of the system

$$(26) \qquad \frac{dx}{dt} = \tilde{f}_0(x), \tilde{f}_0(x) \triangleq f_0(x) + \sum_{i=1}^{m} \tilde{u}_i(x) f_i(x) x(0) = x_0 \in V \subseteq \mathcal{O}$$

fulfils an extra condition, such a stability for example. In the following we shall consider the disturbance decoupling problem for a fixed Lipschitz continuous control $\tilde{u}(x) \in R^m, x \in \mathcal{O}$, defining the vector field $\tilde{f}(\cdot)$ as in (26). Denote as $\tilde{\varphi}_t(y)$, $y \in V \subseteq \mathcal{O}, t \in I_a = (-a, a)$ the local flow generated by the Lipschitz continuous vector field $\tilde{f}$. It is readily seen that the composition of flows $\tilde{z}_t(t; x_0) \triangleq \tilde{\varphi}_t(\Psi(t, x_0)), t \in [0, T]$, is an absolutelly continuous function provided the input disturbance $\Psi(t, x_0)$ is the solution of the system

$$(27) \qquad\qquad \frac{dy}{dt} = \sum_{j=1}^{d} p_j(t) g_j(y) \quad y(0) = x_0$$

which contains the unknown perturbations $p_j(\cdot), j = 1, \ldots, d$. It makes sense to speak about the derivative with respect to the variable $t \in [0, T]$ of the function $\tilde{z}_t(t; x_0)$ but in this case the differential mapping $\tilde{\varphi}(t)_*$ acting on vector fields has no meaning. It is why the procedure adopted in Proposition 2 for the disturbance decoupling problem is not suitable for our purpose and it will be of interest to replace the usual differential mapping using a gradient system solution associated with the Lie algebra $L(f_0, f_1, \ldots, f_m)$ of controlled vector fields. The following Proposition is obtained as a direct consequence of the algebraic representation in a finite dimensional Lie algebra (see Theorem 2.1.2). Let $I_0(f_1, \ldots, f_m)$ be the ideal generated by $f_0$ to into $L(f_0, \ldots, f_m)$.

**Proposition 3.**
    Let $f_j \in C^\infty(\mathcal{O}; R^n), j = 0, 1, \ldots, m$, be given such that the corresponding Lie algebra $\Lambda = I_0(f_1, \ldots, f_m)$ is finite-dimensional. Let $\{Y_1, \ldots, Y_M\} \subseteq \Lambda$ be a fixed basis and consider the mapping $G(t, p; y) \triangleq F_0(t)_0 G_1(t_1) \circ \ldots \circ G_M(t_M)(y)$, for $p \in D_M \triangleq \prod_{1}^{M}(-a_i, a_i), t \in (a_0, a_0)$ and $y \in V \subseteq \mathcal{O}$, where

$G_i(\tau)(x), F_0(t)(x)$ *are the local flows generated by the vector field $Y_i$ in the fixed basis and $f_0$ respectively. Then there exist $q_j \in C^\omega(R^{M+1}, R^N)j = 1, 2, \ldots, m$, such that*

(c)
$$\frac{\partial G}{\partial p}(t, p; y)q_j(t, p) = f_j(G(t, p; y)),$$

*for any $t \in (-a_0, a_0)p \in D_M, y \in V \subseteq \mathcal{O}$ and all $j = 1, 2, \ldots, m$.*

The mapping $G(t, p; y)$ and the conclusion (c) of the above given statement are relevant for a smooth representation of the Lipschitz continuous local flow $\widetilde{\varphi}(t, y)$ taken along smooth curves $y = \Psi(t, x_0), \ x_0 \in V$, of the perturbation.

**Proposition 4.**

Let $f_j \in C^\infty(\mathcal{O}; R^n)j = 0, 1, \ldots, m$, *be given such that the Lie algebra $I_0(f_1, \ldots, f_m)$ is finite–dimensional. Define the mapping $G(t, p; y), t \in (-a_0, a_0), p \in D_M, y \in V \subseteq \mathcal{O}$, and $g_j \in C^\omega(R^{M+1}; R^M)$ fulfilling (c) of Proposition 3. Let $\widetilde{\varphi}_t(y), \ t \in I_0, \ y \in V$, be the local flow generated by the Lipschitz continuous vector field $\widetilde{f}$ in (26), and consider $y = \Psi(t, x_0), \ t \in I_\alpha \subseteq I_a$, a solution in (27). Then the following representation holds:*

(∗)
$$\widetilde{z}_s(t; x_0) \triangleq \widetilde{\varphi}_s(\Psi(t; x_0)) = G(s, \widetilde{p}(s; t, x_0); \Psi(t, x_0)),$$

*for $t \in I_\alpha$, where $\widetilde{p}(s; t, x_0), \ s \in [0, t]$, is the unique solution of*

(∗∗)
$$\frac{dp}{ds} = \sum_{i=1}^{m} \widetilde{u}_i(s, p; t, x_0)q_i(s, p)p(0) = 0, \quad s \in [0, t].$$

**Proof.**

For each $y \in V$ fixed we notice that $\widetilde{\varphi}_t(y), t \in I_\alpha$, can be represented as

(28)
$$\widetilde{\varphi}_t(y) = G(t, p(t; y); y)$$

where $p(t; y), t \in I_\alpha$, is the unique solution of

(29)
$$\frac{dp}{dt} = \sum_{i=1}^{m} \widetilde{u}_i(G(t, p; y))q_i(t, p), \qquad p(0) = 0$$

and $q_j(\cdot), \widetilde{u}(\cdot)$ are given by hypothesis.

To this end we notice that $y = \widetilde{\varphi}_0(y) = G(0; y)$, and computing the derivative

$$\frac{d}{dt}G(t, p(t; y); y) = \frac{\partial G}{\partial p}(t, p(t; y); y)\frac{dp}{dt}(t; y) + \frac{\partial G}{\partial t}(t, p(t, y); y), \text{ for } t \in I_\alpha$$

one can see easily that $\xi(t; y) \stackrel{\triangle}{=} G(t, p(t; y); y), t \in I_\alpha$, is a solution of the differential equation (see (26))

$$(30) \qquad \frac{d\xi}{dt} = f_0(\xi) + \sum_1^m \tilde{u}_i(\xi) f_i(\xi) \stackrel{\triangle}{=} \tilde{f}(\xi), \qquad \xi(0) = y$$

Therefore $G(t, p(t; y); y) = \tilde{\varphi}_t(y),; t \in I_\alpha, \ y \in V$. Now, a direct substitution of $y = \psi(t, x_0)$ into $\tilde{\varphi}_t(y)$ and the computation of $p(t; \psi(t, x_0)) \stackrel{\triangle}{=} \tilde{p}(t; t; x_0)$ as the final value at time $s = t$ of the solution $\tilde{p}(s; t, x_0)$ in $(*, *)$ imply

$$(31) \qquad G(t, \tilde{p}(t; t, x_0); \psi(t, x_0)) = \tilde{\varphi}_t(\psi(t; x_0)), \qquad \text{foreach t} \in I_\alpha,$$

and $G(s, \tilde{p}(s; t, x_0); \psi(t, x_0)) = \tilde{\varphi}_s(\psi(t; x_0))$ for any $s \in [0, t]$.
      The proof is complete.

### Remark 4.
      *The dependence of the composite function* $\tilde{z}_t(t; x_0) \stackrel{\triangle}{=} \tilde{\varphi}_t(\psi(t, x_0))$ *on* $\psi(t, x_0)$ *is restated (see Proposition 4) in both ways, i.e. the smooth dependence on* $\psi(t, x_0)$ *is reflected in the smoothness of the mapping* $G(p; y)$ *with respect to the variable* $y$ *and the nonsmooth dependence on* $\psi(t, x_0)$ *is separated into the solution* $\tilde{p}(s; t, x_0), s \in [0, t]$, *of the auxiliary system* $(*, *)$. *We shall proceed further by giving an explicit expression for each of the above mentioned dependence.*
      As far as the smooth dependence part is concerned it is a routine matter to compute its derivative

$$(32) \qquad \begin{aligned} E(t) &\stackrel{\triangle}{=} \frac{\partial G}{\partial y}(t; \tilde{p}(t; t, x_0)); \psi(t, x_0)) \frac{d\psi}{dt}(t, x_0) \\ &= \sum_{j=1}^d p_j(t) \frac{\partial G}{\partial y}(t, \tilde{p}(t; t, x_0); \psi(t, x_0)) g_j(\psi(t, x_0)). \end{aligned}$$

and one may hope to recognize in the right hand side of (32) a differential mapping applied to the vector fields $g_j$ taken at the point $\tilde{z}_t(t, x_0) = \tilde{\varphi}_t(\psi(t, x_0))$. This is the case, provided we notice that $G(t, \tilde{p}(t; t, x_0); y)$, for $y$ in a neighbourhood of $\psi(t, x_0)$, is a diffeomorphism and its inverse $\tilde{G}(t, \tilde{p}(t; t, x_0); z)$ meets

$$(33) \qquad G(t, \tilde{p}(t; t, x_0); \tilde{G}(t, \tilde{p}(t; t, x_0); z)) = z$$

for $z$ in a neighbourhood of $\tilde{z}_t(t, x_0)$. Actually, the inverse mapping $\tilde{G}(t, p; z)$ associated to $G(t, p; y)$ has an explicit form (see $G(t, p; y)$ in Proposition 3)
(34)
$$\tilde{G}(t, p; z) = G_M(-t_M) \circ \ldots \circ G_1(-t_1) {}_\circ F_0(-t)(z), p = (t_1, \ldots, t_M) \in D_M,$$

and from (33) we obtain

(35)  $$\frac{\partial G}{\partial y}(t, \widetilde{p}; \widetilde{G}(t, \widetilde{p}; z)) \cdot \frac{\partial \widetilde{G}}{\partial z}(t, \widetilde{p}; z) = I_n \text{ (identity matrix)}$$

Therefore, there holds

(36)  $$\frac{\partial \widetilde{G}}{\partial z}(t, \widetilde{p}; \widetilde{z}))^{-1} = \frac{\partial G}{\partial y}(t, \widetilde{p}; \widetilde{G}(t, \widetilde{p}; \widetilde{z})) \qquad \text{and} \qquad \widetilde{G}(t, \widetilde{p}; \widetilde{z}) = \psi(t, x_0)$$

provided

$$\widetilde{z} = \widetilde{z}_t(t, x_0), \qquad \widetilde{p} = \widetilde{p}(t; t, x_0).$$

Using (36), the mapping in the right hand side of (32) can be rewriten
(37)
$$\frac{\partial G}{\partial y}(t, \widetilde{p}; \psi(t, x_0))g_j(\psi(t, x_0)) = \left(\frac{\partial \widetilde{G}}{\partial z}(t, \widetilde{p}; \widetilde{z})\right)^{-1} g_j(\widetilde{G}(t, \widetilde{p}; \widetilde{z})),$$

$$j = 1, \dots, d.$$

Now one may see easily that what we are looking for holds true and assimilate the expression in (37) with the differential mapping $G(t, \widetilde{p})_*$ applied
to the vector fields $g_j$ taken at the point $\widetilde{z}$, i.e.,

(38)
$$\frac{\partial G}{\partial y}(t, \widetilde{p}; \psi(t, x_0))g_j(\psi(t, x_0)) = (G(t, \widetilde{p})_* g_j)(\widetilde{z})$$
$$\overset{\triangle}{=} \left(\frac{\partial \widetilde{G}}{\partial z}(t, \widetilde{p}; \widetilde{z})\right)^{-1} g_j(\widetilde{G}(t, \widetilde{p}; \widetilde{z})),$$

$$j = 1, \dots, d.$$

As far as the Lie algebra $\Lambda = L(g_1, \dots, g_d)$ is of finite type we can represent algebraically the vector fields in (38) with respect to a fixed system of
generators $\{Z_1, \dots, Z_N\} \subseteq \Lambda$, and we obtain:

(39)  $$G(t, p)_* g_j(z) = \{Z_1(z), \dots, Z_N(z)\}A(t, p; z)b_j(t, p; z)$$

for

$$(t, p) \in D_{M+1} \overset{\triangle}{=} \prod_{i=0}^{M}(-a_i, a_i), z \in V \subseteq \mathcal{O},$$

where the $(N \times N)$ matrix $A(t, p; z)$ and the vector function $b_j(t, p; z) \in R^N$
are defined by smooth scalar functions in $C^\infty(D_{M+1} \times V; R)$. The computations involved in proving (39) are similar to those given in the proof

of Theorem 3.2.2 (see (52) there in) and they are based on the controlled invariance of the Lie algebra $\Lambda = L(g_1, \ldots, g_d)$ expressed by

(40)     $\mathrm{ad} f(g_j) \in \Lambda$     for any $j = 1, \ldots, d$, and $f \in \{f_0, f_1, \ldots, f_m\}$

It is left to the reader to convince himself that it can be done in a satisfactory way and finally, the expression (32) is rewriten as:

(41)
$$E(t) = \sum_{j=1}^{d} p_j(t) \frac{\partial G}{\partial y}(t, \widetilde{p}; \psi(t, x_0)) g_j(\psi(t, x_0))$$
$$= \{Z_1(\widetilde{z}), \ldots, Z_N(\widetilde{z})\} A(t, \widetilde{p}; \widetilde{z}) \left[ \sum_{j=1}^{d} p_j(t) b_j(t, \widetilde{p}; \widetilde{z}) \right],$$

if $\widetilde{p} = \widetilde{p}(t; t, x_0)$, $\widetilde{z} = z_t(t, x_0)$, and $\{Z_1, \ldots, Z_N\} \subseteq \Lambda$ is a fixed system of generators. In the case where the whole admissible class of bounded piecewise continuous controls in used $(u(\cdot) \in \mathcal{A}_T)$ then the auxiliary solutions $p^u(s) = \widetilde{p}$, $s \in [0, T]$, do not depend on $\psi(t, x_0)$ and from (41) one may conclude:

**Remark 5.**
    *Assume $\mathrm{ad} f(g_j) \in \Lambda \stackrel{\Delta}{=} L(g_1, \ldots, g_d)$ forany $j = 1, \ldots, d$ andf $\in \{f_0, f_1, \ldots, f_m\}$. For each control $u(\cdot) \in \mathcal{A}_T$ we obtain a local flow $\varphi^u(t; y)$ fulfilling*

(42)
$$\frac{dx}{dt} = f_0(x) + \sum_{i=1}^{m} u_i(t) f_i(x),$$

$$x(0) = y, t \in [0, T], y \in V \subseteq \mathcal{O}, \text{ and } z^u(t) \stackrel{\Delta}{=} \varphi^u(t; \psi(t, \mathbf{x}_0)).$$

*Then*
$$\frac{\partial \varphi^u}{\partial y}(t, \psi(t, x_0)) \frac{d\psi}{dt}(t, x_0) = E^u(t),$$

*where $E^u(t)$ is defined in (41) and $E^u(t) \in \ker dh(z^u(t))$ provided $g_j(x) \in \ker dh(x)$, $\forall x \in \mathcal{O}$, $j = 1, \ldots, d$.*

Generally, the non–smooth part of the composite function $\widetilde{z}_s(t, x_0) \stackrel{\Delta}{=} G(s, \widetilde{p}(s; t, x_0); \psi(t, x_0))$ is determined by the solution $\widetilde{p}(s; t, x_0), 0 \leq s \leq t$, of the system (**) in Proposition 4. As far as a Lipschitz continuous control $\widetilde{u}(x), x \in \mathcal{O}$, is viewed as a limit of smooth ones, we shall start computing $\frac{d}{dt}\widetilde{\varphi}_s(\psi(t, x_0)) = \partial \widetilde{z}_s(t, x_0)/\partial t$ for the smooth case. In other words, the algo-rithm of computing $\frac{d}{dt}\varphi_s(\psi(t, x_0))$ is based on considering $\widetilde{u}(\cdot) \in C^1(\mathcal{O}; R^m)$

and the corresponding derivative

$$
\frac{d}{dt}\varphi_s(\psi(t,x_0)) = \frac{\partial\varphi_s}{\partial y}(\psi(t,x_0))\frac{d\psi}{dt}(t,x_0)
$$

(43)

$$
= \sum_{j=1}^{d} p_j(t)\frac{\partial\varphi_s}{\partial y}(y)g_j(y), \text{ for } y = \psi(t,x_0)
$$

Assume that both Lie algebras $L(g_1,\ldots,g_d)$ and $L(f_1,\ldots,f_m)$ are of the finite type. Let $\{Z_1,\ldots,Z_N\} \subseteq L(g_1,\ldots,g_d)$ and $\{Y_1,\ldots,Y_M\} \subseteq L(f_1,\ldots,f_m)$ be some fixed generators. We need to consider the composed system of generators

$$
B = \{Z_1,\ldots,Z_N,Y_1,\ldots,Y_M\} \text{ for } \Lambda = L(g_1,\ldots,g_d) \cup L(f_1,\ldots,f_m).
$$

As far as

(44)     $\operatorname{ad} f_i(g) \in \Lambda$,     forany $i = 0,1,\ldots,m$, and all $g \in L(g_1,\ldots,g_d)$

is assumed, we may and do compute (43) through the algebraic representation of the $(n \times (M+N))$ matrix

(45)                 $N(s,z) = H(s,z)B(\varphi_{-s}(z))\ 0 \leq s \leq t$

where $H(s,z) \triangleq \left(\dfrac{\partial\varphi_{-s}}{\partial z}(z)\right)^{-1}$ and $\varphi_s(y) = z$ has the unique solution $y = \varphi_{-s}(z)$.

With these notations, one may recognize that $\dfrac{\partial\varphi_s}{\partial y}(y)g_j(y)$,     $y \in \mathcal{O}$, is equal to differential mapping $(\varphi_s)_*$ acting on the vector field $g_j$ and for $a^j(s,z) \in C([0,a] \times V; R^{M+N})$ fixed representing $g_j(\varphi_{-s}((z))$ with respect to $B$ we rewrite (43) as

(46)     $\dfrac{d}{dt}\varphi_s(\psi(t)x_0) = \displaystyle\sum_{j=1}^{d} p_j(t)N(s;z)a^j(s,z)$     for $z = \varphi_s(\psi(t,x_0))$

By definition, $H(s,z)$ meets

(47)             $\dfrac{dH}{ds} = H\dfrac{\partial\widetilde{f}}{\partial x}(\varphi_{-s}(z)), H(0) = I_n, 0 \leq s \leq t,$

where $\widetilde{f}(x) = f_0(x) + \displaystyle\sum_{i=1}^{m}\widetilde{u}_i(x)f_i(x)$, and the algebraic representation of $N(t,z)$ we are looking for will be

(48)                 $N(t,z) = B(z)Z(t;z),$

Here the $(M + N) \times (M + N)$ matrix $Z$ is the solution of a linear matrix equation

(49)
$$\frac{dZ}{ds} = ZA(s; z), 0 \leq s \leq t, \ Z(0) = I_{M+N},$$

where $A(s; z), 0 \leq s \leq t, \ z \in V \subseteq \mathcal{O}$, is a continuous matrix valued mapping. The Leibnitz–Newton formula allow one to write

(50)
$$\begin{aligned} N(t, z) &= N(0, z) + \int_0^t \tfrac{d}{ds} N(s, z) ds \\ &= B(z) + \int_0^t H(s, z) ad\widetilde{f}(B)(\varphi_{-s}(z)) ds \end{aligned}$$

where the $C^1$–vector field $\widetilde{f}(x) = f_0(x) + \sum_{i=1}^m \widetilde{u}_i(x) f_i(x), x \in \mathcal{O}$, is determined by a $C^1$–control $\widetilde{u}(\cdot) \in C^1(\mathcal{O}; R^m)$. This is why the action of $ad\widetilde{f}$ on $B$ must be decomposed and to restrict ourselves considering only first partial derivatives of $\widetilde{u}(\cdot)$. A direct computation shows

(51)
$$\begin{aligned} ad\widetilde{f}(B)(\varphi_{-s}(z)) =\ & ad f_0(B)(\varphi_{-s}(z)) \\ & + \sum_{i=1}^m \widetilde{u}_i(\varphi_{-s}(z)) ad f_i(B)(\varphi_{-s}(z)) \\ & + \sum_{i=1}^m \left\langle \frac{\partial \widetilde{u}_i}{\partial x}(\varphi_{-s}(z)), B(\varphi_{-s}(z)) \right\rangle f_i(\varphi_{-s}(z)), \end{aligned}$$

where the fixed system $B$ is denoted $B = (b_1, \ldots, b_{M+N})$ and
(52)
$$\begin{aligned} \langle \frac{\partial \widetilde{u}_i}{\partial x} \ & (\varphi_{-s}(z)), B(\varphi_{-s}(z)) \rangle \\ & \triangleq \left( \langle \frac{\partial \widetilde{u}_i}{\partial x}(\varphi_{-s}(z)) b_1(\varphi_{-s}(z)) \rangle, \ldots, \langle \frac{\partial \widetilde{u}_i}{\partial x}(\varphi_{-s}(z)), b_{M+N}(\varphi_{-s}(z)) \rangle \right). \end{aligned}$$

Let $A_i(s, z)$ be the $(M + N) \times (M + N)$ matrix associated with

(53)    $$ad f_i(B)(\varphi_{-s}(z)) = B(\varphi_{-s}(z)) A_i(s, z), \qquad i = 0, 1, \ldots, m$$

and the vector function $a_i(s, z) \in R^{M+N}$ fixed such that

(54)    $$f_i(\varphi_{-s}(z)) = B(\varphi_{-s}(z)) a_i(s, z), \qquad i = 1, \ldots, m.$$

According to the above notations we may rewrite

(55)
$$\begin{aligned} ad\widetilde{f}(B)(\varphi_{-s}(z)) &= B(\varphi_{-s}(z))[A_0(s, z) + \sum_{i=1}^m \widetilde{u}_i(s, z) A_i(s, z) \\ & + \sum_{i=1}^m a_i(s, z) \langle \frac{\partial \widetilde{u}_i}{\partial x}, B \rangle (\varphi_{-s}(z))] \\ &= B(\varphi_{-s}(z)) A(s, z) \end{aligned}$$

where $A(s, z)$ is a continuous matrix–valued mapping. Therefore applying (55) to (50), we see easily that:

$$(56) \qquad N(t, z) = B(z) + \int_0^t N(s, z)A(s, z)ds$$

and the approximation sequence $\{N_k(t, z)\}_{k \geq 0}$ associated to (56) fulfils

$$(57) \qquad N_k(t, z) = B(z)Z_k(t, z), \qquad k \geq 0, \qquad \text{forall } z \in V \subseteq \mathcal{O},$$

where $\{Z_k\}_{k \geq 0}$ is the corresponding approximation sequence for the matrix equation (49). The uniform convergence of the sequences $\{Z_k\}_{k \geq 0}$ and $\{N_k\}_{k \geq 0}$ allow one to obtain continuous matrices $Z(t, z)$, $N(t, z)$, $t \in [0, a]$, $z \in V \subseteq \mathcal{O}$, such that

$$(58) \qquad N(t, z) = B(z)Z(t, z), \qquad \text{for all } t \in [0, a], z \in V \subseteq \mathcal{O}.$$

and using (58) in (46) we obtain

$$(59) \qquad \begin{aligned} \frac{d}{dt}\varphi_s(\psi(t, x_0)) &= \sum_{j=1}^d p_j(t)N(s, z)a^j(s, z), a \\ &= B(z)\sum_{j=1}^d p_j(t)Z(s, z)a^j(s, z), \\ &\qquad 0 \leq s \leq t, \text{ for } z = \varphi_s(\psi(t, \mathrm{x}_0)). \end{aligned}$$

The latter expression allows one to deduce on the disturbance decoupling for a Lipschitz continuous control $\tilde{u}(x), x \in \mathcal{O}$, provided we notice that in this case we obtain a set of continuous matrices $\tilde{Z}(s, z)$, $s \in [0, t]$, $z \in V \subseteq \mathcal{O}$, expressing the limit points in $C([0, t]; R^n)$ of the corresponding sequence $\{\frac{d}{dt}\varphi_s^k(\psi(t, x_0)), 0 \leq s \leq t\}_{k \geq 0}$ defined by smooth controls $u^k(\cdot) \in C^1(\mathcal{O}; R^n)$. The abovegiven computation can be summarized as follows:

**Proposition 5.**
 *Let $f_i, g_j \in C^\infty(\mathcal{O}; R^n)$, $i = 0, 1, \ldots, m$, $j = 1, \ldots, d$, be given such that the Lie algebras $L(g_1, \ldots, g_d)$ and $L(f_1, \ldots, f_m)$ are of finite type and $adf(g) \in \Lambda \overset{\triangle}{=} L(g_1, \ldots, g_d) \cup L(f_1, \ldots, f_m)$ for any $f \in \{f_0, f_1, \ldots, f_m\}$ and $g \in \{g_1, \ldots, g_d\}$. Let $\tilde{u}(\cdot) \in C(\mathcal{O}; R^m)$ be a fixed Lipschitz continuous control and consider the controlled vector field $\tilde{f}(x) = f_0(x) + \sum_{i=1}^m \tilde{u}_i(x)f_i(x), x \in \mathcal{O}$, generating a local flow $\tilde{\varphi}_t(y), t \in (-a, a), y \in V \subseteq \mathcal{O}$. Then for each*

$$y = \psi(t, x_0) \in V, \ t \in [0, a),$$

*fulfilling* $\dfrac{dy}{dt} = \displaystyle\sum_{j=1}^{d} p_j(t) g_j(y) y(0) = x_0$, *one an absolutely continuous function*

$\widetilde{\varphi}_s(\psi(t, x_0)), t \in [0, a)$, *such that*

$(c_1)$ $$\frac{d}{dt}\widetilde{\varphi}_s(\psi(t, x_0)) \doteq B(z) \sum_{j=1}^{d} p_j(t) \widetilde{Z}(s, z) a^j(s, z),$$

*for* $z = \widetilde{\varphi}_s(\psi(t, x_0))$, *where* $B = \{Z_1, \ldots, Z_N\} \cup \{Y_1, \ldots, Y_M\}$ *is a fixed system of generators for* $\Lambda$,
$(c_2)$ $\widetilde{Z}(s, z), a^j(s, z)$ *are continuous matrix valued mapping and respectively* $R^{M+N}$ *vector valued function, for* $z = \widetilde{\varphi}_s(\psi(t, x_0))$ *and* $0 \le s \le t$.

**Proof.**

The representation stated in $(c_1)$ and $(c_2)$ is obtained in (59) for a smooth control $\widetilde{u}(\cdot) \in C^1(\mathcal{O}; R^m)$ with bounded partial derivatives $\left\|\dfrac{\partial \widetilde{u}_i}{\partial x}(x)\right\| \le C, x \in \mathcal{O}$, for $i = 1, \ldots, m$, and $C > 0$ some fixed constant. In the general case, $|\widetilde{u}_i(x'') - \widetilde{u}_i(x')| \le C\|x'' - x'\|$, for any $x', x'' \in \mathcal{O}$ and $i = 1, \ldots, m$, where $C > 0$ is a fixed constant. Approximating a Lipschitz continuous function with a sequence $\{u^k(\cdot)\}_{k \ge 1} \subseteq C^1(\mathcal{O}; R^M)$ of smooth and bounded partial derivatives functions we obtain a corresponding sequence of continuous matrix valued and vector valued mappings $\{Z_k(s, z), a_k^j(s, z)\}_{k \ge 0}$ for $z = \widetilde{\varphi}_s^k(\psi(t, x_0)), 0 \ge s \ge t$. The Ascoli–Arzela Lemma allow one to substract uniformly convergent subsequences and denote $\{\widetilde{Z}(s, z), a^j(s, z)\}$ for $z = \widetilde{\varphi}_s(\psi(t, x_0)), 0 \le s \le t$, a limit mapping. By definition, any limit mapping is a continuous one for $z = \widetilde{\varphi}_s(\psi(t, x_0)), 0 \le s \le t$, and using the algebraic representation in (59) we obtain the conclusions.

The proof is complete.

**Remark 6.**

*Under the hypotheses in Proposition 5, the disturbance decoupling problem obtains a positive answer provided* $g_j(x), f_i(x) \in \ker dh(x)$, *for any* $j = 1, \ldots, d, i = 1, \ldots, m$ *and all* $x \in \mathcal{O}$.

## 5.2 Stochastic disturbance decoupling

We have admitted that a deterministic disturbance input $\psi(\cdot)$ doesn't affect the output $h(\cdot) \in C^1(\mathcal{O}; R^l)$ of the control system

(1) $$\frac{dx}{dt} = f_0(x) + \sum_{i=1}^{m} u_i f_i(x) \overset{\triangle}{=} f^u(x)$$

provided the local flow $\varphi_t^u(y), t \in (-a, a) = I_a, y \in V \subseteq \mathcal{O}$, generated by the vector field $f^u(\cdot) \in C^\infty(\mathcal{O}; R^n)$ fulfils

$$(2) \qquad \frac{\partial \varphi_t^u}{\partial y}(\psi(t, x))\frac{d\psi}{dt}(t, x_0) \in \ker dh(\varphi_t^u(\psi(t, x_0))) \text{ for some } t \in [0, \alpha] \subseteq I_a$$

and for each constant control $u \in R^m$.

As far as $\psi(t, x_0), t \in [0, \infty] \subseteq I_a$, is a solution of a differential equation.

$$(3) \qquad \frac{dy}{dt} = \sum_{j=1}^d p_j(t)g_j(y), y(0) = x_0,$$

there is no obstruction in using (2) for defining the disturbance decoupling problem. The expression in (2) is meaningless for stochastic disturbance input fulfilling a stochastic differential equation

$$(4) \qquad dy = \sum_{j=1}^d g_j(y)_0 dw_j(t) \ y(0) = x_0, \ t \geq 0$$

where $w(\cdot) \stackrel{\triangle}{=} (w_j(\cdot), \ldots, w_d(\cdot))$ is a standard Wiener process on a probability space $\{\Omega, \mathcal{F}, P\}$ (see a.2) and "0" stands for Fisk–Stratonovich integral. Let $\mathcal{F}_t \subseteq \mathcal{F}$ be generated by $\{w(s) : 0 \leq s \leq t\}$. By definition, a solution $\psi(t, \omega), t \in [0, T], \omega \in \Omega$, of (4) is a measurable and $\mathcal{F}_t$–adapted $\psi : [0, T \times \Omega \to R^n$, with continuous trajectories $\psi(\cdot, \omega)$, for $\omega \in \Omega$, and fulfilling the integral equation.

$$(5) \qquad \psi(t, \omega) = x_0 + \sum_{j=1}^d \int_0^t g_j(\psi(s, \omega)) \circ dw_j(s), \text{ for any } t \in [0, T]$$

We may and do avoid the nonsmoothness of the trajectories $\psi(\cdot, \omega), \omega \in \Omega$, approximating the original Wiener process $w(\cdot)$ in differential equations (4) with smooth $\mathcal{F}_t$–adapted process $v_\varepsilon(t, \omega), v_\varepsilon(\cdot, \omega) \in C^1([0, T]; R^m)$ for any $\omega \in \Omega$ (see a.2 and Langevin's approximation there in) and consider that the input disturbance $\psi_\varepsilon(t, \omega), t \in [0, T(\omega)]$ is obeying to an ordinary differential equation

$$(6) \qquad \frac{dy}{dt} = \sum_{j=1}^d \frac{dv_\varepsilon^j(t, \omega)}{dt} g_j(y), \ y(0) = x_0 \in \mathcal{O}$$

Now, assimilate the continuous and $\mathcal{F}_t$-adapted process $p_j^\varepsilon(t, \omega) \overset{\triangle}{=} \frac{dv_\varepsilon^j(t, \omega)}{dt}, t \in [0, T], j = 1, \ldots, d$, as some unknown perturbations for each $\varepsilon 0$ fixed. With this interpretation, one can use the previous definition in (2) for each $\varepsilon > 0$ and applying Proposition 2 for deterministic disturbance decoupling we obtain

**Remark 1.**

*Let $f_i, g_j \in C^\infty(\mathcal{O}; R^n) i = 0, 1, \ldots, m, j = 1, \ldots, d$, be given such that the Lie algebra $\Lambda = L(g_1, \ldots, g_d)$ is of finite type and $adf(g_j) \in \Lambda$ for $j = 1, \ldots, d$, and any $f \in \{f_0, f_1, \ldots, f_m\}$. Then, for each $\varepsilon > 0$ fixed, the disturbance decoupling*

$$(2_\varepsilon) \qquad \begin{array}{l} \dfrac{\partial \varphi_t^u}{\partial y}(\psi_\varepsilon(t, x_0)) \dfrac{d\psi_\varepsilon}{dt}(t, x_0) \in \ker dh(\varphi_t^u(\psi_\varepsilon(t, x_0))) \\ \text{for } t \in [0, T(\omega)], \end{array}$$

*holds true for each $u \in R^m$ and any local solution in (6), provided $g_j(x) \in \ker dh(x)$, for $j = 1, \ldots, d$, and all $x \in \mathcal{O}$.*

The definition $(2_\varepsilon)$ in Remark 1 might be taken as a possible choice for the stochastic disturbance decoupling problem which can be replaced by some sufficient conditions totally independent of the parameter $\varepsilon > 0$. It is possible to give a direct definition using stochastic original input disturbance but sufficient conditions in Remark 1
remain the same. Another type of deterministic disturbance decoupling appeared in Proposition 5 where a Lipschitz continuous control $\tilde{u}(\cdot) \in LC(\mathcal{O}; R^m)$ was used. A straight forward implication of the mentioned result will allow one to deduce the corresponding alternative for the input–solution of (b) for each $\varepsilon > 0$ fixed. It is left to the reader to conclude this remark. As far as the disturbance decoupling definition given in (2) allow one to write an equivalent integral form

$$(7) \qquad h(z^u(t, x_0)) = h(x_0) + \int_0^t \langle \frac{\partial h}{\partial x}(z^u(s, x_0)), f^u(z^u(s, x_0)) \rangle ds, \ t \in [0, \alpha]$$

where $z^u(t, x_0) \overset{\triangle}{=} \varphi_t^u(\psi(t, x_0))$ is the solution of the ordinary differential equation.

$$(8) \qquad \frac{dz}{dt} = f^u(z) + \sum_{j=1}^d p_j(t)(\varphi_t^u)_* g_j(z), z(0) = x_0, t \in [0, \alpha],$$

it is obvious that the same integral form (7) can be used to give a precise meaning for a stochastic disturbance decoupling. Namely, in stochastic case, the composite mapping $z^u(t, x_0) \overset{\triangle}{=} \varphi_t^u(\psi(t, x_0))$ fulfils, at least locally for $t \in [0, T(\omega)]$, a stochastic differential equation in integral form

$$(9) \quad z^u(t, x_0) = x_0 + \int_0^t f^u(z^u(s, x_0)) ds + \sum_{j=1}^d \int_0^t (\varphi_s^u)_* g_j(z^u(s, x_0)) \circ dw_j(s)$$

and the output $h(x), x \in \mathcal{O}$, is not affected by the stochastic disturbance defined as Fisk–Stratonovich integral in (9) provided
(10)

$$h(z^u(t, x_0)) = h(x_0) + \int_0^t \left\langle \frac{\partial h}{\partial x}(z^u(s, x_0)), f^u(z^u(s, x_0)) \right\rangle ds \; t \in [0, T(\omega)]$$

for each $u \in R^m$.

The last equation in (10) stands for the stochastic disturbance decoupling problem and a smooth approximation of the original Wiener process by $v_\varepsilon(t, \omega)$ as in (6) allow one to replace (10) by $(2_\varepsilon)$ for each $\varepsilon > 0$.

## 5.6 Stochastic differential equations

Usually, singularly perturbed deterministic equations involve a slow variable $x \in R^n$ which is obeying to an ordinary differential system

$$(1) \qquad \frac{dx}{dt} = f(x, y), x_{(0)} = x_0 \in R^n. \, t \in [0, T],$$

depending on a fast variable $y \in R^m$ which is evolving according to a singularly perturbed differential system

$$(2) \qquad \varepsilon \frac{dy}{dt} = g_0(x, y), \; y(0) = y_0 \in R^m, t \in [0, T],$$

The singularity of the system (2) is measured by the small parameter $\varepsilon > 0$ multiplying the derivative and the limit behaviour of the solutions $(x^\varepsilon(\cdot), y^\varepsilon(\cdot)), \varepsilon > 0$, of (1) and (2) for $\varepsilon \downarrow 0$ may and do involve two approaches. It is a classical problem and one way of solving the problem is based on the algebraic equation

$$(3) \qquad\qquad g_0(x, y) = 0$$

which is solved with respect to $y = \varphi(x)$ allowing one to obtain the limit solution of $(x^\varepsilon(\cdot), y^\varepsilon(\cdot)), \varepsilon > 0$, as $y^0(t) = \varphi(x^0(t))$ with $\dfrac{dx^0}{dt}(t) = f(x^0(t), \varphi(x^0(t)), x^0(0) = x_0$, provided some stability of the vector fields $g_0(x, \cdot)$ is assumed for any $x \in B \subseteq R^n$ in a bounded set. It may occur that the solution of (3) is not unique and even if $\{x^\varepsilon(\cdot), y^\varepsilon(\cdot)\}, \varepsilon > 0$, is uniformly bounded in $C([0, T]; R^n)$ we are led to take into consideration the set of all possible limit points. A more natural interpretation is to look at $y^\varepsilon(t)$, for each $t \in (0, T]$, as the probability masure $P_t^\varepsilon \in \mathcal{P}(R^m)$ on $R^m$ concentrated at the point $y^\varepsilon(t)$ and we may rewrite the solution in (1) as

$$(4) \qquad x^\varepsilon(t) = x_0 + \int_0^t \left[ \int_{R^m} f(x^\varepsilon(s), y) P_s^\varepsilon(dy) \right] ds, \text{ for t} \in [0, T].$$

The space of probability measures on $R^m$, $\mathcal{P}(R^m)$ is endowed with the weak topology generated by the basic space of bounded and continuous functions on $R^m$ denoted by $C_b(R^m)$ and a sequence $\{Q_k\}_k \subseteq \mathcal{P}(R^m)$ converges weakly to a $Q_0 \in \mathcal{P}(R^m)$ if $\lim_{k \to \infty} \int_{R^m} h(y) Q_k(dy) = \int_{R^m} h(y) Q_0(dy)$ for any $h \in C_b(R^m)$.

Under a uniform boundedness condition $|y^\varepsilon|(t) \leq C$, for any $\varepsilon \in (0, 1], t \in (0, T]$, we obtain that the corresponding set $\{P_t^\varepsilon, \varepsilon > 0\} \subseteq \mathcal{P}(R^m)$ is weakly compact (see Billingsley (1)). In addition, any limit point $P_t^0 \in \mathcal{P}(R^m)$ will replace the algebraic solution. It explains the basic of the second approach we may consider and a possible limit solution could be a pair $\{(x^0(t), P_t^0), t \in [0, T]\} \subseteq C([0, T]; R^n) \times \mathcal{P}(R^m)$ such that

$$(5) \qquad x^0(t) = x_0 + \int_0^t \left( \int_{R^m} f(x^0(s), y) P_s^0(dy) \right) ds, \text{ for any t} \in [0, T].$$

### 5.6.1  Singularly perturbed equations viewed as controlled equations

As far as $P_t^0$ is a weak limit point in $\mathcal{P}(R^m)$ of the given sequence $\{P_t^\varepsilon\}, \varepsilon > 0$, using a stability assumption on $g_0(x, y)$ we obtain an admissible class for the limit points $P_t^0$ as the stationary (invariant) probabilities associated with the deterministic flow $\varphi_s(y; x^0(t)), s \geq 0$, generated by $\dfrac{dy}{ds} = g_0(x^0(t), y), s \geq 0$ for each $t \in (0, T]$ fixed. Denote $\Gamma(x) \subseteq \mathcal{P}(R^m)$ the set consisting of all probabilities which are invariant with respect to the flow $\varphi(y; x), s \geq 0$, generated by

(6)
$$\frac{dy}{ds} = g_0(x, y), y(0) = y, s \geq 0, \text{for x} \in R^n \text{ fixed, i.e.}$$

(7)
$$\Gamma(x) = \{P \in \mathcal{P}(R^m) : \int_{R^m} h(\varphi_s(y; x))P(dy) =$$
$$= \int_{R^m} h(y)P(dy), \text{ forany h} \in C_b(R^m), s \geq 0\}$$

The definition of an invariant probability $P \in \mathcal{P}(R^m)$ with respect to the flow $\varphi_s(y; x)$ contained in (7) can be restated as $P_s(A) = P(A)$, for any Borelian $A \subseteq R^m$ and $s \geq 0$ where $\varphi_s(y; x)$ generates the probability $P_s \in \mathcal{P}(R^m)$ by $P_s(A) \triangleq P\{y \in R^m : \varphi_s(y; x) \in A\} = P\{\varphi_{-s}(A; x)\}$. Since a probability measure concentrated at a point $y^0(t)$ belongs to $\Gamma(x^0(t))$ provided $g(x^0(t), y^0(t)) = 0$, it shows that algebraic solutions of $g(x^0(t), y) = 0$ are naturally included in the admissible set $\Gamma(x^0(t))$. With these remarks, one may reconsider the original singularly perturbed problem starting with the relaxed integral equation

(*)
$$x(t) = x_0 + \int_0^t \left( \int_{R^m} f(x(s), y)P_s(dy) \right) ds, \ t \in [0, T],$$

where the set $M \subseteq \{P_t \in \mathcal{P}(K), t \in [0, T]\}$, of generalized controls consists of measurable mappings, i.e. $t \to \int_K h(y)P_t(dy)$ is measurable from $[0, T]$ to $R$ for each $h \in C(K; R)$, where $K \subseteq R^m$ is a fixed compact set. Supposing that the component $y^\varepsilon(t), t \in [0, T]$, of the solution $(x^\varepsilon(t), y^\varepsilon(t)), t \in [0., T]$, of (1) and (2), fulfils a boundedness condition $y^\varepsilon(t) \in K$, for any $\varepsilon \in [0, 1], t \in [0, T]$, and some compact $K \subseteq R^m$, we obtain a particular set of generalized controls $\{D_t^\varepsilon, t \in [0, T]\} \in M$ for any $\varepsilon \in (0, 1]$, which are generated by $y^\varepsilon(t), t \in [0, T]$. Now, the problem is to characterize the set of limit points associated to the relatively compact set $\{\{(x^\varepsilon(t), P_t^\varepsilon), t \in [0, T]\}; \varepsilon > 0\}$, in the space $C([0, T]; R^n) \times M$.

This interpretation allow one to obtain dynamical limit solutions which are not necessarily involved in algebraic solutions of the equation $g_0(x, y) = 0$. On the other hand, the same approach could be used for interpreting the singularly perturbed stochastic differential equation which are defined by a deterministic equation for the slow variable $x(\cdot) \in C([0,T]; R^n)$

$(\alpha)$  $$x(t) = x_0 + \int_0^t \left( \int_{R^m} f(x(s), y) P_s(dy) \right) ds, \ t \in [0,T],$$

and the admissible class $\widehat{M}$ of generalized control functions $\prod = \{P_t \in \mathcal{P}(R^m), \ t \in [0,T]\}$ consists of measurable mappings, i.e. $t \to \int_{R^m} h(y) P_t(dy)$ is measurable from $[0,T]$ to $R$ for each $h \in C(R^m; R)$ fulfilling $|h(y)| \le C_h(1 + |y|^2)$. The singularly perturbed stochastic differential equation is defined

$(\beta)$  $$\varepsilon dy = g_0(x(t), y)dt + \sqrt{\varepsilon} \sum_{i=1}^k g_i(x(t), y) dw_i(t), \ t \in [0,T], \ \varepsilon > 0$$

where $w(\cdot)$ is a standard $k$–dimensional Wiener process on the probability space $\{\Omega, \mathcal{F}, P\}$ and $x(\cdot) \in C([0,T]; R^m)$ is arbitrarily fixed.

The admissible class $\widehat{M}$ is too large for defining solutions in $(\alpha)$ and we shall restrict ourselves to the bounded generalized controls in $\widehat{M}_0 \subseteq \widehat{M}$ defined as

$(\alpha_1)$

$$\widehat{M}_0 = \left\{ \prod \in \widehat{M} : \left| \int_{R^m} |y|^2 P_t(dy) \right| \le C_0, t \in [0,T], \text{ for some constant } C_0 \right\}$$

It is a routine matter to define a unique solution of the equation $(\alpha)$ for each $\prod \in \widehat{M}_0$ provided the following conditions are fulfilled

$(\alpha_2)$

$(i) f \in C(R^n \times R^m; R^n), \|f(x, y)\| \le C_1(1 + \|x\|)(1 + \|y\|^2) \ (\forall) \ x \in R^n, y \in R^m$

$(ii)$
$\|f(x_2, y) - f(x_1, y)\| \le L_K(1 + \|y\|^2)\|x_2 - x_1\| \ (\forall) x_1, x_2 \in K \subset R^n \text{and } y \in R^m,$

for some constants $C_1 L_k > 0$, where the compact $K$ is arbitrarily fixed.

$(iii)$  $$\frac{\partial f}{\partial y}(x, y)\| \le C_1(1 + \|x\|) \ (\forall) \ x \in R^n, y \in R^m.$$

**Remark.**

A bounded set $\widehat{B}_0 \subseteq C([0, T]; R^n)$ of solution in $(\alpha)$ is defined provided the bounded generalized controls in $\widehat{M}_0$ are used. Namely, let $x(t), t \in [0, T]$, be a solution in $(\alpha)$ corresponding to an arbitrarily fixed $\prod \in \widehat{M}_0$. Then, using ( $\alpha_2$, (ii)), we obtain $\|x(t)\| \leq \|x_0\| + C_2 \int_0^t (1 + \|x(s)\|) ds$, where $C_2 = C_1(1 + C_0)$, and applying Gronwall's Lemma we obtain

$$1 + \|x(t)\| \leq (1 + \|x_0\|) \exp C_2 T \text{ and } \|x(t)\| \leq (1 + \|x_0\|) \exp C_2 T - 1,$$

for any $t \in [0, T]$. Therefore $\widehat{B}_0 \subseteq \{x(\cdot) \in C([0, T]; R^n) : \|x(\cdot)\| \leq \widehat{C}_0\}$, where $\widehat{C}_0 \triangleq (1 + \|x_0\|)(\exp C_1(1 + C_0)T) - 1$, and $C_0, C_1$ are given in $(\alpha_1), (\alpha_2)$,

## 5.6.2 Bounded solutions for singularly perturbed system $(\alpha), (\beta)$

The main obstruction for defining bounded solutions of $(\alpha), (\beta)$ comes from the singularlty perturbed diffusion equation (3) which involves a large parameter $N = \frac{1}{\varepsilon}$ in the coeficients of the right hand side. As far as a solution of $(\beta)$ is accepted provided it generates a bounded generalized control for each $\varepsilon \in [0, 1]$ we shall notice that it can be fulffiled imposing the usual linear growth and Lipschitz continuity condition as follows

$(\beta_1)$
$$i) \ \|g_j(x, y)\|^2 \leq C(1 + \|y\|^2),$$
$$\|g_j(x, y_2) - g_j(x, y_1)\|^2 \leq L\|y_2 - y_1\|^2,$$
$$\text{forany } x \in R^n \text{ and } y_1, y_2, y \in R^m$$

$(ii)$
$$\|g_j(x_2, y) - g_j(x_1, y)\|^2 \leq C(1 + \|y\|^2)\|x_2 - x_1\|^2,$$
$$(\forall) \ y \in R^m \text{ and } x_1, x_2 \in R^n$$

for some constants $C, L > 0$ and all $j = 0, 1, \ldots, k$.

**Remark 2.**
Assuming the conditions $(\beta_1)$ fulfilled we may obtain bounded solutions of $(\beta)$ for each $\varepsilon \in (0, 1]$ fixed. Namely, for $x(\cdot) \in ([0, T]; R^n), \varepsilon \in (0, 1]$, fixed, the solution $y^\varepsilon(t), t \in [0, T]$, fulfils

$$y^\varepsilon(t) - y_0 = \frac{1}{\varepsilon} \int_0^t g_0(x(s), y^\varepsilon(s)) ds +$$

$$\frac{1}{\sqrt{\varepsilon}} \sum_{j=1}^k \int_0^t g_j(x(s), y^\varepsilon(s)) dw_j(s), t \in [0, T].$$

*and the linear growth condition allow one to estimate the norm $L_2$–norm $(E\|y^\varepsilon(t)\|^2)^{1/2}$ of $y^\varepsilon(t)$ as follows*

$$E\|y^\varepsilon(t)\|^2 \leq A \left[ \|y_0\|^2 + \frac{C}{\varepsilon} \int_0^t (1 + E\|y(s)\|^2)ds + \frac{Ck}{2} \int_0^t (1 + E\|y(s)\|^2)ds \right]$$

*for all $t \in [0,T]$, and $A > 0$ some constant.*
*Applying Gronwall's Lemma for $\varphi(t) = 1 + E\|y^\varepsilon(t)\|^2, t \in [0,T]$, where "$E$" stands for the expectation taken with respect to the original probability $P$ of $\{\Omega, \mathcal{F}, P\}$, we obtain*

(∗)                      $E\|y^\varepsilon(t)\|^2 \leq C_1^\varepsilon$ for all $t \in [0,T]$,

*and some constant $C_1^\varepsilon > 0$ independent of $x(\cdot) \in C([0,T]; R^n)$. In addition, a similar direct computation allow one to see*

(∗∗)     $E\|y_2^\varepsilon(t) - y_1^\varepsilon(t)\|^2 \leq C_2^\varepsilon (\max_s \|x_2(s) - x_1(s)\|)^2$ for all $t \in [0,T]$,

*and some constant $C_2^\varepsilon > 0$ independent of $x_i(\cdot)$, where $y_i^\varepsilon(\cdot)$ is the solution of $(\beta)$ corresponding to $x_i(\cdot) \in C([0,T]; R^n)$.*
    For each $\varepsilon \in (0,1]$ fixed, the solution $y^\varepsilon(t), t \in [0,T]$, of $(\beta)$, depends on $x(\cdot) \in C([0,T]; R^n)$ and is uniformly bounded as it is expressed in (∗) of Remark 2. It generates a special bounded generalized control $\overset{\varepsilon}{\prod}(x(\cdot)) \in \widehat{M}_\varepsilon$, where the set $\widehat{M}_\varepsilon$ is defined as in $(\alpha_1)$ with $C_1^\varepsilon$ replacing the constant $C_0$. In addition, the dependence on $x(\cdot) \in C([0,T]; R^n)$, of the solution $y^\varepsilon(t), t \in [0,T]$, is stipulated in the above given property (∗∗) and it insures a specific behaviour of $\overset{\varepsilon}{\prod}(x(\cdot)) \in \widehat{M}_\varepsilon$ as a functional of $x(\cdot) \in C([0,T]; R^n)$. The conclusions in Remark 2 can be restated for $\overset{\varepsilon}{\prod}(x(\cdot)) \overset{\triangle}{=} \{P_t^\varepsilon(x(\cdot)), t \in [0,T]\}$ as follows
($\gamma$)

1)$|\int_{R^m} h(y)P_t^\varepsilon(x(\cdot))(dy)| \leq C_h^\varepsilon$, forany $t \in [0,T], x(\cdot) \in C([0,T]; R^n)$ and
$h \in C(R^m; R)$ fulfilling $|h(y)| \leq C_h(1 + \|y\|^2)$

$$|\int_{R^m} h(y)[P_t^\varepsilon(x_2(\cdot)) - P_t^\varepsilon(x_1(\cdot))](dy)| \leq$$
$$\leq C_h^\varepsilon \max_{s \leq t} \|x_2(s) - x_1(s)\|, t \in [0,T],$$

for any $x_1(\cdot), x_2(\cdot) \in C([0,T];R^n)$ and $h \in C^1(R^m;R)$ fulfilling

$$|h(y)| \le C_h(1 + \|y\|^2), \left\|\frac{\partial h}{\partial y}(y)\right\|^2 \le C_h(1 + \|y\|^2) \ \forall y \in R^m, \text{ where } C_h > 0$$

is some constant depending on $h$.

**Remark 3.**

It is readily seen that the conclusions $(\gamma)$ written for a generalized control $\prod^\varepsilon(x(\cdot))$ generated by the solution $y^\varepsilon(\cdot)$ of $(\beta)$ allow one to define a unique solution $x^\varepsilon(\cdot) \in C([0,T];R^n)$ of the deterministic equation $(\alpha)$ obeying to the conditions $(\alpha_2)$. In addition, the corresponding generalized control $\prod^\varepsilon(x^\varepsilon(\cdot)) \triangleq \{P^\varepsilon_t(x^\varepsilon(\cdot)), t \in [0,T]\} \in \widehat{M}_\varepsilon$ is generated by the solution $y^\varepsilon(t), t \in [0,T]$, of $(\beta)$ choosing $x(\cdot) = x^\varepsilon(\cdot)$, and for each $\varepsilon \in (0,1]$ fixed the pair $(x^\varepsilon(t), y^\varepsilon(t)), t \in [0,T]$, fulfils the singularly perturbed system $(\alpha)$ and $(\beta)$.

The limit behaviour for the solution $(x^\varepsilon(t), y^\varepsilon(t)), t \in [0,T]$, of $(\alpha)$ and $(\beta)$ will be analyzed with respect to $\varepsilon \downarrow 0$ and the following uniform boundedness condition is assumed

$$(H_1) \qquad E\|y^\varepsilon(t)\|^2 \le C_0, \text{ for any } \varepsilon \in (0,1], \ t \in [0,T].$$

**Remark 4.**

Assuming $(H_1)$ fulfilled we obtain a generalized control $\prod^\varepsilon(x(\cdot)) \triangleq \{P^\varepsilon_t(x(\cdot)), t \in [0,T]\} \in \widehat{M}_0$, for all $\varepsilon \in (0,1]$ and any $x(\cdot) \in C([0,T]\ R^n)$, where $\widehat{M}_0$ is defined in $(\alpha_1)$ using the fixed constant $c_0$ assumed in $(H_1)$. The conclusions in Remark 3 applied to a generalized control $\pi^\varepsilon(x(\cdot)) \in \widehat{M}_0$ for any $x(\cdot) \in C([0,T];R^n)$ allow one to obtain a unique solution $x^\varepsilon(\cdot) \in C([0,T]\ R^n)$ of the deterministic equation $(\alpha)$ and a direct computation similar to that performed in Remark 1 will conclude the uniform boundedness property of determinstic solutions $x^\varepsilon(\cdot)$, i.e. $\|x^\varepsilon(\cdot)\| \le \widehat{C}_0$, for any $\varepsilon \in (0,1]$, and some constant $\widehat{C}_0 > 0$.

One may wander about the realizability of the hypothesis $(H_1)$ and it will be nice to give a simple sufficient condition ensuring it.

**Example.**

It is considered a particular singularly perturbed system for which the corresponding diffusion equation has the following form

$$(\beta_2) \qquad \varepsilon dy = -ydt + g_0(x)dt + \sqrt{\varepsilon}\sum_{j=1}^k g_j(x)(x)dw_j(t), y(0) = y(0) \in R^m,$$

where $g_j \in C_b(R^n; R^m), j = 0, 1, \ldots, k$, meet $(\beta_1)$.  A straight compu-
tation allow one to represent solutions in $(\beta_2)$ as $y^\varepsilon(t) = (\exp -\frac{1}{\varepsilon}t)z^\varepsilon(t)$,
where $(E\|z^\varepsilon(t)\|^2)^{1/2} \le \widetilde{C}[(\exp \frac{1}{\varepsilon}) + 1]$ for some constant $\widetilde{C} > 0$, and any
$\varepsilon \in (0, 1], t \ge 0$. It shows that the hypothesis $(H_1)$ is not so difficult to be
proved provided a strong stability of the drift term is assumed. The above
given remarks and elementary computations can be summarized in the following.

**Proposition 1.**
   *Let $f \in C(R^n \times R^m; R^n)$ and $g_j \in C(R^n \times R^m; R^m), j = 0, 1, \ldots, k$,
be given such that the conditions $(\alpha_2)$ and $(\beta_1)$ are fulfilled. Assume that
the solution $y^\varepsilon(\cdot), \varepsilon \in (0, 1]$, of the singularly perturbed diffusion equation $(\beta)$ obeys to the hypothesis $(H_1)$. Then there exists a unique solution
$(x^\varepsilon(t), P_t^\varepsilon, t \in [0, T]), \varepsilon \in (0, 1]$ of the system $(\alpha)$ and $(\beta)$ which is uniformly
bounded, i.e. $\|x^\varepsilon(t)\| \le \widehat{C}_0$ and $\int\limits_{R^m} \|y\|^2 P_t^\varepsilon(dy) \le \widehat{C}_0$ for any $t \in [0, T]$ and
$\varepsilon \in (0, 1]$.*

## 5.6.3   Characterization of the dynamical limits

The integer $p = 2$ used in $(H_1)$ allow one to obtain a strong convergence
in $L_1$ norm provided the boundedness condition in $L_2$ is fulfilled. It gives
the motivation for replacing the original probability space $\{\Omega, \mathcal{F}, P\}$ by a
new one and to consider the following unperturbed stochastic differential
equation.

$$(8) \qquad dy = g_0(x, y)ds + \sum_{i=1}^{k} g_i(x, y)dw_i(s), s \ge 0,$$

for each $x \in R^n$, where $g_j, j = 0, 1, \ldots, k$, are the mappings fulfilling the
conditions in $(\beta_1)$.
   The basic probability space $\{\Omega, \mathcal{F}, P\}$ is replaced by a new one $\{\widetilde{\Omega}, \widetilde{\mathcal{F}}, \widetilde{P}\}$,
where $\widetilde{\Omega} = \Omega \times \overline{\Omega}, \widetilde{\mathcal{F}} = F \otimes \overline{\mathcal{F}}, \widetilde{P} = P \otimes \overline{P}$, are completely determined
specifying $\overline{\Omega} = [0, 1], \overline{\mathcal{F}}$ is the $\sigma$-algebra of Borelian sets in $[0, 1]$, and $\overline{P}$
stands for the Lebesque measure on $\overline{\Omega}$. With these notations, $w(s), s \ge$
$0$, is a $k$-dimensional standard Wiener process on $\{\widetilde{\Omega}, \widetilde{\mathcal{F}}, \widetilde{P}\}$ provided it is
so on the original probability space $\{\Omega, \mathcal{F}, P\}$. Choosing initial condition
$y(0) = \overline{y}(\cdot) \in L_2(\overline{\Omega}; \overline{P})$ in (8) we obtain that the unique solution $y(s, \cdot)$
is in $L_2(\widetilde{\Omega}, \widetilde{P})$ for each $s \ge 0$. Write $\varphi(s, \overline{y}(\cdot); x), s \ge 0$, for the unique

global solution of (8) corresponding to a fixed $x \in R^n, \overline{y}(\cdot) L_2(\overline{\Omega}; \overline{P})$. It makes sense to consider perturbations $x + p(s), s \geq 0$, and $\overline{y}(\cdot) + q(\cdot)$, with $\sup \|p(s)\| = \|p(\cdot)\| < \infty, q(\cdot) \in L_1(\overline{\Omega}, \overline{P}), \|q(\cdot)\|_1 = \overline{E}\|q(\cdot)\|$ and the solution $\varphi(s; x + p(\cdot), \overline{y}(\cdot) + q(\cdot)), s \geq 0$ in (8) is well defined. A dynamical limit for the singularly perturbed stochastic differential equations $(\alpha)$ and $(\beta)$ could be obtained provided a stability condition is assumed

$(H_2)$ $$\sup_{s \geq 0} \widetilde{E}\|\varphi(s, \overline{y}(\cdot); x) - \varphi(s, \overline{y}(\cdot); x)\| \to 0$$

uniformly in $\|x\| \leq \widetilde{C}_0, \|\overline{y}(\cdot)\|_2 \leq \widetilde{C}_0$ if $\|p(\cdot)\| + \|q(\cdot)\|_1 \to 0$.

Now we are in position to state some auxiliary results. Let $x^0(\cdot) \in C([0,T]; R^n)$ with $x^0(0) = x_0 \in R^n$ be fixed and the constant $\widetilde{c}_0$ is as in Proposition 1 and $(H_2)$. Define a bounded set $B(x^0(t)) \subseteq R^n$ (not necessarily closed)

$$(9) \quad B(x^0(t)) = \{x \in R^n : x = \lim \frac{1}{T_n} \int_0^{T_n} \widetilde{E}f(x^0(t), \varphi(s, \overline{y}(\cdot); x^0(t)))ds$$
$$\text{forsome } \{T_n\} \uparrow \infty \text{ and } \overline{y}(\cdot) \in L_2(\overline{P}), \|\overline{y}(\cdot)\|_2 \leq \widehat{C}_0\}$$

Take a function $h_\varepsilon : (0,1] \to [0, \infty)$, fulfilling

$$(10) \quad \lim_{\varepsilon \to 0} h(\varepsilon) = 0, \lim_{\varepsilon \to 0} \frac{h(\varepsilon)}{\varepsilon} =$$
$$= \infty(h(\varepsilon) = \sqrt{\varepsilon}, h(\varepsilon) = -\varepsilon ln\varepsilon)$$

Let $(x^\varepsilon(t), y^\varepsilon(t)), t \in [0,T], \varepsilon \in (0,1]$, be the solution in $(\alpha)$ and $(\beta)$ obeying to the hypothesis $(H_1)$. Write

$$(11) \quad \widehat{x}^\varepsilon(t) = \frac{1}{h(\varepsilon)}[x^\varepsilon(t + h(\varepsilon)) - x^\varepsilon(t)]$$

$$f^\varepsilon(t) = \frac{1}{h(\varepsilon)} \int_t^{t+h(\varepsilon)} \mathcal{U}_\varepsilon(x_\varepsilon(\cdot))(s)ds, \text{ where the mapping associated with}$$

$$\prod^\varepsilon(x(\cdot)) \text{ generated by } y^\varepsilon(\cdot), \mathcal{U}_\varepsilon : C([0,T]; R^n) \to C([0,T]; R^n) \text{ is as follows}$$

$$(12) \quad \mathcal{U}_\varepsilon(x(\cdot))(t) = \int_{R^m} f(x(t), y)P_t^\varepsilon(x(\cdot))(dy), t \in (0,T]$$

considering $x^\varepsilon(t) = x^\varepsilon(T), y^\varepsilon(t) = y^\varepsilon(T)$, for $t \in [T, T + h(\varepsilon)]$.

**Lemma 1.**

Let the hypothesis $(H_1)$ and $(H_2)$ be fulfilled for $(\alpha)$ and $(\beta)$. Let $\{\varepsilon_n\} \downarrow 0, x^0(\cdot) \in C([0,T]; R^n) f^0(\cdot) \in L_\infty([0,T]; R^n)$ be such that $\lim_{n\to\infty} x^{\varepsilon_n}(\cdot) = x^0(\cdot)$ in $C([0,T]; R^n)$ and $\lim_{n\to\infty} f^{\varepsilon_n}(\cdot) = f^0$ in $L_1([0,T]; R^n)$ with $L_1^*$ topology, i.e

$$\lim_{n\to\infty} \int_0^T \langle f^{\varepsilon_n}(t),$$

$$m(s)\rangle ds = \int_0^T \langle f^0(s), m(s)\rangle ds \text{ for any } m(\cdot) \in L_\infty([0,T]; R^n). \text{ Then}$$

a) $x^0(t) = x_0 + \displaystyle\int_0^t f^0(s)ds$, for any $t \in [0,T]$,

b) $f^0(t) \in \overline{\text{conv}}B(x^0(t))$ for any $t \in [0,T]$,

where the bounded set $B(x^0(t))$ is defined in (9).

The set $B(x^0(t)) \subseteq R^n$ is related to the invariant probability measures of the stochastic flow $\varphi(s, y; x), s \geq 0$, generated by the unperturbed stochastic differential equation (8) and it is the purpose of the next Lemma.

**Definition.**

Let $\varphi(s, y; x), s \geq 0, y \in R^m$, be the stochastic flow generated by diffusion equation (8) for each $x \in R^n$ fixed. We say that a probability measure $Q$ on $R^m, Q \in \mathcal{P}(R^m)$, is invariant (stationary) with respect to $\varphi(s; x, y), s \geq 0$,

$$\text{if } \int_{R^m} \psi(y)Q(dy) = \int_{R^m} \left( \int_{R^m} \psi(z)Q(s, y; x)(dz) \right) Q(dy) \text{ for any } s \geq 0 \text{ and}$$

$\psi \in C_b(R^m)$, where $Q(s, y; x) \in \mathcal{P}(R^m)$ is generated by the stochastic flow $\varphi(s, y; x)$

Let $M > 0, x \in R^n$ be fixed, and define the compact and convex sets $A_M \subseteq \mathcal{P}(R^m), \Gamma(x) \subseteq A_M$

(13) $$A_M = \{Q \in \mathcal{P}(R^m) : \int_{R^m} \|y\|^2 Q(dy) \leq M\}$$

(14)
$$\Gamma(x) = \{Q \in A_M : Q \text{ is invariant with respect to } \varphi(s, y; x), s \geq 0, y \in R^m\}$$

**Remark 5.**

The convexity property of $A_M$ and $\Gamma(x)$ is readily seen but the compactness of a set $S \subseteq \mathcal{P}(R^m)$ is obtained provided the topology induced by $C_b(R^m)$

*on $\mathcal{P}(R^m)$ is used (see Billingsley (1)). In particular, the same topology allow one to define measurability of a map $t \to P_t \in \mathcal{P}(R^m)$ provided each scalar mapping $t \to \int_{R^m} h(y)P_t(dy)$ from $[0,T]$ to $R$ is measurable for an arbitrarily fixed $h \in C_b(R^m)$.*

**Lemma 2.**
*Let the hypotheses of Lemma 1 be fulfilled and $x^0(\cdot) \in C([0,T];R^n), f^0(\cdot) \in L_\infty([0,T];R^n)$ agree with (a) and (b). Consider $B(x^0(t)) \subseteq R^n$ defined in (9). Then*

(♮)
$$B(x^0(t)) \subseteq \{x \in R^n : x = \int_{R^m} f(x^0(t), y)Q_t(dy),$$
$$Q_t \in \Gamma(x^0(t))\} \overset{\triangle}{=} C(x^0(t))$$

(♮♮)
$$f^0(t) = \int_{R^m} f(x^0(t), y)Q_t(dy), \ for \ some Q_t \in \Gamma(x^0(t))$$
$$\text{and any } t \in [0,T]$$

*We sall state and prove the main result*

**Theorem.**
*Let $(H_1)$ and $(H_2)$ be fulfilled for $(\alpha)$ and $(\beta)$. Assume $x^0(\cdot) = \lim x^{\varepsilon_n}(\cdot)$ in $C([0,T];R^n)$ for some $\{\varepsilon_n\} \downarrow 0$. Then there exists a measurable map $t \to P_t^0$ from $[0,T]$ to $\mathcal{P}(R^m)$ such that $P_t^0 \in \Gamma(x^0(t))$ for each $t \in [0,T]$ and $(x^0(t)), P_t^0$ is the solution of $(\alpha)$ fulfilling*

$(c_1)$
$$x^0(t) = x_0 + \int_0^t ds \int_{R^m} f(x^0(s), y)P_s^0(dy), t \in [0,T],$$

$(c_2)$
$$\text{the map } t \to \int_{R^m} f(x^0(t), y)P_t^0(dy) \text{ is in } L_\infty([0,T];R^n)$$

**Proof.**
*By hypotheses, the conclusions in Lemmas 1 and 2 are fulfilled and let $f^0(\cdot) \in L_\infty([0,T];R^n)$ be a weak limit in $L_1$ of the sequence $\{f^{\varepsilon_n}(\cdot)\}_n$ such that*

(15)
$$x^0(t) = x_0 + \int_0^t f^0(s)ds \text{ for all } t \in [0,T].$$

Using again Lemma 2 we obtain

$$g(t) = f^0(t) = \int_{R^m} f(x^0(t), y)Q(dy) \overset{\triangle}{=} h(t, Q) \text{ for some } Q \in \Gamma(x^0(t))$$

and for each $t \in [0, T]$. Now, the multivalued mapping $t \to \Gamma(x^0(t))$ from $[0, T]$ to $A_M \overset{\triangle}{=} \{Q \in \mathcal{P}(R^m) : \int_{R^m} \|y\|^2 Q(dy) \leq M\}$ is upper semicontinuous and applying Castaing–Fillippov's theorem (sec Warga (3)) we obtain a measurable univalent mapping $t \to P_t^0 \in \Gamma(x^0(t))$ such that

(16)        $$f^0(t) = \int_{R^m} f(x^0(t), y)P_t^0(dy), \text{ for any } t \in [0, T].$$

Replacing $f^0(t)$ represented in (16) into (15) we obtain the conclusions and the proof is complete.

The analysis will be completed pointing out the influence of the stability assumption $(H_2)$ which is contained in the elementary proofs of Lemmas 1 and 2.

**Proof of Lemma 1**

Write $\widehat{x}^\varepsilon(t) = \dfrac{1}{h(\varepsilon)}[x^\varepsilon(t + h(\varepsilon)) - x^\varepsilon(t)],$

$$f^\varepsilon(t) = \frac{1}{h(\varepsilon)} \int_t^{t+h(\varepsilon)} \mathcal{U}_\varepsilon(x^\varepsilon(\cdot))(s)ds, \text{ for } t \in [0, T]$$

where $\mathcal{U}_\varepsilon(x(\cdot))$ is defined in (12).

By a direct computation we convince ourselves that

(17)

$$\int_0^t \widehat{x}^\varepsilon(s)ds = \frac{1}{h(\varepsilon)}\left[\int_0^t x^\varepsilon(s + h(\varepsilon))ds - \int_0^t x^\varepsilon(s)ds\right]$$

$$= \frac{1}{h(\varepsilon)}\left[\int_{h(\varepsilon}^{t+h(\varepsilon)} x^\varepsilon(\sigma)d\sigma - \int_0^t x^\varepsilon(s)ds\right]$$

$$= \frac{1}{h(\varepsilon)}\left[\int_l^{t+h(\varepsilon)} x^\varepsilon(\sigma)d\sigma - \int_0^{h(\varepsilon)} x^\varepsilon(s)ds\right]$$

Using $\lim\limits_{n\to\infty} x_n^\varepsilon(\cdot) = x^0(\cdot)$ and $\lim\limits_{n\to\infty} h(\varepsilon_n) = 0$ from (17) we obtain

$$(18) \qquad \int_0^t \widehat{x}^\varepsilon(s)ds = x^\varepsilon(t) - x_0 + \eta(\varepsilon)$$

where $\|\eta(\varepsilon)\| \le \max\limits_{t\in[0,T]} \|x^\varepsilon(t) - x^0(t)\| = \|x^\varepsilon(\cdot) - x^0(\cdot)\|$ obeys to

$$(19) \qquad \lim_{n\to\infty} \eta(\varepsilon_n) = 0$$

On the other hand, by definition

$$(20) \qquad \int_0^t \widehat{x}^\varepsilon(s)ds = \int_0^t f^\varepsilon(s)ds$$

and using $(H_1)$ we obtain $\{f^\varepsilon(\cdot)\}_{\varepsilon>0} \subseteq C([0,T]; R^n)$ uniformly bounded and therefore weakly compact in $L_1([0,T]; R^n)$. For $\{\varepsilon_n\} \downarrow 0$, let $x^0(\cdot) \in C([0,T]; R^n)$ and $f^0(\cdot) \in L_\infty([0,T]; R^n)$ such that

$$(21) \qquad \lim_{n\to\infty} x^{\varepsilon_n}(\cdot) = x^0(\cdot), \text{ in } C([0,T]; R^n) \text{ and } \lim_{n\to\infty} f^{\varepsilon_n}(\cdot) = f^0(\cdot)$$

weakly in $L_1([0,T]; R^n)$

Passing to the limit in (18) and (20) we obtain the representation

$$(22) \qquad x^0(t) = x_0 + \int_0^t f^0(s)ds \text{ for all } t \in [0,T],$$

and the conclusion $(*)$ holds.

In addition, from (21) one may see easily that $f^{\varepsilon_n}(\cdot)$ and $f^0(\cdot)$ are obeying to

$$(23) \qquad \liminf_n \langle \lambda, f^{\varepsilon_n}(t) \rangle \le \langle \lambda, f^0(t) \rangle \le \limsup_n \langle \lambda, f^{\varepsilon_n}(t) \rangle$$

for any $\lambda \in R^n$ and almost every where respect to $t \in [0,T]$. Applying a separation theorem for convex sets in $R^n$, from (23) we obtain

$$(24) \qquad f^0(t) \in conv\, K(t), \text{ for any } t \in [0,T]\backslash N,$$

where the Lebesgue measure of $N$ is zero, and the compact set $K(t) \subseteq R^n$ is collecting limit points of $\{f^{\varepsilon_n}(t)\}$, i.e.

$$(25) \qquad \begin{aligned} K(t) &\triangleq \{x \in R^n\} : x = \lim_{k\to\infty} f^{\varepsilon_k}(t) \\ &\text{for some } \{\varepsilon_k\}_k \subseteq \{\varepsilon_n\}_n, \{\varepsilon_k\} \downarrow 0\}. \end{aligned}$$

for each $t \in [0, T]$.

Redefine $f^0(t)$ on the set $N$ such that

(26)                           $f^0(t) \in \text{ conv } K(t)$ for any $t \in [0, T]$,

and from now on $f^0(\cdot)$, fulfilling (26), is taken as the solution in the equation ($*$).

We may and do rewrite the function $f^\varepsilon(t)$ as

(27)
$$f^0(t) = \frac{1}{h(\varepsilon)} \int_t^{t+h(\varepsilon)} (Ef(x^\varepsilon(s), y^\varepsilon(s)))ds =$$
$$\frac{1}{h(\varepsilon)} \int_t^{t+h(\varepsilon)} (Ef(x^0(s), y^\varepsilon(s)))ds + \eta_1(\varepsilon)$$

where $\lim_{n \to \infty} \eta_1(\varepsilon_n) = 0$

A time change $\sigma = t + \varepsilon s$ allow one to see

(28)      $$f^\varepsilon(t) = \frac{1}{T(\varepsilon)} \int_0^{T(\varepsilon)} (Ef(x^0(s), y^\varepsilon(t + \varepsilon s)))ds + \eta_2(\varepsilon)$$

where $T(\varepsilon) \triangleq \dfrac{h(\varepsilon)}{\varepsilon} \uparrow \infty$ for $\varepsilon \downarrow 0$ and $\lim_{n \to \infty} \eta_2(\varepsilon_n) = 0$. Write $y_t^\varepsilon(s) \triangleq y^\varepsilon(t + \varepsilon s), x_t^\varepsilon(s) = x^\varepsilon(t + \varepsilon s)$, for $0 \leq s \leq T(\varepsilon)$, and $y_t^\varepsilon(s)$ becomes the solution for the following diffusion equation

(29)      $$dy = g_0(x_t^\varepsilon(s), y)ds + \sum_{j=1}^k g_j(x_t^\varepsilon(s), y)dw_j(s), s \geq 0,$$

for $y(0) = y^\varepsilon(t)$, and where $w(s, \omega) : [0, \infty) \times \Omega \to R^k$ is a new Wiener process independent of $y^\varepsilon(t)$. The behaviour of $x_t^\varepsilon(s), s \geq 0$, is determined by definition as

(30)                           $x_t^\varepsilon(s) = x^0(t) + p_\varepsilon(s),$

where $\max_{0 \leq s \leq T(\varepsilon)} \|p_\varepsilon(s)\| \triangleq \eta_3(\varepsilon)$ fulfils $\lim_{n \to \infty} \eta_3(\varepsilon_n) = 0$. The property (30) of the coefficients in diffusion equation (29) is not enough for obtainting a limit point of (28) when $\{\varepsilon_n\} \downarrow 0$. The uniform boundedness condition imposed on $\{y^\varepsilon(t)\}_{\varepsilon>0}$ in $(H_1)$ allow one to obtain weak compactness of the sequence $\{Q_t^{\varepsilon_n} \subseteq \mathcal{P}(R^m)$ generated by $y^{\varepsilon_n}(t)$ and to obtain $Q_t^0 \in \mathcal{P}(R^m)$ such that

$$
\begin{aligned}
&\lim_{k\to\infty} Q_t^{\varepsilon_k}(\psi) = Q_t^0(\psi), \text{ for any } \psi \in C_b(R^m) \\
&\text{and some } \{\varepsilon_k\}_k \leq \{\varepsilon_n\}_n, \varepsilon_k \downarrow 0
\end{aligned}
$$
(31)

The weak limit $Q_t^0$ is still unconnected with the invariant probability measures we are looking for and a real help may come from the basic Skorohod's theorem (see Ikeda, Watanabe (2)).

The probabilities $Q_t^{\varepsilon_k}, Q_t^0 \in \mathcal{P}(R^m)$ can be viewed as being generated by some random vectors $\overline{y}_k; \overline{y}_0$ on the standard probability space $\{\overline{\Omega}, \overline{\mathcal{F}}, \overline{P}\}$ such that

(32) $$ E\|\overline{y}_k\| \leq C, E\|\overline{y}_0\| \leq C, \ k \geq 1, \text{ and} $$

$\lim_{k\to\infty} \overline{y}_k(\overline{\omega}) = \overline{y}_0(\overline{\omega})$ on $\overline{\Omega} = [0,1]$ except a null Lebesgue measure. Using (32) we obtain the uniform integrability of the sequence $\{\overline{y}_k(\cdot)\}_k$ in $L_1(\overline{\Omega}; \overline{P})$ and

(33) $$ \lim_{k\to\infty} \overline{E}\|\overline{y}_k(\cdot) - \overline{y}_0(\cdot)\| = 0 $$

Now reconsider the diffusion equation (29) on the new probability space $\{\widetilde{\Omega}, \widetilde{\mathcal{F}}, \widetilde{P}\}$

$$
\begin{aligned}
dy &= g_0(x^0(t) + p_\varepsilon(s), y)ds+ \\
&+ \sum_{j=1}^{k} g_j(x^0(t) + p_\varepsilon(s), y)dw_j(s) \\
y(0) &= \overline{y}_0(\cdot) + q_\varepsilon(\cdot)
\end{aligned}
$$
(34)

where $\overline{y}_0(\cdot)$ fulfils (33), $p_\varepsilon(\cdot)$ is defined in (30), and $q_\varepsilon \in L_2(\overline{\Omega}; \overline{p})$ fulfils

(35) $$ \lim_{k\to\infty} \|q_{\varepsilon_k}(\cdot)\|_1 = 0, \text{ for some } \{\varepsilon_k\}_k \subseteq \{\varepsilon_n\}_n, \ \{\varepsilon_k\} \downarrow 0 $$

The solution $\widetilde{y}_t^\varepsilon(s)$, of the diffusion equation (34) on $\{\widetilde{\Omega}, \widetilde{\mathcal{F}}, \widetilde{P}\}$ generates the same probability measure on $R^m$ as $y_t^\varepsilon(s)$, in (29), but for (34) the stability hypothesis $(H_2)$ can be applied and we obtain

$$
\begin{aligned}
&\sup_{s\geq 0}(\widetilde{E})\|\widetilde{y}_t^\varepsilon(s) - \varphi(s; x^0(t), \overline{y})\| = \\
&\sup_{s\geq 0} \|\varphi(s; x^0(t) + p_\varepsilon, \overline{y}_0 + q_\varepsilon) - \varphi(s; x^0(t), \overline{y}_0)\| \longrightarrow 0
\end{aligned}
$$
(36)

provided $\|p_\varepsilon(\cdot)\| + \|q_\varepsilon(\cdot)\|_1 \longrightarrow \infty$

The expression (28) can be simplified provided (36) is used and we obtain

$$(37) \qquad f^{\varepsilon}(t) = \frac{1}{T(\varepsilon)} \int_0^{T(\varepsilon)} \widetilde{E} f(x^0(t), \varphi(s; x_0(t), \overline{y}_0))) ds + \eta_3(\varepsilon)$$

where $\overline{y}_0 \in L_2(\overline{\Omega}; \overline{P})$, $\lim_{k \to \infty} \eta_3(\varepsilon_k) = 0$, for some $\{\varepsilon_k\}_k \subseteq \{\varepsilon_n\}_n$ with $\varepsilon_k \downarrow 0$ and $\lim_{\varepsilon \to 0} T(\varepsilon) = \infty$

Using (37) in (28) we obtain that the limit points in the compact set $K(t)$ (see (25)) is contained in $\overline{\text{conv}}\, B(x^0(t))$, for each $t \in [0, T]$, where $B(x^0(t))$ is defined in (9). Therefore (25) can be rewriten

$$(38) \qquad f^0(t) \in \overline{\text{conv}} B(x^0(t)) \text{ for all } t \in [0, T]$$

and the proof is complete.

**Proof of Lemma 2**

Let $\{\varepsilon_n\}_n \downarrow 0, x^0(\cdot) \in C([0, T]; R^n)$ and $f^0(\cdot) \in L_{\infty}([0, T]; R^n)$ defined as in Lemma 1. The proof is concluded provided $B(x^0(t)) \subseteq \Gamma(x^0(t))$ for any $t \in [0, T]$, is proved. Let $x^* \in B(x^0(t))$ and by definition (see (9))

$$(39) \qquad x^* = \lim_{k \to \infty} \frac{1}{T_k} \int_0^{T_k} (\widetilde{E} f(x^0(t)), \varphi(s; x_0(t), \overline{y}(\cdot))) ds$$

for some $\{T_k\} \uparrow \infty$ and $\overline{y}(\cdot) \in L_1(\overline{\Omega}; \overline{P})$. Denote $Q_s \in \mathcal{P}(R^m)$ the probability generated by $\varphi(s; x^0(t), \overline{y}(\cdot))$ on $R^m$ and we obtain

$$(40) \qquad \widetilde{E} f(x^0(t), \varphi(s, x^0(t), \overline{y}(\cdot))) = \int f(x^0(t), y) Q_s(dy)$$

The hypothesis $(H_2)$ allow one to obtain

$$(41) \qquad \int \|y\|^2 Q_s(dy) \le C \text{ for any } s \ge 0$$

In addition

$$(42) \qquad Q_k \overset{\Delta}{=} \frac{1}{T_k} \int_0^{T_k} Q_s ds \in \mathcal{P}(R^m) \text{ and}$$

$$(43) \qquad \int_{R^m} \|y\|^2 Q_k(dy) \le C \text{ for any } k \ge 1$$

Therefore $\{Q_k\}_{k\geq 1} \subseteq A_M$ (compact) and let $Q_0 \in A_M$ be a limit point. We obtain $\lim\limits_{k} \int\limits_{R^m} \psi(y)Q_k(dy) = \int\limits_{R^m} \psi(y)Q_0(dy)$, for any $\psi(\cdot) \in C_b(R^m)$ and using (43) we may apply again Skorohod's theorem. It allow onw to obtain $\bar{y}_k(\cdot), \bar{y}_0(\cdot) \in L_2(\bar{\Omega}, \bar{P})$ such that

$$(44) \qquad\qquad\qquad \|\bar{y}_k(\cdot) - \bar{y}_0(\cdot)\|_1 \to 0$$

By hypothesis, $\|\frac{\partial f}{\partial y}(x, y)\| \leq C_1(1 + \|x\|)$ (see $(\alpha_2$, (iii)) and from (44) we obtain

$$(45) \qquad x^* = \int\limits_{R^m} f(x^0(t), y)Q_0(dy) = \lim\limits_{k\to\infty} \int f(x^0(t), y)Q_k(dy)$$

and $Q_0 \in A_M$. It remains to check that $Q_0$ is an invariant probability for the stochastic flow $\varphi(s; x^0(t), y), s \geq 0$. Indeed, the ergodic type limit defining $Q_0$ allow one to see easily that

$$(46)$$

$$\int (E\psi(\varphi(\tilde{s}; x_0(t), y)))Q_0(dy) = \lim\limits_{T_k\uparrow\infty} \frac{1}{T_k} \int\limits_0^{T_k} (\tilde{E}\psi(\varphi(s + \tilde{s}; x^0(t), \bar{y}(\cdot)))ds =$$

$$= \lim\limits_{k\to\infty} \int\limits_{R^m} \psi(y)Q_k(dy) = \int\limits_{R^m} \psi(y)Q_0(dy) \text{ for any } \psi \in C_b(R^m)$$

Therefore $Q_0 \in \Gamma(x^0(t)) \subseteq A_M$ and the conclusion ($\sharp$) is verified. By definition, the convex and compact set $C(x^0(t)) \subseteq R^n$ is obeying to $\overline{conv}B(x^0(t)) \subseteq C(x^0(t))$ for any $t \in [0, T]$. Using $(**)$ of Lemma 1 we obtain finally $f^0(t) \in C(x^0(t))$ for any $t \in [0, T]$ and the proof is complete

### Remark 6.

*The main result stated in the above given Theorem is based on two inconvenient assumptions which are expressed implicitely in the hypothesis $(H_2)$ and in condition $(\alpha_2, (iii))$. Both are generated by the boundedness assumption which is stated in the space $L_2$ contained in the hypothesis $(H_1)$. If $(H_1)$ is required to be fulfilled for some $p > 2$ replacing $p = 2$, then $(H_2)$ and $(\alpha, (iii))$ could be replaced by more adequate conditions. Namely the stability hypothesis $(H_2)$ could be stated in $L_2$ norm replacing $L_1$ norm and $(\alpha_2)$ (iii)) changes into*

$$\left\|\frac{\partial f}{\partial y}(x, y)\right\| \leq c_1(1 + \|x\|)(1 + \|y\|) \text{ for any } x \in R^m$$

## Comment on computation of invariant probabilities measures

The main hypotheses $(H_1)$ and $(H_2)$ of the Theorem involve a diffusion equation $(\beta)$ and its unperturbed form in (8) for which $\varphi(s, y; x)$ is the corresponding stochastic flow. The fulfillment of $(H_1)$ and $(H_2)$ insures that the set $\Gamma(x)$, in (14) of invariant probabilities measures associated with the stochastic flow $\varphi(s, y; x)$ of (8) is nonempty. Using smooth functions $\psi \in C_b^2(R^m)$ we may rewrite the invariance property in a differential form and denote

$$
(a) \qquad
\begin{aligned}
\mathcal{U}(s, y) &= \int_{R^m} \psi(z) Q(s, y; x)(dz) \\
&= E\psi(\varphi(s, y; x)), \ s \geq 0, y \in R^m,
\end{aligned}
$$

for an arbitrarily fixed $x \in R^n$.

The function $\mathcal{U}(s, y)$ is the solution for a Cauchy problem of a parabolic equation

$$
(b) \qquad
\begin{aligned}
\frac{\partial \mathcal{U}}{\partial s}(s, y) &= \left\langle \frac{\partial \mathcal{U}}{\partial y}(s, y), g_0(x, y) \right\rangle + \frac{1}{2} Tr \left( \frac{\partial^2 U}{\partial y^2}(s, y) G(x, y) \right), \\
\mathcal{U}(0, y) &= \psi(y),
\end{aligned}
$$

where the $(m \times m)$ matrix $G(x, y) \overset{\Delta}{=} \sum_{l=1}^{k} g_l(x, y) g_l^*(x, y)$ and $g_j(x, y), j = 0, 1, \ldots, k$, define the unperturbed diffusion equation in (8).

An invariant probability measure $P_x \in \Gamma(x)$ fulfils

$$
\int_{R^m} \mathcal{U}(s, y) P_x(dy) = \int_{R^m} \psi(y) P_x(dy) \text{ for any } s \geq 0
$$

and therefore

$$
(c) \qquad
\int_{R^m} \frac{\partial \mathcal{U}}{\partial s}(s, y) P_x(dy) = 0 \text{ for any } s \geq 0 \text{ and} \psi \in C_b^2(R^m)
$$

In particular for $s = 0$ and using (b) in (c) we obtain

$$
(d) \qquad
\int_{R^m} (D(x, y)\psi(y)) P_x(dy) = 0 \text{ for any } \psi \in C_b^2(R^m)
$$

where the elliptic partial differential operator $D$ meet

$$
D(x, y) \overset{\Delta}{=} \sum_{j=1}^{m} g_0^j(x, y) \frac{\partial}{\partial y^j} + \frac{1}{2} \sum_{i,j=1}^{m} G_{ij}(x, y) \frac{\partial^2}{\partial y_i \partial y_j}
$$

$$g_0 = (g_0^1, \ldots, g_0^m), \quad G = (G_{ij})_{i,j}$$

A probability measure $P_x \in \Gamma(x)$ fulfilling (d) can be asimilated as a solution in distributions (a generalized solution) of the adjoint elliptic equation associated to $D(x, y)\psi = 0$

It shows the complexity of describing the set $\Gamma(x)$ of invariant probabilities measures. Nevertheless, it may occur that a limit point $\lim_{s\uparrow\infty} \varphi(s, y; x) = \varphi_0(x)$ exists, in a weak sense and independent of $y \in R^m$, which enables one to obtain the set $\Gamma(x)$ as a singleton $P_x \in \mathcal{P}(R^m)$ generated by the random vector $\varphi_0(x)$ fulfilling $E\|\varphi_0(x)\|^2 \le \hat{C}_0$. It is the case with the example we have considered earlier where the corresponding stochastic flow $\varphi(s, y; x)$ is obeying to the following diffusion equation

$$(e) \qquad dy = (-y + g_0(x))ds + \sum_{i=1}^{k} g_i(x)dw_i(s), s \ge 0, y(0) = y \in R^m$$

Since $\varphi(s, y; x)$ is a Gaussian random vector for each $s \ge 0$, we may and do restrict ourselves considering only Gaussian limit points for $\lim_{s\uparrow\infty} \varphi(s, y; x)$. The explicit representation of the solution of (e) allow one to see the mean value and covariance matrix are expressed as follows

$$(f) \qquad E\varphi(s, y; x) = (\exp -s)y + [1 - \exp -s]g_0(x) \triangleq \widehat{\varphi_s}(x)$$

$$R_s(x) \triangleq E[\varphi(s, y; x) - \widehat{\varphi_s}(x)][\varphi(s, y; x) - \widehat{\varphi_s}(x)]^* =$$
$$\frac{1}{2}[1 - \exp -2s] \sum_{i=1}^{k} g_i(x)g_i^*(x),$$

Both mean value and correlation matrix depend on initial vector $y \in R^m$ and of the parameter $x \in R^n$ but the limit of them for $s \uparrow \infty$ are independent of $y \in R^m$ and we obtain

$$(g) \qquad m_0(x) = \lim_{s\uparrow\infty} \widehat{\varphi_s}(x) = g_0(x)$$

$$\mathbb{R}_0(x) = \lim_{s\uparrow\infty} R_s(x) = \frac{1}{2} \sum_{i=1}^{k} g_i(x)g_i^*(x).$$

Any Gaussian random vector $z(\cdot) \in L_2(\Omega, P)$ obeying to $Ez(\cdot) = m_0(x)$, $Ez(\cdot)z^*(\cdot) = R_0(x)$ can be taken as a probabilistic realization of $(m_0(x), R_0(x))$ but the probability $\mathbf{P}_x \in \mathcal{P}(R^m)$ generated on $R^m$ is uniquely determined by the given vectors $g_j(x) \in R^m$, $j = 0, 1, \ldots, m$. In conclusion, for each $x \in R^n$ fixed the set $\Gamma(x)$ consists of the unique probability

$P_x \in \mathcal{P}(R^m)$ determined by a Gaussian random vector for which the mean value and the corresponding matrix are defined in (9).


### Bibliographical notes

The problems included in this chapter are labelled in a standard way but the solution for stabilization of affine control system in the lack of full rank condition had to be reconsidered entirely. The algorithm for periodical defined control and solution stabilizing affine control systems originates in a work of Henry Hermes (see SIAM J. Control and Optimization, vol. 18, No. 4, 1980).

The problem of disturbance decoupling (see Isidori [8]) is enlarged to Lipschitz continuous controls and considering stochastic perturbations in addition. The last section can be taken as a slight challenge of using a Lie structure for singularly perturbed problems and what is included here reflects a dynamical point of view which belongs to Zvi Arstein. It was extended to singularly perturbed stochastic differential equations in a recent paper (see Vasilache, Vârsan, *Singularly perturbed stochastic differential equations with dynamical limits*, Stability, control and singular perturbations, Qualitative problems of differential eq. and theory of control systems, World Scientific, Singapore–New Jersey–London–Hong Kong (1995)).

# Bibliography

[1] P. Billingsley, *Convergence of probability measures*, Wiley, 1968.

[2] N. Ikeda, S. Watanabe, *Stochastic Differential Equations and Diffusion Processes*, North–Holland, 1981.

[3] J. Warga, *Optimal Control of Differential and Functional Equations*, Academic Press, 1975.

# Appendix

**(a.1) Lemma 1.**

    *Assume that the $(n \times n)$ continuous matrices $Z_i(p), p \in D_M =$*

$\prod_1^M (-a_j, a_j), i = 1, \cdots, n-1,$ *are given such that $Z_i(0) = I_n$. For a*

*nonsingular $(n \times n)$ matrix $A$ denote $Z_i^a(p) \triangleq A^{-1} Z_i(p) A$ and $M^a(p) \triangleq$*
*$(e_1, Z_1^a(p)e_2, \cdots, Z_{n-1}^a(p)e_n)$, where $e_1, \ldots, e_n \in R^n$ is the canonical basis.*
*Then*

    *a) $\det M^a(p) \geq m_1(p) \cdots m_{n-1}(p) + 0(p)$ if $m_i(p) \geq 0$, where*

    *b) $m_i(p) \triangleq \min \left\{ \lambda \in \text{ spectral values of } \dfrac{Z_i(p) + Z_i^T(p)}{2} \right\}$*

*$m_i(0) = 1, \ i = 1, \ldots, n-1,$ and*
*$|0(p)| \leq K \max\{\|Z_i(p) - I_n\|, \ i = 1, \ldots, n-1\}$, for some constant $K > 0$.*

**Proof.**

It is easily seen

$$
\begin{aligned}
e_n^T Z_{n-1}^a(p) e_n &= Tr Z_{n-1}^a(p) e_n e_n^T \\
&= Tr e_n e_n^T A^{-1} Z_{n-1}(p) A \\
&= Tr A e_n e_n^T A^{-1} Z_{n-1}(p),
\end{aligned}
$$

and for $A_n \triangleq A e_n e_n^T A^T$, $S \triangleq (A^T)^{-1}$ we obtain $A e_n e_n^T A^{-1} = A_n S$, where
$A_n \geq 0, S > 0$ can be diagonalized by a unitary matrix $\mathcal{U}$. More precisely,

(1)     $A e_n e_n^T A^T = A_n S = \mathcal{U} \text{ diag } (k_1, \cdots, k_n) \mathcal{U}^T$ where $k_i \geq 0$.

197

Using (1) we obtain

$$e_n^T Z_{n-1}^a(p)e_n \; = Tr A_n S Z_{n-1}(p)$$

(2)
$$= \sum_{i=1}^{n} k_i e_i^T \mathcal{U}^T Z_{n-1}(p)\mathcal{U}e_i$$

$$= \sum_{i=1}^{n} k_i e_i^T \mathcal{U}^T \left( \frac{Z_{n-1}(p) + Z_{n-1}^T(p)}{2} \right) \mathcal{U}e_i$$

see $e_i^T \mathcal{U}^T \left( \dfrac{Z_{n-1}(p) - Z_{n-1}^T(p)}{2} \right) \mathcal{U}e_i = 0, \; i = 1, \dots, n)$ and

(3)     $e_n^T Z_{n-1}^a(p)e_n \geq \left( \displaystyle\sum_{1}^{n} k_i \right) m_{n-1}(p) = m_{n-1}(p)$ provided $m_{n-1}(p) \geq 0$

see $\displaystyle\sum_{1}^{n} k_i = T_r A_n S = T_r A e_n e_n^T A^{-1} = 1)$. Similarly, it is easily seen that

(4)     $e_i^T Z_{i-1}^a(p)e_i \geq m_{i-1}(p), \qquad i = 2, \dots, n,$ provided $m_j(p) \geq 0$

For $\det M^a(p)$ we rewrite

(5)                       $\det M^a(p) = \det\{C_{ij}(p), i, j \in \{2, \dots, n\}\},$

where $C_{ij}(p) = e_i^T Z^a(p)e_j$, for some $Z^a(p) \in \{Z_1^a(p), \dots, Z_{n-1}^a(p)\}$ fixed provided $i, j \in \{2, \dots, n\}$ is fixed.

By definition $C_{ij}(0) = 0$ for $i \neq j$ and notice
(6)
$$C_{ij}(p) = \tilde{C}_{ij}(p) = e_i^T(Z^a(p) - I_n)e_j = e_i^T A^{-1}(Z(p) - I_n)Ae_j, \text{ for i} \neq \text{j,}$$
$$C_{ii}(p) = \tilde{C}_{ii}(p) \triangleq e_i^T Z_{i-1}^a(p)e_i, \; i = 2, \dots, n.$$

Using (6) we rewrite (5) as :

(7)                       $\det M^a(p) = \det\{\tilde{C}_{ij}(p) : i, j \in \{2, \dots, n\}\},$

where $\tilde{C}_{ij}(p)$ are given in (6) and $\tilde{C}_{ii}(p), i = 2, \dots, n,$ fulfil (4).
By definition,

(8)     $\det M^a(p) = \displaystyle\sum_{k \in P} \underline{\underline{\sigma}}(k)\tilde{C}_{2i_2}(p) \times \cdots \times \tilde{C}_{ni_n}(p) = \sum_{k \in P} \underline{\underline{\sigma}}(k)C_k(p),$

where $k \triangleq \{i_2, \dots, i_n\} \in P$ is a permutation of $\{2, \dots, n\}$ and $\underline{\underline{\sigma}}(k) \triangleq$ signature of $k$.

For $k_0 = \{2,\dots,n\}$ we obtainthe scalar function $\tilde{C}_{22}(p) \times \cdots \tilde{C}_{nn}(p) = C_{k_0}(p)$ fulfilling (see (4))

(9) $\;C_{k_0}(p) \geq m_1(p) \times \cdots \times m_{n-1}(p)$, provided $m_j(p) \geq 0, j = 1, \dots, n-1$.

For any $k \in P$, $k \neq k_0$, we obtaina corresponding scalar function $C_k(p)$ which can be written as

(10)
$$C_k(p) = \tilde{C}(p) \prod_{j=2}^{n_1} \tilde{C}_{i_j i_j}(p),$$

where $\tilde{C}(p)$ is a cyclic element of the form $\tilde{C}(p) = \tilde{C}_{k_1 k_2}(p)\tilde{C}_{k_2 k_3}(p)\cdots\tilde{C}_{k_m k_1}(p)$, and $\{k_1,\dots,k_m\}$ is a permutation in $k\backslash\{i_2,\dots,i_{n_1}\}$ with $k_i \neq k_j$, $k \overset{\triangle}{=} \{i_2,\dots,i_n\}$

It is the goal of the next computation to estimate a ciclic element $\tilde{C}(p)$ with $k_i \neq k_j$.

In this respect (see (6))

(11) $\qquad \tilde{C}_{k_1 k_2} \overset{\triangle}{=} e_{k_1}^T A^{-1}(Z(p) - I_n)Ae_{k_2}, \; Z \in \{Z_1,\dots,Z_{n-1}\}$

and notice $e_{k_1}^T e_{k_2} = 0$ for $k_1 \neq k_2$.

Therefore $\tilde{C}(p)$ in (10) can be written
(12)
$$\tilde{C}(p) = (e_{k_1}^T A^{-1}(Z^1(p) - I_n)Ae_{k_2})(e_{k_2}^T A^{-1}(Z^2(p) - I_n)Ae_{k_3})\cdots$$
$$(e_{k_m}^T A^{-1}(Z^m(p) - I_n)Ae_{k_1})$$

with $Z^i \in \{Z_1,\dots,Z_{n-1}\}$. We obtain
(13)
$$\tilde{C}(p)\; = T_r(\qquad\qquad Ae_{k_1}e_{k_1}^T A^{-1})[Z^1(p) - I_n](Ae_{k_2}e_{k_2}^T A^{-1})\cdots$$
$$(Z^{m-1}(p) - I_n)(Ae_{k_m}e_{k_m}^T A^{-1})(Z^m(p - I_n)]$$
$$= T_r A_1 W_1(p) A_2 W_2(p) \cdots A_m W_m(p)$$

where $A_i \overset{\triangle}{=} Ae_{k_i}e_{k_i}^T A^{-1}$, $W_i(p) = (Z^i(p) - I_n)$, $i = 1,\dots,m$.

We may and do rewrite

$$A_i = (Ae_{k_i}e_{k_i}^T A^T)S, \; \text{for } S = (AA^T)^{-1} > 0,$$

and by a diagonalization with an orthogonal matrix $\mathcal{U}_i$ we obtain

$$A_i = \mathcal{U}_i^T \text{ diag } (r_1^i,\dots,r_n^i)\mathcal{U}_i, r_j^i \geq 0, \;\; \sum_{j=1}^n r_j^i = 1$$

and

$$
\begin{aligned}
\widetilde{C}(p) &= \sum_{j=1}^{n} r_j^i e_j^T \mathcal{U}_1 W_1(p) A_2 W_2(p) \times \cdots \times A_m W_m(p) \mathcal{U}_1^T e_j \\
&= \sum_{j=1}^{n} r_j^1 \widetilde{C}_j(p).
\end{aligned}
$$

(14)

By definition,

$$
\begin{aligned}
\widetilde{C}_j(p) &= \underline{e}_j^T \mathcal{U}_1 W_1(p) A_2 W_2(p) \cdots A_m W_m(p) \mathcal{U}_1^T e_j \\
&= T_r A_2 W_2(p) \cdots A_m W_m(p) \mathcal{U}_1^T e_j e_j^T \mathcal{U}_1 W_1(p),
\end{aligned}
$$

(15)

and repeating the above algorithm for $A_2$ we obtain $A_2 = \mathcal{U}_2^T \, \text{diag} \, (r_1^2, \ldots, r_n^2) \mathcal{U}_2$, with $r_j^2 \geq 0, \sum_{j=1}^{n} r_j^2 = 1$, and $\widetilde{C}_j(p)$ in (15) becomes

$$
\begin{aligned}
\widetilde{C}(p) &= \sum_{t=1}^{n} r_t^2 e_t^T \mathcal{U}_2 W_2(p) A_3 \\
&\qquad \cdots \times A_m W_m(p) (\mathcal{U}_1^T e_j e_j^T U_1) W_1(p) \mathcal{U}_2^T e_t \\
&= \sum_{t=1}^{n} r_t^2 d_t^j(p)
\end{aligned}
$$

(16)

Here

$$
\begin{aligned}
d_t^j(p) = T_r A_3 \quad & W_3(p) A_4 \cdots \times A_m W_m(p) (\mathcal{U}_1^T e_j e_j^T \mathcal{U}_1) \\
& \times W_1(p) (\mathcal{U}_2^T e_t e_t^T \mathcal{U}_2) W_2(p)
\end{aligned}
$$

(17)

Finally, we obtain

$$
\widetilde{C}(p) = \sum_{i_1=1}^{n} \cdots \sum_{i_m=1}^{n} (r_{i_1} \cdots r_{i_m}) T_r [\widetilde{W}_{i_1} \widetilde{W}_{i_2}(p) \cdots \widetilde{W}_{i_m}(p)],
$$

(18)

where $\sum_{j=1}^{M} r_{ij} = 1$, $r_{ij} \geq 0$, $\widetilde{W}_{i_j}(p) \triangleq (\mathcal{U}_j^T e_{i_j} e_{i_j}^T \mathcal{U}_j) W_j(p)$, and $W_j(p) = Z^j(p) - I_n$ fulfils $W_j(0) = 0$ (null matrix).

It is a routine matter to obtain

$$
\widetilde{C}(p) \geq -\|\widetilde{W}_{i_1}(p)\| \times \cdots \times \|\widetilde{W}_{i_m}(p)\| = 0(p)
$$

(19)

and $\|(0(p))\| \leq \max\{\|Z_i(p) - I_n\|, i = 1, \dots, n - 1\}$

Therefore, for any $k \in P$, $k \neq k_0$, the $C_k(p)$ in (8) is estimated by

(20) $\qquad C_k(p) \geq 0(p)$ provided $\widetilde{C}_{jj}(p) \geq m_j(p) \geq 0$.

Using (9) and (20) in (8) we obtain

(21) $\qquad \det M^a(p) \geq \prod_{i=1}^{n-1} m_i(p) + 0(p), \quad$ provided $m_i(p) \geq 0$,

where $\lim_{|p| \to 0} 0(p) = 0$ and $m_i(0) = 1$.

The proof is complete.


### a.2 Proof of Proposition 2.4.2 and some auxiliary results

Using Theorem 2.2.1, a global gradient system

(1) $\qquad \dfrac{\partial y}{\partial t_j} = X_j(p; y), \quad j = 1, \cdots M, \ y(0; x) = x, \ p \in R^M.$

is defined, where $X_j : R^M \times R^n \to R^n$ are $C^\infty$-functions fulfilling

$$X(p; y) \triangleq \{X_1(p; y), \dots, X_M(p; y)\}$$
(2)
$$= \{Y_1(y), \dots, Y_M(y)\} A(p), \quad p \in R^M$$

with $A(p)$ a nonsingular analytical $(M \times M)$ matrix verifying $A(0) = I_M$. Let $h_j \in \text{Vect}(R^M)$ be such that

(3) $\qquad A(p)h_j(p) = e_j, \ j = 1, \dots, M, \ h_i = q_i, i = 1, \dots, m,$

Using (2) and (3) we obtain $\dfrac{\partial y}{\partial p}(p; x)h_j(p) = Y_j(x), \ j = 1, \dots, M$, and the homomorphism $\lambda : L(g_1, \dots, g_m) \to L(q_1, \dots, q_m)$ (see $g_i = Y_i, \ i = 1, \dots, m$) is defined such that $\lambda(g) = h$ iff $\dfrac{\partial y}{\partial p}(p; x)h(p) = g(x)$ with $y(p; x)$ a local solution of (1). Approximations of the Chow type for controllable system

(4) $\qquad \dfrac{dp}{dt} = \sum_{i=1}^{m} u_i q_i(p), \ p(0) = 0, \ p \in R^M$

is the purpose of the next Lemma. Everywhere in what follows $T > 0, C > 0$ and $\varepsilon_0 > 0$ are fixed. Let $g_{ij}$ be the first vector field in $Y_j$ and $a_{ij} = \lambda(g_{ij})$.

**Lemma 1.**

Assume $L(g_1, \ldots, g_m)$ finitely generated over $R$. Let $q_1, \ldots, q_m \in$ Vect$(R^M)$ and $h_j \in L(q_1, \ldots, q_m)$, $j = 1, \ldots, M$, fulfilling (12). For $\varepsilon \in [0, \varepsilon_0]$ and $v \in R^M, \|v\| \leq C$ consider $\widetilde{p}(t, v), t \in (0, T)$, the solution of

$$(*) \qquad \frac{dp}{dt} = \varepsilon \sum_{j=1}^{M} v_j h_j(p), \quad t \in [0, T], \quad p(0) = 0$$

Then for each $h \stackrel{\triangle}{=} T/N$ there exist $C'$ and periodic controls $u_{ij}(t, h)$, $u_i(t, h)$ : $[0, \infty) \to R$ with a common period $T$ such that the solution $p_h(t, v), t \in [0, T]$, in

$$(**) \qquad \frac{dp}{dt} = \varepsilon \sum_{j=1}^{M} v_j u_{ij}(t, h) q_{ij}(p) + \sum_{i=1}^{m} u_i(t, h) q_i(p), \quad p(0) = 0$$

exists. In addition, the following properties are fulfilled:

a) $p_h(T, v) = \varepsilon T v + 0(\sqrt{h}; v)$ with $0(\sqrt{h}; 0 = 0$;

b) $p_h(t, v) = \widetilde{p}(t, v) + 0(\sqrt{h}; v, t)$;

c) $-u_i(T - t; h) = u_i(t, h)$, $i = 1, \ldots, m$, $t \in [0, T].$;

d) $\lim_{\delta \downarrow 0} \dfrac{0(\sqrt{h}; v, t)}{\delta} = 0$, uniformly in $\|v\| \leq C, t \in [0, T]$, provided $\delta = h^r, r \in (0, 1/2)$.

The controls $u_{ij}(t, h), u_i(t, h)$ in Lemma 1 are defined as in [6] for $t \in [0, T/2]$ and they are redefined for $t \in [T/2, T]$ as follows $u_{ij}(t, h) = 0, j = m+1, \ldots, M$, $t \in [T/2, T]$,

$$u_i(T/2 + t) = -u_i(T/2 - t), \ i = 1, \ldots, m, \ t \in [0, T/2].$$

We obtain $u_i(t, h) : [0, T] \to R$ with $u_i(T - t, h) = -u_i(t, h)$, $i = 1, \ldots, m$, and the limit equations has to be changed slightly by multiplying the vector field $h_j$ with a scalar function $s_j(t) = 1$, $t \in [0, T/2]$, $s_j(t) = 0$, $t \in [T/2, T]$, for $j = m+1, \ldots, M$. It doesn not change the expression of $\widetilde{p}(T)$, where $v_j$ has to be replaced with $\dfrac{1}{2} v_j$ for $j = m+1, \ldots, M$, and this modification in the limit equations is omitted.

**Lemma 2.**

Assume that $L(g_1, \ldots, g_m)$ is finitely generated over $R$ with $\{Y_1, \ldots, Y_M\}$, $Y_i = g_i$, $i = 1, \ldots, m$, a fixed system of generators and denote $h_j =$

$\lambda(Y_j)$, $j = 1, \ldots, M$, where $h_j$ fulfil (12). Then any solution $\widetilde{p}(t)$ in (13) will generate a solution $\widetilde{x}(t; x_0)$, in (14). In particular, any solutions $\widetilde{p}(t, v)$ and $p_h(t, v)$ in Lemma 1 will generate the corresponding solutions $\widetilde{x}(t, v; x_0) \overset{\triangle}{=} G(\widetilde{p}(t, v); x_0)$ in (14) and $x_h(t, v; x_0) \overset{\triangle}{=} G(p_h(t, v); x_0)$ a solution in

$$(**) \qquad \frac{dx}{dt} = \varepsilon \sum_{j=1}^{M} v_j, u_{ij}(t, h)g_{ij}(x) + \sum_{i=1}^{m} u_i(t, h)g_i(x), \quad x(0) = x_0$$

where $G(p; x_0) = G_1(t_1) \circ \cdots \circ G_M(t_M)(x_0)$.

The periodic solution of $\dfrac{dp}{dt} = \displaystyle\sum_{i=1}^{m} u_i(t, h)q_i(p)$ is useful for obtainting a new representation of the solution $p_h(t, v)$ in Lemma 1, and it is the purpose of the next Lemma.

**Lemma 3.**

Assume the hypotheses in Lemma 1 are fulfilled and let $p_h(t, v) \overset{\triangle}{=} p_h(t)$ be the solution in (**). Then

$$p_h(T) = \varepsilon \sum_{j=1}^{M} v_j \int_0^T u_{ij}(t, h) \frac{\partial \widetilde{H}}{\partial p}(t, p_h(t)) q_{ij}(p_h(t)) dt$$

for some $\widetilde{H}(t, p)$ periodically defined in $t$ and $C^\infty$ in $p \in V(0)$ fulfilling

$$(*) \qquad \frac{\partial \widetilde{H}}{\partial t} + \sum_{i=1}^{m} u_i(t, h) \left\langle \frac{\partial \widetilde{H}}{\partial p}(t, p), q_i(p) \right\rangle = 0,$$

$$\widetilde{H}(0, p) = \widetilde{H}(T, p) = p$$

**Proof.**

The function $\widetilde{H}(t, p) : [0, \infty) \times V(0) \longrightarrow R^M$ in Lemma 3 is defined via a local solution $H(r; p)$ in a gradient system of partial differential equations,

$$(5) \qquad \frac{\partial H}{\partial t_j}(r; p) + \left\langle \frac{\partial H}{\partial p}(r; p), X_j(r, p) \right\rangle = 0, \quad j = 1, \ldots, M$$

$r = (t_1, \ldots, t_M) \in R^M, p \in R^M$, where the vector fields $X_j$ fulfil the representation

$$(6) \qquad X(r; p) \overset{\triangle}{=} \{X_1(r; p), \ldots, X_M(r; p)\} = \{h_1(p), \ldots, h_M(p)\} A(r)$$

with the same nonsingular and analytical $(M \times M)$ matrix $A(r)$ as in (2).

A local solution for (5) is

(7) $$H(r; p) = H_M(-t_M) \circ \cdots \circ H_1(-t_1)(p)$$

for $r = (t_1, \ldots, t_M) \in \mathcal{U}(0) \subset R^M$ and $p \in V(0) \subset R^M$, where $H_j(t)(p)$ is the local flow generated by the vector field $h_j \in \text{Vect}(R^M)$.

Denote $p_0(t) : [0, \infty) \longrightarrow R^M$ the periodic solution in

(8) $$\frac{dp}{dt} = \sum_{i=1}^{m} u_i(t, h) q_i(p), \quad p(0) = p(T) = 0$$

which obeys $\|p_0(t)\| \le \eta$, $t \in [0, T]$, for $\eta$ sufficiently small. Define $r_0(t)$, $t \ge 0$, as the unique periodic function fulfilling

(9) $$H(r_0(t); 0) = p_0(t)$$

and we obtainthat $r_0(\cdot)$ is the solution of

(10) $$\frac{dr}{dt} = \sum_{i=1}^{m} u_i(t, h) l_i(r), \quad r(0) = 0 = r(T)$$

where $A(r) l_i(r) = e_i$, $i = 1, \ldots, m$, and $\{e_1, \ldots, e_M\}$ is the canonical base in $R^M$. Denote $\tilde{H}(t, p) = H(r_0(t); p)$ and using (5) and

(11) $$X(r; p) l_i(r) = h_i(p) = q_i(p), \quad i = 1, \ldots, m$$

we obtain by straight computation that $\tilde{H}(t, p)$ fulfils the equation (*). Now using (*) and the equation for $p_h(t)$ we obtaine asily that

(12) $$\frac{d}{dt} \tilde{H}(t, p_h(t)) = \varepsilon \sum_{j=1}^{M} v_j u_{ij}(t, h) \frac{\partial \tilde{H}}{\partial p}(t, p_h(t)) q_{ij}(p_h(t))$$

By definition

$$p_h(T) = \tilde{H}(T, p_h(T)) \ = \tilde{H}(0, p_h(0)) + \int_0^T \left( \frac{d}{dt} \tilde{H}(t, p_h(t)) \right) dt =$$

$$= \int_0^T \frac{d}{dt} (\tilde{H}(t, p_h(t)) dt$$

and the proof is complete.

**Lemma 4.**

*Assume the conditions in Lemma 3 are fulfilled and let $p_h(t,v)$ be the solution in Lemma 1 fulfilling the representation given in Lemma 3. Then for $\delta = h^{1/3}$, $\varepsilon = \delta$, we obtain*

$$p_h(T,v) = \delta T(E_M + \sum(\delta,v))v$$

*where* $\lim_{\delta\downarrow 0} \sum(\delta,v) = \bigcirc$ *(null matrix) uniformly in* $\|v\| \leq c$.

**Proof.**

By hypotheses, the conditions in Lemma 1 are fulfilled and using (a) and (d) we obtain $\lim_{\delta\downarrow 0}\frac{1}{\delta T}p_h(T,v) = v$ for any $v \in R^M, \|v\| \leq C$. Therefore, the $(M \times M)$ matrix $S(\delta;v)$,

$$S(\delta;v) \triangleq \left(\int\limits_0^T u_{ij}(t,h)\frac{\partial\widetilde{H}}{\partial p}(t,p_h(t))q_{ij}(p_h(t))dt, \; j = 1,\dots,M\right)$$

converges to the identity matrix $E_M$ provided $\delta \to 0$. We obtain

$$p_h(T,v) = \delta T(E_M + \sum(\delta;v))v,$$

where $\sum(\delta,v) \triangleq S(\delta;v) - E_M$ fulfils the conclusion in Lemma.
The proof is complete.

**Proof of the Proposition 2.4.2** By hypothesis, the conditions in Lemmas 1–4 are fulfilled and let

$$x_h(t,v;x_0) \triangleq G(p_h(t,v);x_0), \; t \in [0,T],$$

be defined as in Lemma 2 where $p_h(t,v)$ is the solution of (**) in Lemma 1. Then $x_h(t,v;x_0)$, $t \in [0,T]$, is a solution of (1) (see (*) in Lemma 2) and using Lemma 4 we obtain that $x_h(T,v;x_0) \triangleq G(p_h(T,v);x_0)$ fulfils

(13)
$$\|x_h(\;T,v;x_0)\|^2$$
$$= \|x_0\|^2 + \delta T G(0;x_0)\frac{\partial G}{\partial p}(0,x_0)\sum_{j=1}^M v_j h_j(0) + 0(\delta T)_{\|v\|^2}$$
$$= \|x_0\|^2 + \delta T \sum_{j=1}^M \langle x_0, Y_j(x_0)\rangle \, v_j + o(\delta T) \, \|v\|^2$$

with $\lim_{r\to 0} 0(r)/r = 0$ and where $G(0; x_0) = x_0, \dfrac{\partial G}{\partial p}(0; x_0)h_j(0) = Y_j(x_0), j = 1, \dots, M$. Define

(14)                     $v_j(x_0) = -\langle x_0, Y_j(x_0) \rangle, \quad j = 1, \dots, M$

and from (13) follows

(15)                     $\|x_h(T, v(x_0); x_0)\|^2 \le \|x_0\|^2 - \dfrac{1}{2}\delta T \|v(x_0)\|^2$

for any $\|x_0\| \le C_0$ if $\delta \triangleq h^{1/3}$ is sufficiently small. The system (**) in Lemma 2 and its periodically defined solution $x_h(t, v(x_0); x_0)$, $t \ge 0$, doesn not leave invariant the set $V$. It can be adjusted by multiplying $g_{ij}(x)$ and $g_i(x)$ by $a(x_0) \triangleq \|v(x_0)\|^2$ and we obtain (see(6))

(16),       $\dfrac{dx}{dt} = a(x_0) \left( \displaystyle\sum_{j=1}^{M} \varepsilon u_{ij}(t, h)v_j g_{ij}(x) + \sum_{i=1}^{m} u_i(t, h)g_i(x) \right)$

$x(0) = x_0$, where $u_{ij}(t, h)$, $u_i(t, h)$ are the same as in Lemma 2. The system (16) leaves invariant the set $V$ (see $x_0 \in V \iff a(x_0) = 0$) and in addition its solutions $\widetilde{x}_h(t, v; x_0)$, $t \in [0, T]$, fulfils a similar estimate as in (13).

Namely, the corresponding auxiliary system for $\widetilde{p}_h(t, v)$, $t \in [0, T]$, defining $\widetilde{x}_h(t, v; x_0)$ is

(17)    $\dfrac{dp}{dt} = a(x_0) \left( \displaystyle\sum_{j=1}^{M} \varepsilon v_j u_{ij}(t, h)q_{ij}(p) + \sum_{i=1}^{m} u_i(t, h)q_i(p) \right), \quad p(0) = 0,$

and taking $\widetilde{q}_i(p) \triangleq a(x_0)q_i(p)$, $I = 1, \dots, m$, as the new vector fields defining the auxiliary controllable system we obtain

(18)                     $\widetilde{h}_j(p) = a^{l_j}(x_0)h_j(p), \quad j = 1, \dots, M.$

Lemmas 1-4 remain unchanged if we replace $q_i, q_{ij}, h_j$ by the corresponding $\widetilde{q}_i, \widetilde{q}_{ij}$ and $\widetilde{h}_j$, $j = 1, \dots, M$, $i = 1, \dots, m$, except that in Lemma 4 the identity matrix $I_M$ must be replaced by a diagonal matrix

(19)                     $D_M = \text{diag}\,(a^{l_j}(x_0)), \quad j = 1, \dots, M.$

It changes the estimate in (13) as follows:

$$\|\widetilde{x}_h(T, v; x_0)\| = \|x_0\|^2 + \delta T \sum_{j=1}^{M} a^{l_j}(x_0)\langle x_0, \ Y_j(x_0)\rangle \ v_j + o(\delta T)_{\|v\|^2}$$

and choosing

(20)        $v_j = \widetilde{v}_j(x_0) = -a^{l_j}(x_0)\langle x_0, Y_j(x_0)\rangle, \quad j = 1,\ldots, M.$

we obtain

$$\|\widetilde{x}_h(T, x; x_0)\|^2 \leq \|x_0\|^2 - \frac{1}{2}\delta T \|\widetilde{v}(x_0)\|^2 \quad \text{forany } \|x_0\| \leq C_0$$

if $\delta \stackrel{\triangle}{=} h^{1/3}$ is sufficiently small. Replace $v_j$ in (16) by $\widetilde{v}_j(x_0)$ in (20) and denote $\widetilde{x}_h(t; x_0)$, $t \geq 0$, the periodically defined solution in (16). We obtain that $\widetilde{x}_k = \widetilde{x}_h(kT; x_0) \stackrel{\triangle}{=} \widetilde{x}_h(kT; \widetilde{x}_{k-1})$ fulfils

$$\|\widetilde{x}_k\|^2 \leq \|\widetilde{x}_{k-1}\|^2 - \frac{1}{2}\delta T \|\widetilde{v}(x_{k-1})\|^2, \quad k = 1, 2, \ldots,$$

and finally

$$\|\widetilde{x}_{k+1}\|^2 \leq \|x_0\|^2 - \frac{1}{2}\delta T \left(\|\widetilde{v}(x_0)\|^2 + \|\widetilde{v}(\widetilde{x}_1)\|^2 + \cdots + \|\widetilde{v}(\widetilde{x}_k)\|^2\right),$$

$$\lim_{k \to \infty} \widetilde{v}(\widetilde{x}_k) = 0$$

Any sequence $(t_N) \uparrow \infty$ is rewriten $t_N = k_N T + \widetilde{t}_N$ with $0 \leq \widetilde{t}_N < T$ and by continuous dependence of a solution on parameters we obtain $\lim_{t_N \to \infty} \widetilde{x}_h(t_N; x_o) \in V \cap \{x : \|x\| \leq C_0\}$ and

$$\lim_{k_N \to \infty} \widetilde{x}_h(k_N T; x_0) \in V \cap \{x : \|x\| \leq C_0\}.$$

The proof is complete.

### a.3 Properties of continuous semi–martingales and stochastic integrals

Let a complete probability space $\{\Omega, F, P\}$ be given. Assume that a family of sub $\sigma$-fields $F_t \subseteq F, t \in [0, T]$, is given such that the following properties are fulfilled:

i) Each $F_t$ contains all null sets of $F$. ii) $(F_t)$ is increasing, i.e. $F_t \subseteq F_s$ if $t \geq s$. iii) $(F_t)$ is right continuous, i.e $\bigcap_{\varepsilon > 0} F_{t+\varepsilon} = F_t$ for any $t < T$.

Let $X(t)$, $t \in [0, T]$, be a measurable stochastic process with values in $R$.

We will assume, unless otherwise mentioned, that it is $(F_t)$-adapted, i.e, $X(t)$ is $F_t$-measurable for any $t \in [0, T]$.

The process $X(t)$ is called continuous if $X(t; w)$ is a continuous function of $t$ for almost all $\omega \in \Omega$. Let $L_c$ be the linear space consisting of all continuous stochastic processes. We introduce the metric $\rho$ by $|X - Y| = \rho(X, Y) = E[\sup_t |X(t) - Y(t)|^2 / (1 + \sup_t |X(t) - Y(t)|^2]^{1/2}$. It is equivalent to the topology of the uniform convergence in probability. A sequence $\{X^n\}$ of $L_c$ is a Cauchy sequence iff for any $\varepsilon > 0$, $P(\sup_t |X^n(t) - X^m(t)| > C) \to 0$ if $n, m \to \infty$. Obviously $L_c$ is a complete metric space.

We introduce the norm $\| \cdot \|$ by $\|X\| = E[\sup_t |X(t)|^2]^{1/2}$ and denote by $L_c^2$ the set of all elements in $L_c$ with finite norms. We may say that the topology of $L_c^2$ is the uniform convergence in $L_2$. Since $\rho(X) \le \|X\|$, the topology by $\| \, \|$ is stronger than that by $\rho$ and it is easy to see that $L_c^2$ is a dense subset of $L_c$.

**Definition**

Let $X(t)$, $t \in [0, T]$, be a continuous $F_t$-adapted process. (i) It is called a martingale if $E|X(t)| < \infty$ for any $t$ and satisfies $E[X(t)/F_s] = X(s)$ for any $t > s$. (ii) It is called a local martingale if there exists an increasing sequence of stopping times $(T_n)$ such that $T_n \uparrow \infty$ and each stopped process $X_{(t)}^{T_n} \equiv X(t \wedge T_n)$ is a martingale. (iii) It is called an increasing process if $X(t; \omega)$ is an increasing function of $t$ almost surely (a.s) with respect to $\omega \in \Omega$, i.e. there is a $P$-null set $N \subseteq \Omega$ such that $X(t; \omega)$, $t \in [0, T]$, is an increasing function for any $\omega \in \Omega \setminus \wedge$. (iv) It is called a process of bounded variation if it is written as the difference of two increasing processes. (v) It is called a semi–martingale if it is written as the sum of a local martingale and a process of bounded variation.

We will quote two classical results of Doob concerning martingales without giving proofs

**Theorem** *Let $X(t)$, $t \in [0, T]$, be a martingale. (i)* **Optional sampling theorem**. *Let $S$ and $\mathcal{U}$ be stopping times with values in $[0, T]$. Then $X(S)$ is integrable and satisfies $E[X(S)/F(\mathcal{U})] = X(S \wedge \mathcal{U})$, where $F(\mathcal{U}) = \{A \in F_T : A \cap \{\mathcal{U} \le t\}\} \in F(t)$ for any $t \in [0, T]\}$. (ii)* **Inequality**. *Suppose $E|X(T)|^{P} < \infty$ with $p > 1$. Then*

$$E \sup_t |X(t)|^p \le q^p E|X(T)|^p$$

*where*

$$\frac{1}{p} + \frac{1}{q} = 1$$

**Remark.**

Let $S$ be a stopping time. If $X(t)$ is a martingale the stopped process $X^S(t) \equiv X(t \wedge S)$ is also a martingale. In fact, by Doob's optional sampling theorem we have for $t \geq s E[X^S(t)|F(s)] = X(t \wedge S \wedge s) = X(S \wedge s) = X^S(s)$. Similarly, if $X$ is a local martingale the stopped process $X^S$ is a local martingale. The proofs of the following statements can be found in Kunita (7).

**Theorem.**

Let $X(t)$ be a continuous local martingale. (i) If $E\left[\sup_t |X(t)|\right] < \infty$, then $X$ is a martingale. (ii) Let $p > 1$. Then $X$ is an $L^\infty$-martingale iff $E\left[\sup_t |X(t)|^p\right] < \alpha$.

**Remark.**

Let $X$ be a local martingale. Then there is an increasing sequence of stopping times $S_k \uparrow \infty$ such that each stopped process $X^{S_k}$ is a bounded martingale. In fact, define $S_k$ by $S_k = \inf\{t > 0 : |X(t)| \geq k\}$ $(= \infty$ if $\{\cdots\} = \varphi)$. Then $S_k \uparrow \infty$ and it holds $\sup_t |X^{S_k}(t)|^{\leq k}$, so that each $X^{S_k}$ is a bounded martingale.

Let $M_c$ be the set of all square integrable martingale, $X(t)$ with $X(0) = 0$. Because of Doob's inequality the norm $\|X\|$ is finite for any $X \in M_c$. Hence $M_c$ is a subset of $L_c^2$. We denote $M_c^{loc}$ the set of all continuous local martingales $X(t)$ such that $X(0) = 0$. It is a subset of $L_c$.

**Theorem.**

$M_c$ is a closed subspace of $L_c^2 \cdot M_c^{loc}$ is a closed subspace of $L_c$. Furthe more, $M_c$ is dense in $M_c^{loc}$.

**Remark.**

Denote by $H_t^2$ the set consisting of all random variables $h : \Omega \to R$ which are $F(t)$−measurable and $E|h|^2 < \infty$. Then for each $X \in L_c^2$ there holds $E[X(T)/F(t)] = \hat{h}$ where $\hat{h} \in H_t^2$ is the optional solution for

$$\min_{h \in H_t^2} E|X(T) - h|^2 = E|X(T) - \hat{h}|^2.$$

**Definition** Let $X(t)$ be a continuous stochastic process and $\Delta$ a partition of the interval $[0, T] : \Delta = \{0 = t_0 < \ldots < t_n = T$, and let $|\Delta| = \max(t_{i+1} - t_i)$. Associated with the partition $\Delta$, we define a continous process $\langle X \rangle^\Delta(t)$ as

$$\langle X \rangle^\Delta(t) = \sum_{i=0}^{k-1} (X(t_{i+1}) - X(t_i))^2 + (X(t) - X(t_k))^2$$

where $k$ is the number such that $t_k \le t < t_{k+1}$. We call it the quadratic variation of $X(t)$ associated with the partition $\Delta$.

Now let $\{\Delta\}_m$ be a sequence of partitions such that $|\Delta_m| \to 0$. If the limit of $\langle x \rangle^{\Delta_m}(t)$ exists in probability and it is independent of the choice of sequence $\{\Delta_m\}$ a.s., it is called the quadratic variation of $X(t)$ and is denoted by $\langle X \rangle(t)$. We will see that a natural class of process where quadratic variations are well defined is that of continuous semimartingales.

**Lemma.**

*Let $X$ be a continuous process of bounded variation. Then the quadratic variation exists and is equals to zero a.s.*

**Proof.**

Let $|X|(t; \omega)$ be the total variation of the function $X(s; \omega), 0 \le s \le t$. Then these holds

$$\langle X \rangle^\Delta(t) \le \left( \sum_{j=0}^{k-1} | X(t_{j+1}) - X(t_j)| + |X(t) - X(t_k)| \right)$$
$$x \max_i |X(t_{iM}) - X(t_i)|$$
$$\le |X|(t) \max |X(t_{i+1}) - X(t_i)|.$$

The right hand side converges to 0 as $|\Delta| \to 0$ a.s. .

**Theorem**

*Let $M$ be a bounded continuous martingale. Let $\{\Delta_n\}$ be a sequence of partitions such that $|\Delta_n| \to 0$. Then $(M)^{\Delta_n}(t)$, $t \in [0, T]$, converges uniformly to a continuous increasing process $\langle M \rangle(t)$ in $L^2$ sense, i.e.,*

$$\lim_{n \to \infty} E[\sup_t |\langle M \rangle^{\Delta_n}(t) - \langle M \rangle(t)|^2] = 0$$

*The proof is based on the following two lemmas.*

**Lemma**

*For any $t > s$, there holds*

$$E[\langle M \rangle^\Delta(t)/F(s)] - \langle M \rangle^\Delta(s) = E[(M(t) - M(s))^2|F(s)] - M^2(s).$$

*In particular, $M^2(t) - \langle M \rangle^\Delta(t)$ is a continuous martingale.*

**Lemma**
*It holds* $\lim_{n,m\to\infty} E[|\langle M\rangle^{\Delta_n}(T) - \langle M\rangle^{\Delta_m}(T)|^2] = 0.$

**Theorem.**
*Let $M(t)$ be a continuous local martingale. Then there is a continuous increasing process $\langle M\rangle(t)$ such that $\langle M\rangle^{\Delta}(t)$ converges uniformly to $\langle M\rangle(t)$ in probability.*

**Corollary**
*$M^2(t) - \langle M\rangle(t)$ is a local martingale if $M(t)$ is a continuous local martingale.*

**Theorem**
*Let $X(t)$ be a continuous semi–martingale. Then $\langle X\rangle^{\Delta}(t)$ converges, uniformly to $\langle M\rangle(t)$ in probability as $|\Delta| \to 0$, where $\langle M\rangle(t)$ is the local martingale part of $X(t)$.*

**Stochastic integrals**
Let $M(t)$ be a continuous local martingale and let $f(t)$ be a continuous $F(t)$ adapted process. We will define the stochastic integral of $f(t)$ by the differential dM(t) using the properties of martingales, specially those of quadratic variations.

Let $\Delta = (0 = t_0\langle\ldots\langle t_{n-T})$ be a partition of $[0,T]$. For any $t \in [0,T]$ choose $t_k$ of $\Delta$ such that $t_k \le t < t_{k+1}$ and define

(1) $\qquad L^{\Delta}(t) = \sum_{i=0}^{k-1} f(t_i)(M(t_{i+1} - M(ti) + f(t_k))(M(t) - M(t_k))$

It is easily seen that $L^{\Delta}(t)$ is a continuous local martingale. The quadratic variation is computed directly as

$$
\begin{aligned}
\langle L^{\Delta}\rangle(t) &= \sum_{i=0}^{k-1} f^2(t_i)(\langle M\rangle(t_{iM}) - \langle M\rangle(t_i)) \\
&\quad + f^2(t_k)(\langle M\rangle(t) - \langle M\rangle(t_k) \\
&= \int_0^t |f^{\Delta}(s)|^2 d\,\langle M\rangle(s),
\end{aligned}
$$

(2)

where $f^{\Delta}(s)$ is a step process defined from $f(s)$ by $f^{\Delta}(s) = f(t_k)$ if $t_k \le s < t_{k+1}$. Let $\Delta'$ be another partition of $[0,T]$.

We define $L^{\Delta'}(t)$ similarly using the same $f(s)$ and $M(s)$. Then there holds

$$\langle L^{\Delta} - L^{\Delta'}\rangle(t) = \int_0^t |f_{(s)}^{\Delta} - f_{(s)}^{\Delta'}|^2 d\,\langle M\rangle(s).$$

Now let $\{\Delta_n\}$ be a sequence of partitions of $[0, T]$ such that $|\Delta_n| \to 0$. Then $\langle L^{\Delta_n} - L^{\Delta_m}\rangle(T)$ converges to 0 in probability as $n, m \to \infty$. Hence $\{L^{\Delta_n}\}$ is a Cauchy sequence in $M_c^{loc}$. We denote the limit as $L(t)$.

### Definition
The above $L(t)$ is called the Itô integral of $f(t)$ by $dM(t)$ and is denoted by $L(t) = \int_0^t f(s)dM(s)$.

### Definition
Let $X(t)$ be a continuous semimartingale decomposed to the sum of a continuous local martingale $M(t)$ and a continuous process of bounded variation $A(t)$.

Let $f$ be an $F(t)$–adapted process such that $f \in L^2(\langle M\rangle)$ and $\int_0^T \leftarrow$ $|f(s)|d|A|(s)\langle\infty$. Then the Ito integral of $f(t)$ by $dX(t)$ is defined as

$$\int_0^t f(s)\,dX(s) = \int_0^t f(s)\,dM(s) + \int_0^t f(s)\,dA(s)$$

**Remark.**

*If $f$ is a continuous semimartingale then $\int_0^t f(s)\,dX(s)$ exists. We will define another stochastic integral by the differential*

$$\circ dX(t): \int_0^t f(s) \circ dX(t) = \lim_{|\Delta|\to 0}\left[\sum_{i=0}^{k-1} \frac{1}{2}(f(t_{i+1}) + f(t_i))(X(t_{i+1}) - X(t_i))\right.$$

$$\left. + \frac{1}{2}(f(t) + f(t_k))(X(t) - X(t_k))\right]$$

### Definition

If the above limit exists, it is called the Fisk–Stratonovich integral of $f(s)$ by $dX(s)$.

If $f$ is a continuous semimartingale, the Fisk–Stratonovich integral is well defined and satisfies

$$\int_0^t f(s) \circ dX(s) = \int_0^t f(s)dX(s) + \frac{1}{2}\langle f, X\rangle(t)$$

### Proof

Is easily seen from the relation

$$\sum_{i=0}^{k-1} \frac{1}{2}(\ f(t_{i+1}) + f(t_i))(X(t_{i+1}) - X(t_i)) + \frac{1}{2}(f_(t) + f(t_k))(X(t) - X(t_k))$$

$$= \sum_{i=0}^{k-1} f(t_i)(X(t_{i+1}) - X(t_i)) + f(t_k)(X(t) - X(t_k)) + \frac{1}{2}\langle f, X\rangle^\Delta(t)$$

where the joint quadratic variation

$$\langle f, X\rangle^\Delta(t) = \sum_{i=0}^{k-1}(f(\ t_{i+1} - f(t_i))(X(t_{i+1}) - X(t_i))$$
$$+(f(t) - f(t_k))(X(t) - X(t_k))$$

and $k$ is the number such that $t_k \leq t < t_{k+1}$.

### Theorem

*Let $X$ and $Y$ be continuous semi-martingales. The joint quadratic variation associated with the partition $\Delta$ is defined as before and is written $\langle X, Y\rangle^\Delta$. Then $\langle X, Y\rangle^\Delta$ converges uniformly in probability to a continuous process of bounded variation $\langle X, Y\rangle(t)$. If $M$ and $N$ are local martingale parts of $X$ and $Y$, respectively, then $\langle X, Y\rangle$ coincides with $\langle M, N\rangle$.*

### Remark.

*Let $w(t) = w^1(t), \ldots, w^m(t)$ be an $m$–dimensional $(F_{(t)})$–adapted continuous stochastic process. It is an $(F_{(t)})$–Wiener process iff each $w^i(t)$ is a scalar Wiener process with the joint quadratic variation fulfilling $\langle w_i, w_j\rangle(t) = \delta_{ij}t$ where $\delta_{ij} = 1$ and $\delta_{ij} = 0$, if $i \neq j$.*

**a.4 Approximations of the diffusion equation by smooth ordinary differential equations**

The $m$–dimensional Wiener process is approximated by a smooth process and it allows one to use non anticipative smooth solutions of ordinary differential equations as approximations for solutions of a diffusion equation.

It comes from Langevin's classical procedure of defining a stochastic differential equation. By a standard $m$–dimensional Wiener process we mean a measurable function $w(t;\omega) \in R^m$, $(t;\omega) \in [0,\infty) \times \Omega$ with continuous trajectories $w(t,\omega)$ for each $\omega \in \Omega$ such that $w(0,\omega) = 0$, and

$i_1)$ $E(w(t_2)) - w(t_1)/F_{t_1}) = 0,$

$i_2)$ $E([w(t_2) - w(t_1)][w(t_2) - w(t_1)]^T/\mathcal{F}_{t_1}) = I_m(t_2 - t_1)$ for any $0 \le t_1 \langle t_2,$

where $(\Omega, \mathcal{F}, P)$ is a given complete probability space and $F_t \subseteq F$ is the $\sigma$–algebra generated by $(w(s), s \le t)$.

The quoted Langevin procedure replaces a standard $m$–dimensional Wiener process by a $C'$ non-anticipative process $v_\varepsilon(t)$ (see $v_\varepsilon(t)$ is $F_t$ measurable) as follows

$$v_\varepsilon(t) = w(t) - \int\limits_0^t (\exp -\beta(t - s))dw(s), \qquad \beta = \frac{1}{\varepsilon},\ \varepsilon \downarrow 0$$

where the integral in the right hand side is computed as $w(t) - \beta \int_0^t w(s)(\exp -\beta(t - s))ds$ for each $t > 0$. (the integration by parts formula)

Actually, $v_\varepsilon(t)$, $t \in [0, T]$, is the solution of the following

$$(2) \quad \frac{dv_\varepsilon}{dt}(t) = \beta \int\limits_0^t \exp -\beta(t-s)dw(s) = \beta w(t) - \beta^2 \int\limits_0^t w(s)(\exp -\beta(t-s))ds$$

and by a direct computation we obtain

$$(3) \qquad\qquad E\|v_\varepsilon(t) - w(t)\|^2 \le \varepsilon, \quad t \in [0, T]$$

Rewrite $v_\varepsilon(t)$ in (1) as

$$v_\varepsilon(t) = w(t) - \eta_\varepsilon(t),$$

where

$$(4) \qquad\qquad \eta(t) = \int\limits_0^t \exp -\beta(t - s)dw(s)$$

fulfils $d\eta^i = -\beta\eta^i dt + dw^i(t), \quad i = 1,\ldots,m$, and

(5) $$\frac{dv_\varepsilon}{dt}(t) = \beta\eta_\varepsilon(t), \quad \beta = \frac{1}{\varepsilon}, \quad \varepsilon \downarrow 0.$$

Now, we are given continuous functions

$$f(t,x), g_j(t,x) : [0,T] \times R^n \to R^n, \quad j = 1,\cdots,m,$$

such that

$$\alpha) = \begin{cases} i) f, g_j \;\; \text{arebounded}, & j = 1,\ldots,m \\ ii) \|h(t,x'') - h(t,x')\| \le L\|x'' - x'\| & (\forall)x', x'' \in R^n, \ t \in [0,T], \end{cases}$$

where $L > 0$ is a constant, $h \overset{\Delta}{=} f, g_j$

In addition, we assume that $(g_j \quad \in \quad C_b^{1,2}([0,T] \times R^n)]$

$\beta)\ \dfrac{\partial g_j}{\partial t}, \ \dfrac{\partial^2 g_j}{\partial t \partial x}, \ \dfrac{\partial^2 g_j}{\partial x^2}, \quad j = 1,\ldots,m$, are continuous and bounded

functions. Let $x_0(t)$ and $x_\varepsilon(t), t \in [0,T]$, be the solution in

(6)
$$dx = [f(t,x) + \frac{1}{2}\sum_{i=1}^m \frac{\partial g_i}{\partial x}(t,x)\,g_i(t,x)]dt$$
$$+ \sum_{i=1}^m g_i(t,x)dw_{(t)}^i, \quad x(0) = x_0$$

and

(7) $$\frac{dx}{dt} = f(t,x) + \sum_{i=1}^m g_i(t,x)\frac{dv_\varepsilon^i(t)}{dt}, \quad x(0) = x_0$$

correspondingly, where $v_\varepsilon(t)$ is associated with $w(t)$ in (1) and fulfils (2)–(5).

It is the Fisk–Stratonovich integral (see 0 below) which allow one to rewrite the system (6) as

(6s) $$dx = f(t,x)dt + \sum_{i=1}^m g_i(t,x) \circ dw^i(t), x(0) = x_0$$

**Theorem 1**
*Assume that continuous functions $f(t,x), g_i(t,x), t \in [0,T], x \in R^n$, are given such that $(\alpha)$ and $(\beta)$ are fulfilled. Then $\lim_{\varepsilon \to 0} E\|x_\varepsilon(t) - x_0(t)\|^2 = 0, \quad t \in$*

$[0, T]$, *where $x_0(t)$ and $x_\varepsilon(t)$ are the solutions defined in (6) and, respectively,*
*(7).*

**Proof** Using (5) we rewrite the solution $x_\varepsilon(t)$ as

$$
x_\varepsilon(t) = x_0 + \int_0^t f(s), x_\varepsilon(s))ds
$$

(8)

$$
+ \sum_{i=1}^m \beta \int_0^t g_i(s, x_\varepsilon(s))d\eta_\varepsilon^i(s), \quad t \in [0, T]
$$

and from $F_t$- measurability of $\eta_\varepsilon(t)$ we obtain that $x_\varepsilon(t)$ is $\mathcal{F}_t$-measurable and non-anticipative with respect to $\{\mathcal{F}_t\}$, $t \in [0, T]$. Therefore, the stochastic

integrals $\displaystyle\int_0^t g_i(s, x_\varepsilon(s))$

$dw^i(s)$ and $\displaystyle\int_0^t g_i(s, x_\varepsilon(s))d\eta_\varepsilon^i(s)$ are well defined, and using (4) we obtain

$$
\int_0^t g_i(s, x_\varepsilon(s))d\eta_\varepsilon^i(s) = -\beta \int_0^t g_i(s, x_\varepsilon(s))\eta_\varepsilon^i(s)ds
$$

(9)

$$
+ \int_0^t g_i(s, x_\varepsilon(s))dw^i(s).
$$

Using (9) in (8) there follows

$$
x_\varepsilon(t) = x_0 + \int_0^t f(s, x_\varepsilon(s))ds
$$

(10)

$$
+ \sum_{i=1}^m \int_0^t g_i(s, x_\varepsilon(s))dw^i(s)
$$

$$
- \sum_{i=1}^m \int_0^t g_i(s, x_\varepsilon(s))d\eta_\varepsilon^i(s), \quad t \in [0, T]
$$

In what follows it will be proved that

(11) $\displaystyle -\int_0^t g_i(s, x_\varepsilon(s))d\eta_\varepsilon^i(s) = \frac{1}{2}\int_0^t \frac{\partial g_i}{\partial x}(s, x_\varepsilon(s))\, g_i(s, x_\varepsilon(s))ds + O_t(\varepsilon)$

where $E\|O_t(\varepsilon)\|^2 \le c_1\varepsilon$, $(\forall)\ t \in [0,T]$, for some constant $c_1 > 0$, and using (11) in (10) we rewrite (10) as

(12)
$$x_\varepsilon(t) = x_0 + \int_0^t \left[ f(s, x_\varepsilon(s)) + \frac{1}{2} \sum_{i=1}^m \frac{\partial g_i}{\partial x}(s, x_\varepsilon(s))\, g_i(s, x_\varepsilon(s)) \right] ds$$
$$+ \sum_{i=1}^m \int_0^t g_i(s, x_\varepsilon(s))\, dw^i(s) + O_t(\varepsilon),$$

where $E\|O_t(\varepsilon)\|^2 \le c_1\varepsilon$.

The hypotheses $(\alpha)$ and $(\beta)$ allow one to check that

$$\widetilde{f}(t,x) \triangleq f(t,x) + \frac{1}{2} \sum_{i=1}^m \frac{\partial g_i}{\partial x}(t,x) g_i(t,x)$$

and $g_i(t,x)$ fulfil $(\alpha)$ also but with a new Lipschitz constant $\widetilde{L}$.

The proof will be complete noticing that

(13) $$E\|x_\varepsilon(t) - x_0(t)\|^2 \le c_2 \int_0^t E\|x_\varepsilon(s) - x_0(s)\|^2 ds + c_3\varepsilon, \quad t \in [0,T]$$

for some constants $c_2, c_3 > 0$ and Gronwall's lemma applied to (13) implies

$$E\|x_\varepsilon(t) - x_0(t)\|^2 \le \varepsilon c_3(\exp Tc_2) = \varepsilon c_4$$

which proves the conclusion.

To obtain (11) fulfilled we use an ordinary calculus as integration by parts formula (see $x_\varepsilon(t)$ is a $C^1$ function in $t \in [0,T]$). More precisely
(15)
$$-\int_0^t g_i(s, x_\varepsilon(s)) d\eta_\varepsilon^i(s) = -g_i(t, x_\varepsilon(t)\eta_\varepsilon^i(t) + \int_0^t \eta_q^i(s) \left( \frac{d}{ds} g_i(s, x_\varepsilon(s)) \right) ds$$
$$= \sum_{j=1}^m \int_0^t \frac{\partial g_i}{\partial x}(s, x_\varepsilon(s)) g_j(s, x_\varepsilon(s)) \frac{dv_\varepsilon^j(s)}{ds} \eta_\varepsilon^i(s) ds + \vartheta_t^1(\varepsilon)$$
$$\triangleq \sum_{j=1}^m T_{ij}(t) + O_t^1(\varepsilon),$$

where

$$O_t^1(\varepsilon) \triangleq \int_0^t \eta_\varepsilon^i(s) \left[ \frac{\partial g_i}{\partial t}(s, x_\varepsilon(s)) + \frac{\partial g_i}{\partial x}(s.x_\varepsilon(s)) f(s, x_\varepsilon(s)) \right] ds$$
$$- g_i(t, x_\varepsilon(t)) \eta_\varepsilon^i(t)$$

satisfies

(16) $$E\|O_t^1(\varepsilon)\|^2 \le k\varepsilon, \ \ t \in [0, T],$$

(see $E\|\eta_\varepsilon^i(t)\|^2 \le \varepsilon$ in (4) and $f, g_i, \dfrac{\partial g_i}{\partial x}; \dfrac{\partial g_i}{\partial t}$ are bounded). On the other hand, $\dfrac{dv_\varepsilon^j}{dt} = \beta\eta_\varepsilon^t(t)$ (see (5)) and using $\beta\eta_\varepsilon^j(t)dt = dw^j(t) - d\eta_\varepsilon^j(t)$ (see (4) we obtain that $T_{ij}$ in (15) fulfils

(17) $$T_{ij}(t) = -\int\limits_0^t g_{ij}(s, x_\varepsilon(s))\eta_\varepsilon^i(s)d\eta_\varepsilon^j(s) + \vartheta_t^2(\varepsilon),$$

wehre $g_{ij}(t, x) \triangleq \dfrac{\partial g_i}{\partial x}(t, x) \ g_j(t, x)$ and

$$\vartheta_t^2(\varepsilon) \triangleq \int\limits_0^t g_{ij}(s, x_\varepsilon(s))\eta_\varepsilon^i(s)dw^j(s)$$

obeys

$$E\|\vartheta_t^2(\varepsilon)\|^2 \le k_2\varepsilon, \ \ t \in [0, T] \le (18)$$

(see $g_{ij}$ bounded and $E\|\eta_\varepsilon^i(t)\|^2 \le \varepsilon$).

For $i = j$ we use the formula

(18) $$(\eta_\varepsilon^i(t))^2 = 2\int\limits_0^t \eta_\varepsilon^i(s)d\eta_\varepsilon^i(s) + \int\limits_0^t ds$$

and

$$
\begin{aligned}
T_{ii}(t) \;=\;& \frac{1}{2}\int_0^t g_{ii}(s, x_\varepsilon(s))ds \\
& -\frac{1}{2}\int_0^t g_{ii}(s, x_\varepsilon)d(\eta_\varepsilon^i(t))^2 + \vartheta_t^2(\varepsilon) \\
=\;& \frac{1}{2}\int_0^t g_{ii}(s, x_\varepsilon(s))ds \\
& -\frac{1}{2}g_{ii}(t, x_\varepsilon(t))(\eta_\varepsilon^i(t))^2 \\
& +\frac{1}{2}\int_0^t \left[\frac{d}{ds}g_{ii}(s, x_\varepsilon(s))\right](\eta_\varepsilon^i(s))^2 ds + \vartheta_t^2(\varepsilon) \\
=\;& \frac{1}{2}\int_0^t g_{ii}(s, x_\varepsilon(s))ds + \tilde{O}_t^2(\varepsilon).
\end{aligned}
$$

(19)

Here

$$
\begin{aligned}
\tilde{O}_t^2(\varepsilon) \;\triangleq\;& O_t^2(\varepsilon) - \frac{1}{2}g_{ii}(t, x_\varepsilon(t))(\eta_\varepsilon^i(t))^2 \\
& +\frac{1}{2}\int_0^t \left[\frac{\partial}{\partial t}g_{ii}(s, x_\varepsilon(s)) + \frac{\partial g_{ii}}{\partial x}(s, x_\varepsilon(s))f(s, x_\varepsilon(s))\right. \\
& \left. +\beta\sum_{j=1}^m \frac{\partial g_{ii}}{\partial x}(s, x_\varepsilon(s))g_j(s, x_\varepsilon(s))\eta_\varepsilon^j(s)\right](\eta_\varepsilon^i(s))^2 ds
\end{aligned}
$$

fulfils $E\|\tilde{O}_t^2(\varepsilon)\|^2 \leq \tilde{K}_2\varepsilon$ taking into account that

$$(20)\qquad E(\eta_\varepsilon^i(t))^4 = -\int_0^t 4\beta\, E(\eta_\varepsilon^i(s))^4 ds + 6\int_0^t E(\eta_\varepsilon^i(s))^2 ds$$

implies

$$(21)\qquad E(\eta_\varepsilon^i(t))^4 = 6(\exp -4\beta t)\int_0^t (\exp\, 4\beta s)\, E(\eta_\varepsilon^i(s))^2 ds$$

and $E(\eta_\varepsilon^i(t))^4 \leq 6\varepsilon^2$

For $i \neq j$, there holds

$$\eta_\varepsilon^i(t)\eta_\varepsilon^j(t) = \int_0^t \eta_\varepsilon^i(s)d\eta_\varepsilon^j(s) + \int_0^t \eta_\varepsilon^j(s)d\eta_\varepsilon^i(s)$$

$$= 2\int_0^t \eta_\varepsilon^i(s)d\eta_\varepsilon^j(s) + \int_0^t \eta_\varepsilon^j(s)dw^i - \int_0^t \eta_\varepsilon^i(s)dw^j$$

and

(22)
$$\int_0^t \eta_\varepsilon^i(s)d\eta_\varepsilon^j(s) = \frac{1}{2}(\eta_\varepsilon^i(t)\eta_\varepsilon^j(t))$$

$$-\frac{1}{2}\left[\int_0^t \eta_\varepsilon^j(s)dw^i(s) - \int_0^t \eta_\varepsilon^i(s)dw^j(s)\right].$$

Using (22) in (17) for $i \neq j$ we obtain

(23)     $$T_{ij}(t) = -\frac{1}{2}\int_0^t g_{ij}(s, x_\varepsilon(s))\, d(\eta_\varepsilon^i(s)\eta_\varepsilon^j(s)) + \tilde{O}_t(\varepsilon) + O_t^2(\varepsilon),$$

where

$$\tilde{O}_t(\varepsilon) \triangleq \frac{1}{2}\int_0^t g_{ij}(s, x_\varepsilon(s))[\eta_\varepsilon^j(s)dw^i(s) - \eta_\varepsilon^i(s)dw^j(s)]$$

satisfies

(24)                                   $$E\|\tilde{O}_t(\varepsilon)\|^2 \leq \tilde{k}\varepsilon$$

Integrating by parts the first term in (23) and using

(25)                  $$E|\eta_\varepsilon^i(s)\eta_\varepsilon^j(s)|^2 = E|\eta_\varepsilon^i(s)|^2|\eta_\varepsilon^j(s)|^2$$
$$\leq \varepsilon^2 \ (i \neq j)$$

we finally obtain

(26)              $$T_{ij}(t) = \tilde{O}_t^2(\varepsilon) \text{ with } E\|\tilde{O}_t^2(\varepsilon)\|^2 \leq \tilde{C}\varepsilon, \ t \in [0, T]$$

and using (19), (26) in (15) we find (11) fulfilled.
    The proof is complete.

**Remark 1.**

*Under the conditions in Theorem 1 it might be useful to notice that the computations remain unchanged if a stopping time $\tau : \Omega \to [0, \infty)$, $\{\omega : \tau \geq t\} \in \mathcal{F}_t$, $(\forall)\ t \in [0, T]$, is used. Namely*

$$\lim_{\varepsilon \downarrow 0} E\|x_\varepsilon(t \wedge \tau) - x_0(t \wedge \tau)\|^2 = 0\ (\forall)\ t \in [0, T],$$

*if the random variable $\tau : \Omega \to [0, \infty)$ is adapted to $\{\mathcal{F}_t\}$ i.e. $\{\omega : \tau \geq t\} \in \mathcal{F}_t$ for $t \in [0, T]$.*

*By definition*

$$x_\varepsilon(t \wedge \tau) = x_0 + \int_0^{t \wedge \tau} f(s, x_\varepsilon(s)) ds + \sum_{j=1}^m \int_0^{t \wedge \tau} g_j(s, x_\varepsilon(s)) \frac{dv_\varepsilon^j}{ds}(s),$$

$$x_0(t \wedge \tau) = x_0 + \int_0^{t \wedge \tau} \tilde{f}(t, x_0(s)) ds + \sum_{j=1}^m \int_0^{t \wedge \tau} g_j(s, x_0(s) dw^j(s),\ t \in [0, T],$$

*where*

$$\tilde{f}(t, x) \triangleq f(t, x) + \frac{1}{2} \sum_{j=1}^m \frac{\partial g_j}{\partial x}(t, x)\ g_j(t, x).$$

*Using the characteristic function*

$$\mathcal{X}(t, \omega) = \begin{cases} 1 & \text{if } \tau(\omega) \geq t \\ 0 & \text{if } \tau(\omega) < t \end{cases}$$

*which is a non-anticipative function, we rewrite $y_\varepsilon(t) \triangleq x_\varepsilon(t \wedge \tau)$ and $y_0(t) \triangleq x_0(t \wedge \tau)$ as:*

$$(*)\quad y_\varepsilon(t) = x_0 + \int_0^t x(s)\ f(s, y_\varepsilon(s)) ds + \sum_{j=1}^m \int_0^t \mathcal{X}(s) g_j(s, y_\varepsilon(s)) \frac{dv_\varepsilon^j(s)}{ds},$$

$$(*)\quad y_0(t) = x_0 + \int_0^t \mathcal{X}(s)\ \tilde{f}(s, y_0(s)) ds + \sum_{j=1}^m \int_0^t \mathcal{X}(s) g_j(s, y_0(s)) dw^j(s).$$

*Now the computations in Theorem 1 repeated for $(*)$ and $(**)$ allow one to obtain the conclusion.*

**Remark 2**

*Using Remark 1 we may remove the boundedness assumption on $f, g_j$ in the hypothesis $(\alpha)$ of Theorem 1. That is to say, the solutions $x_\varepsilon(t), x_0(t), t \in [0, T]$, exist assuming only the hypotheses $(\alpha, (ii))$ and $(\beta)$, and to obtain the conclusion we multiply $f, g_j$ by a $C^\infty$ scalar function $0 \le \alpha_N(x) \le 1$ such that $\alpha_N(x) = 1$ if $x \in S_N(0)$, $\alpha_N(x) = 0$ if $x \in R^n \backslash S_{2N}(0)$ where $S_\rho(0) \le \varepsilon R^n$ is the ball of radius $\rho$ and centered at origin. We obtain new bounded functions*

$$f^N(t, x) = f(t, x)\, \alpha^N(x), \quad g_j^N(t, x) = g_j(t, x)\, \alpha^N(x)$$

*fulfilling $(\alpha)$ and $(\beta)$ of Theorem 1, and therefore $\lim\limits_{\varepsilon \downarrow 0} E\|x_\varepsilon^N(t) - x_0^N(t)\|^2 = 0 \ (\forall)\ t \in [0, T]$, where $x_\varepsilon^N(t)$ and $x_0^N(t)$ are the corresponding solutions. On the hand, using a stopping time*

$$\tau_N(\omega) = \inf\{t \ge 0 : \ x_0(t, \omega) \in S_N(0)\}$$

*we obtain $x_0(t \wedge \tau_N) = x_0^N(t \wedge \tau_n), \ t \in [0, T]$, (see Friedmann) where $x_0(t), \ t \in [0, T]$ is the solution of the equation (6) with $f, g_j$ fulfilling $(\alpha, ii)$ and $(\beta)$.*

*Finally we obtain: c) $\lim\limits_{\varepsilon \to 0} E\|x_\varepsilon^N(t \wedge \tau_N) - x_0(t \wedge \tau_N)\|^2 = 0$ for any $t \in [0, T]$,*

*for arbitrarily fixed $N > 0$ and the conclusion (c) represent the approximation of the solution in (6) under the hypotheses $(\alpha, ii)$ and $(\beta)$.*

**Remark 3.**

*The nonanticipative process $v_\varepsilon(t), \ t \in [0, T]$, used in Theorem 1 is only of the class $C^1$ with respect to $t \in [0, T]$, but a minor change in the approximating equations as follows $(\beta = \dfrac{1}{\varepsilon}, \varepsilon \downarrow 0)$ $\dfrac{dv_\varepsilon}{dt}(t) = y_1, \ \varepsilon\dfrac{dy_1}{dt} =$*

*$-y_1 + y_2, \dots, \quad \varepsilon\dfrac{dy_{k-1}}{dt} = -y_{k-1} + y_k. \ \varepsilon dy_k = -y_k \ dt + dw(t), \ t \in [0, T], \ y_j(0) = 0, \ j = 1, \dots k, v_\varepsilon(0) = 0$ will allow one to obtain a nonanticipative $v_\varepsilon(t), \ t \in [0, T]$ of the class $C^k$, for an arbitrarily fixed $k$.*

### (a.5) Some elementary notions of smooth manifolds

A locally Euclidian space $E$ of dimension $n$ is a topological space such that, for each $x \in E$, there exists a homeomorphism $h$ mapping some open neighbourhood of $x$ onto open set in $R^n$.

**Definition**

A manifold $M$ of dimension $n$ is a topological space which is locally Euclidian of dimension $n$, is Hausdorff and has a countable basis

The dimension of a locally Euclidian space is a well defined object considering Brouwer's theorem on invariance of domain (an open subset $\mathcal{U} \subseteq R^n$ can not be homeomorphic to an open subset $V \subseteq R^m$ if $n \neq m$).

A coordinate chart on a manifold $M$ is a pair $(\mathcal{U}, h)$ where $\mathcal{U}$ is an open set of $M$ and $h$ a homeomorphism of $\mathcal{U}$ onto an open set of $R^n$. Sometimes $h$ is represented as a set $(h_1, \ldots, h_n)$ and $h_i : U \to R$ is called the $i$-th coordinate function. If $x \in \mathcal{U}$, the $n$-tulpe of real numbers $(h_1(x), \ldots, h_n(x))$ is called the set of local coordinates of $x$ in the coordinate chart $(\mathcal{U}, h)$.

A coordinate chart $(\mathcal{U}, h)$ is called a cubic coordinate chart if $h(U)$ is an open cube about the origin in $R^n$. If $x \in U$ and $h(x) = 0$ then the coordinate chart is said to be centred at $x$.

Let $(\mathcal{U}_1, h^1)$ and $(\mathcal{U}_2, h^2)$ be two coordinate charts on a manifold $M$ with $\mathcal{U}_1 \cap \mathcal{U}_2 \neq \varphi$. Let $(h_1^2, \ldots, h_n^2)$ be the set of coordinate functions associated with the mapping $h^2$

The homeomorphism

$$h^2 \circ (h^1)^{-1} : h^1(\mathcal{U}_1 \cap \mathcal{U}_2) \to h^2(\mathcal{U}_1 \cap \mathcal{U}_2)$$

taking for each $x \in \mathcal{U}_1 \cap \mathcal{U}_2$, the set of local coordinates $(h_1^1(x), \ldots, h_n^1(x))$, is called a coordinates transformation on $U_1 \cap U_2$. Clearly $h_0^1(h^2)^{-1}$ gives the inverse mapping, which expresses $(h_1^2(x), \ldots, h_n^2(x))$ in terms of $(h_1^1(x), \ldots, h_n^1(x))$. Frequently, the set $(h_1(x), \ldots, h_n(x))$ is represented as an $n$-vector $\mathrm{col}\,(\mathrm{x}_1, \ldots, \mathrm{x}_n)$ and consistently the coordinates transformation $h^1 \circ (h^2)^{-1}$ can be represented in the form

$$x^1 = \mathrm{col}\,(\mathrm{x}_1^1, \ldots, \mathrm{x}_n^1) = \mathrm{col}\,(\mathrm{x}_1^1(\mathrm{x}_1^2, \ldots, \mathrm{x}_n^2), \ldots, \mathrm{x}_n^1(\mathrm{x}_1^2, \ldots, \mathrm{x}_n^2)) = \mathrm{x}^1(x)$$

and the inverse transformation $h_0^2(h^1)^{-1}$ in the form.

$$x^2 = x^2(x^1)$$

Two coordinate charts $(\mathcal{U}, h^1)$ and $(\mathcal{U}_2, h_2)$ are $C^\infty$-compatible if, whenever $\mathcal{U}_1 \cap \mathcal{U}_2 \neq \emptyset$, the coordinates transformation $h_0^1(h^2)^{-1}$ is a diffeomorphism, i.e., if $x^1(x^2)$ and $x^2(x^1)$ are both $C^\infty$ maps.

A $C^\infty$ atlas on a manifold $M$ is collection $C = \{(\mathcal{U}_i; h^i) : i \in I\}$ of pairwise $C^\infty$-compatible coordinate charts with the property $\underset{i \in I}{U}\, U_i = M$. An atlas is complete, if not properly contained in any other atlas.

**Definition**

A smooth or $C^\infty$ manifold is a manifold equipped with a complete $C^\infty$ atlas.

**Remark.**

*If $\mathcal{A}$ is any $C^\infty$ atlas on a manifold $M$, there exists a unique complete $C^\infty$ atlas $\mathcal{A}^*$ containing $\mathcal{A}$ and the later is defined as the set of all coordinates, charts $(\mathcal{U}, h)$ which are compatible with every coordinate chart $(\mathcal{U}_i, h_i)$ of $\mathcal{A}$. This set contains $\mathcal{A}$, is a $C^\infty$ atlas, and is complete by construction.*

**Definition**

Let $M_1$ and $M_2$ be smooth manifolds. A mapping $F : M_1 \to M_2$ is a smooth mapping if for each $x \in M_1$ there exist coordinate charts $(\mathcal{U}, h)$ of $M_1$ and $(V, g)$ of $M_2$ with $x \in \mathcal{U}$ and $F(x) \in V$, such that the expression of $F$ in local coordinates is $C^\infty$.

Let $F : M_1 \to M_2$ be a smooth mapping of manifolds.

i) $F$ is an immersion if rank $(F) = \dim M_1$ for all $x \in M_1$ where the rank $(F)$ at a point $x \in M_1$ is the rank of the Jacobian matrix $\dfrac{\partial \widehat{F}}{\partial \widehat{x}}$ at $h(x) = \widehat{x}$, where $\widehat{F} = g_0 F_0 h^{-1}$. ii) $F$ is an univalent immersion if $F$ is an immersion and is injective. iii) $F$ is an embedding if $F$ is an univalent immersion and the topology induced on $F(M_1)$ by the one of $M_1$ coincides with the topology of $F(M_1)$ as a subset of $M_2$.

**Remark**

*The mapping $F$, being smooth, is in particular a continuous mapping of topological spaces and the topology induced on $F(M_1)$ by the one of $M_1$ may properly contain the topology of $F(M_1)$ as a subset of $M_2$.*

**Definition**

The image $F(M_1)$ of an univalent immersion is called an immersed submanifold of $M_2$ and the image of an embedding is called an embedded submanifold of $M_2$.

**Remark**

*Conversely, one may say that a subset $S$ of $M_2$ is an immersed (respectively, embedded) submanifold of $M_2$ if there is another manifold $M_1$ and an univalent immersion (respectively, embedding) $F : M_1 \to M_2$ such that $F(M_1) = S$.*

*Let $M$ be a smooth manifold of dimension $m$ and $(\mathcal{U}, h)$ a cubic coordinate chart. Let $n$ be a integer $0 < n < m$, and $x_0$ a point of $U$. The subset of $\mathcal{U}$*

$$S_{x_0} = \{x \in \mathcal{U} : x_i(x) = x_i(x_0), \ i = n+1, \cdots, m\}$$

*is called an n-dimensional slice of $\mathcal{U}$ passing through $x_0$. In other words, a slice of $\mathcal{U}$ is the locus if all points of $\mathcal{U}$ for which some coordinates are constant.*

### Theorem

*Let $M$ be a smooth manifold of dimension $m$. A subset $S \subseteq M$ is an embedded submanifold of dimension $n < m$ iff for each $y \in S$ there exists a cubic coordinate chart $(\mathcal{U}, h)$ of $M$ with $y \in \mathcal{U}$ such that $\mathcal{U} \cap S$ coincides with an n-dimensional slice of $\mathcal{U}$ passing through $y$.* The difference between immersed

and embedded submanifold can be clarified by the following example.

Let $M_1$ be the open interval $(0, 2\pi)$ of the real line and $M_2 = R^2$. Let $t$ denote a point in $M_1$ and $(x_1, x_2)$ a point in $M_2$.

The mapping $F$ is defined by

$$x_1(t) = \sin\ 2t \quad x_2(t) = \sin\ t$$

This mapping is an immersion because

$$\text{rank}\ (F) = \text{rank} \begin{bmatrix} \dfrac{dx_1(t)}{dt} \\ \dfrac{dx_2}{dt}(t) \end{bmatrix} = \text{rank} \begin{bmatrix} 2\ \cos\ 2t \\ \cos\ t \end{bmatrix} = 1$$

for all $t \in (0, 2\pi)$. It is also univalent because $F(t_1) = F(t_2) \Rightarrow t_1 = t_2$. However, the mapping is not an embedding.

The mapping $F$ takes the open set $(\pi - \varepsilon, \pi + \varepsilon)$ of $M_1$ onto a subset $S$ of $F(M_1)$ which is open by definition in the topology induced by the one of $M_1$, but is not an open set in the topology of $F(M_1)$ as a subset of $M_2$.

This is because $S$ can not be seen as the intersection of $F(M_1)$ with an open set of $R^2$.

### Tangent Vectors

Let $M$ be a smooth manifold of dimension $n$. A real valued function $\varphi$ is said to be smooth in a neighbourhood of $x \in M$, if the domain of $\varphi$ includes an open set $V$ of $M$ containing $\alpha$ and the restriction of $\varphi$ to $V$ is a smooth function. The set of all smooth functions in a neighbourhood of $x$ is denoted by $C^\infty(x)$. Note that $C^\infty(x)$ forms a vector space over $R$ and that any two functions $\varphi_1, \varphi_2 \in C^\infty(x)$ may be multiplied and $\varphi(y) = \varphi_1(y) \cdot \varphi_2(y)$ is defined for all $y$ in a neighbourhood of $x$.

**Definition**
A tangent vector $v$ at $x$ is a map $v : C^\infty(x) \to R$ with the following properties: $(i)$ (linearity): $v(a\varphi_1 + b\varphi_2) = av(\varphi)_1 + bv(\varphi_2)$ for all $\varphi_1, \varphi_2 \in C^\infty(x)$ and $a, b \in R$;

$ii)$ (Leibnitz' rule): $v(\varphi_1\varphi_2) = \varphi_1(x)v(\varphi_2) + \varphi_2(x)v(\varphi_1)$ for all $\varphi_1, \varphi_2 \in C^\infty(x)$.

**Definition**
Let $M$ be a smooth manifold. The tangent space to $M$ at $x$ written $T_xM$, is the set of all tangent vectors at $x$.
A map satisfying the properties $(i)$ and $(ii)$ is also called a derivation.

**Definition**
Let $M_1$ and $M_2$ be a smooth manifolds. Let $F : M_1 \to M_2$ be a smooth mapping. The differential of $F$ at $x \in M_1$ is the map $F_* : T_xM_1 \to T_{F(x)}M_2$ defined as follows.
For $v \in T_xM_1$ and $\varphi \in C^\infty(F(x))$, $F_*(v)(\varphi) = v(\varphi_0F)$.

**Remark.**
$F_*$ is a map of the tangent space of $M_1$ at a point $x$ into the tangent space of $M_2$ at the point $F(x)$. So one has to express the way in which $F_*(v)$ maps the set $C^\infty(F(x))$ into $R$. There is one of such maps for each point $x$ of $M_1$.

**Theorem**
The differential $F_*$ is a linear map and $(G_0F)_* = G_*F_*$ for any two smooth mappings. Let $(U, h)$ be a coordinate chart around $x \in M$ $M = m$. With this coordinate chart one may associate $m$ tangent vectors at $x$, denoted $\left(\dfrac{\partial}{\partial x_1}\right)_x, \dots, \left(\dfrac{\partial}{\partial x_m}\right)_x$ defined in the following way:

$$\left(\frac{\partial}{\partial x_i}\right)_x (\varphi) = \left[\frac{\partial}{\partial x_i}(\varphi_0h^{-1})\right]_{h(x)} \quad \text{for } 1 \leq i \leq m,$$

where $x_i = h_i(x)$, $\varphi \in C^\infty(x)$

**Theorem**
Let $M$ be a smooth manifold of dimension $m$. Let $x$ be any point of $M$. The tangent space $T_xM$ to $M$ at $x$ is an $m$-dimensional vector space over $R$. If $(\mathcal{U}, h)$ is a coordinate chart around $x$, then the tangent vectors $\left(\dfrac{\partial}{\partial x_1}\right)_x, \dots, \left(\dfrac{\partial}{\partial x_m}\right)_x$ form a basis of $T_xM$.

*From the above theorem it is seen that* $v = \sum\limits_{i=1}^{m} v_i \left( \dfrac{\partial}{\partial x_i} \right)_x$ *, where* $v_1, \cdot, v_m$
*are real numbers if* $v$ *is a tangent vector at* $x$. *The* $v_i$ *is coordinate can be*
computed explicitly in the following way. Let $h_i$ be the $i$-th coordinate
function. Clearly $hi \in C^{\infty}(x)$, and

$$v(h_i) = \sum_{j=1}^{m} v_j \left( \frac{\partial}{\partial x_j} \right) (h_i) = \sum_{j=1}^{m} v_j \left[ \frac{\partial}{\partial x_j}(hi \circ h^{-1}) \right]_{h(x)} = v_i,$$

because $hi \circ h^{-1}(x_1, \ldots, x_n) = x_i$.

### Definition
Let $M$ be a smooth manifold, of dimension $m$. A vector field $F$ on $M$ is
a mapping assigning to each $x \in M$ a tangent vector $f(x)$ in $T_x M$. A vector
field is smooth if for each $x \in M$ there exists a coordinate chart $(\mathcal{U}, h)$ about
$x$ and $m$ real-valued smooth functions $f_1, \ldots, f_m$ defined on $\mathcal{U}$ such that for
all $y \in \mathcal{U} f(y) = \sum\limits_{i=1}^{m} f_i(y) \left( \dfrac{\partial}{\partial x_i} \right)_y$.

### Theorem
Let $f$ be a smooth vector field on a manifold $M$. For each $x^0 \in M$ there
exists an open interval $I_0 \subseteq R$, depending on $x^0$, such that $0 \in I_0$ and a
smooth mapping $G : W \to M$ defined on the subset $W$ of $R \times M$, $W =$
$\{(t, x_0) \in R \times M : t \in I_0\}$ with the following properties: i) $G(0; x^0) = x^0$,

ii) $\dfrac{d}{dt} G(t; x_0) = f(G(t; x^0))$, $t \in I_0$. iii) $G(s; G(t; x^0)) = G(s+t; x^0)$

whenever both side are defined. iv) whenever $G(t; x^0)$ is defined, there exists
an open neighbourhood $\mathcal{U}$ of $x^0$ such that the mapping

$$G(t)w : \mathcal{U} \to M \text{ defined by } G(t)(x) = G(t; x)$$

is a diffeomorphism onto its image, and

$$G^{-1}(t) = G(-t)$$

The mapping $G$ is called the flow of $f$.

## Definition

A vector field is complete if, for all $x^0 \in M$, the interval $I_0$ coincide with $R$, i.e., the flow $G$ is defined on the whole cartesian product $R \times M$ and the integral curves $G(t; x^0)(x^0 \in M)$, are defined for all $t \in R$.

## Definition

Let $f$ be a smooth vector field on $M$ and $\varphi$ a smooth real-valued function on $M$. The derivative of $\varphi$ along $f$ is a function $M \to R$, written $L_f\,\varphi$ and defined as $(L_f\varphi)(x) = (f(x))(\varphi)$, i.e., $(L_f\varphi)(x)$ is the value on $\varphi$ of the tangent vector $f(x)$. The function $L_f\varphi$ is a smooth function and in local coordinates $L_f\,\varphi$ is represented by

$$(L_f\,\varphi)(x_1, \cdots, x_m) = \left[ \frac{\partial \varphi}{\partial x_1}, \cdots, \frac{\partial \varphi}{\partial x_m} \right] \begin{bmatrix} f_1 \\ \vdots \\ f_m \end{bmatrix}$$

If $f^1$, $f^2$ are vector fields and $\varphi$ a real-valued function, we denote $L_{f^1}L_{f^2}\varphi = Lf^1(L_{f^2}\varphi)$.

The set of all smooth vector fields on a smooth manifold is denoted $V(M)$. This set is a vector space since if $f, g \in V(M)$ and $a, b \in R$, their linear combination $af + bg$ is a smooth vector field defined by

$$(af + bg)(x) = af(x) + bg(x).$$

The set $V(M)$ forms a Lie algebra with the vector space structure already discussed and a product $[\cdot, \cdot]$ defined in the following way. If $f$ and $g$ are vector fields, $[f, g]$ is a new vector field whose value at $x$, a tangent vector in $T_x M$, maps $C^\infty(x)$ into $R$ according to the rule

$$([f, g](x))(\varphi) = (L_f L_g \varphi)(x) - (L_g L_f \varphi)(x)$$

In other words, $[f, g](x)$ takes $\varphi$ into the real number

$$(L_f L_g \varphi)(x) - (L_g L_f \varphi)(x).$$

### Theorem

$V(M)$ with the product $[f, g]$ thus defined is a Lie algebra. The product $[f, g]$ is called the Lie bracket of the two vector fields $f$ and $g$ and the reader may easily compute the expression of $[f, g]$ in local coordinate as

$$
\begin{bmatrix} \dfrac{\partial g_1}{\partial x_1} & \cdots & \dfrac{\partial g_1}{\partial x_m} \\ \cdots & \cdots & \cdots \\ \dfrac{\partial g_m}{\partial x_1} & \cdots & \dfrac{\partial g_m}{\partial x_m} \end{bmatrix} \begin{bmatrix} f_1 \\ \vdots \\ f_m \end{bmatrix} - \begin{bmatrix} \dfrac{\partial f_1}{\partial x_1} & \cdots & \dfrac{\partial f_1}{\partial x_m} \\ \cdots & \cdots & \cdots \\ \dfrac{\partial f_m}{\partial x_1} & \cdots & \dfrac{\partial f_m}{\partial x_m} \end{bmatrix} \begin{bmatrix} g_1 \\ \vdots \\ g_m \end{bmatrix} = \dfrac{\partial g}{\partial x} f - \dfrac{\partial f}{\partial x} g
$$

$$
= adf(g)
$$

The notion of Lie bracket of vector fields is very much related to its applications in the study of nonlinear control systems, or could be of interest in solving two properties.

### Theorem

Let $N \subseteq M$ be an embedded submanifold of $M$. Let $V$ be an open set of $N$ and $f, g \in V(M)$ such that for all $x \in V f(x) \in T_x N$ and $g(x) \in T_x N$ for all $x \in V$.

### Theorem

Let $f, g$ be two smooth fields on $M$. Let $G(t)$ denote the flow of $f$. For each $x \in M$

$$
\lim_{t \to 0} \frac{1}{t}[G(-t)_* g(G(t; \ x)) - g(x)] = [f, \ g](x).
$$

### Remark

The first term of the expression under bracket is a tangent vector at $x \in M$ obtained in the following may. With $x$, the mapping $G(t); \ x)$ (always defined for sufficiently small $t$) associates a point $y = G(t; \ x)$. The vector field is evaluated at $y$ and the value $g(y) \in T_y M$ is taken back to $t_x M$ via the differential $G(-t)_*$ (which maps the tangent space at $y$ onto the tangent space at $x = G(-t; \ y)$. Let $f$ be a smooth vector field on $M, g$ a smooth vector field on $N$ and $F : M \to N$ a smooth function. The vector fields $f, g$ are said to be $F$-related if $F_* f = g_0 F$. Note that the vector field $G(-t)_* g(G(t; x))$ considered in the above Remark is $G(-t)$ related to $g$.

### Remark

If $\overline{f}$ is $F$-related to $f$ and $\overline{g}$ is $F$-related to $g$, then $[\overline{f}, \overline{g}]$ is $F$-related to $[f, g]$.

**Remark**

*The Lie bracket of $g$ and $f$ may be interpreted as the value at $t = 0$ of the derivative with respect to $t$ of a function defined as $W(t) = G(-t)_* g(G(t; x))$ and it is easily seen that for any $k \geq 0$ $\left( \dfrac{d^k}{dt^k} W(t) \right)_{t=0} = ad^k f(g)(x)$. If $W(t)$ is analytic in a neighbourhood of $t = 0$, then $W(t)$ can be expanded in the form*

$$W(t) = \sum_{k=0}^{\infty} ad^k f(g)(x) \frac{t^k}{k!},$$

*known as the Campbell–Baker–Hausdorff formula. One may define an object which dualizes the notion of a vector field.*

**Definition**

Let $M$ be a smooth manifold of dimension $m$. A covector field (also called one-form) $\omega$ on $M$ is a mapping assigning to each point $x \in M$ a tangent covector $\omega(x) \in T_x^* M$. A covector field $\omega$ is smooth if for each $x \in M$ there exists a coordinate chart $(\mathcal{U}, h)$ about $x$ and $m$ real-valued smooth functions $\omega_1, \ldots, \omega_m$ defined on $\mathcal{U}$, such that, for all $y \in \mathcal{U}$,

$$\omega(y) = \sum_{i=1}^{m} \omega_i(y)(dh_i)_y$$

The expression of a covector field in local coordinates is often given the form of a row vector $\omega = \text{row}\,(\omega_1, \ldots, \omega_m)$ in which the $\omega_i$-s are real-valued function of $x_1, \cdots, x_m$.

If $\omega$ is a covector field and $f$ is a vector field, $\langle \omega, f \rangle$ denotes the smooth real-valued function defined by

$$\langle \omega, f \rangle(x) = \langle \omega(x),\ f(x) \rangle = \sum_{i=1}^{m} \omega_i(x)\, f_i(x)$$

With any smooth function $\varphi : M \to R$ one may associate a covector field by taking at each $x$ the cotangent vector $(d\varphi)x$. The covector field thus defined is usually still represented by the symbol $d\varphi$. However, the converse is not always true.

**Definition**

A covector field $\omega$ is exact if there exists a smooth real-valued function $\varphi : M \to R$ such that $\omega = d\lambda$. The set of all smooth covector fields on a manifold $M$ is denoted by the symbol $V^*(M)$.

One may define the notion of derivative of a covector field $\omega$ along a vector field $f$. Let $x$ be a point of the domain of the flow $G(t)$ determined by $f \in V(M)$. Recall that $G(t)_* : T_x M \to T_{G(t;x)} M$ is a linear mapping and let $G(t)^* : T^*_{G(t;x)} M \to T^*_x M$ denote the dual mapping. With $\omega$ and $G(t)$ we associate a new covector field whose value at a point $y$ in the domain of $G(t)$ is defined by $G(t)^* \omega(G(t;y))$. The covector field thus defined is said to be $G(t)$ related to $\omega$.

**Theorem**

*Let $f$ be a smooth vector field and $\omega$ a smooth covector field on $M$. For each $x \in M$ the derivative $L_f \omega$ of $\omega$ along $f$ exists and*

$$L_f \omega(x) = \lim_{t \to 0} \frac{1}{t}[G(t)^* \omega(G(t;x)) - \omega(x)]$$

*The derivative $L_f \omega$ of $\omega$ along $f$ is covector field on $M$ and its expression in local coordinates is given by the (row) $m$–vector*

$$[f_1, \ldots, f_m] \begin{bmatrix} \dfrac{\partial \omega_1}{\partial x_1} & \cdots & \dfrac{\partial \omega_m}{\partial x_1} \\ \cdots & \cdots & \cdots \\ \dfrac{\partial \omega_1}{\partial x_m} & \cdots & \dfrac{\partial \omega_m}{\partial x_m} \end{bmatrix} + [\omega_1, \ldots, \omega_m] \begin{bmatrix} \dfrac{\partial f_1}{\partial x_1} & \cdots & \dfrac{\partial f_1}{\partial x_m} \\ \cdots & \cdots & \cdots \\ \dfrac{\partial f_m}{\partial x_1} & \cdots & \dfrac{\partial f_m}{\partial x_m} \end{bmatrix}$$

$$= \left[ \frac{\partial \omega^T}{\partial x} f \right] + \omega \frac{\partial f}{\partial x},$$

*where $T$ denotes transpose.*

*Now for a given set of smooth vector fields $g_i \in V(M)$ $i = 1, \ldots, k$, we will associate a gradient system on $M$. For the sake of simplicity we assume that $g_i$ are complete vector fields. Let $x^\circ \in M$ be fixed, and define*

$$G(p; x^\circ) = G_1(t_1) \circ \cdots \circ G_k(t_k)(x^\circ), \quad p \in (t_1, \ldots, t_k) \in R^k,$$

*where $G_i(t; x)$, $t \in R$, $x \in M$, is the flow determined by the complete vector field $g_i$. The partial derivative $\dfrac{\partial G}{\partial t_i}(p; x^\circ)$ exists and equals $\lim_{t \to 0} \dfrac{1}{t}[G(p + te_i; x^\circ) - G(p; x^\circ)]$, where $e_1, \ldots, e_k \in R^k$ is the canonical basis. By definition $\dfrac{\partial G}{\partial t_i}(p; x^\circ)$ is a tangent vector at some point of $M$ and the following statement gives a more explicit form of the partial derivatives $\dfrac{\partial G}{\partial t_i}(p; x^\circ)$. Denote $y_1 = y \in M$, $y_{j+1} = G_j(-t_j, y_j)$, $j = 1, \ldots, k - 1$, and notice that $G_j(t_j)$ takes $y_{j+1}$ to $y_j$. The differential map $G_j(t_j)_*$ of $G_j(t_j)$*

at $x = y_{j+1}$ is a linear mapping of $T_{y_{j+1}}M$ on $T_{y_j}M$. On the other hand $G_j(t_j)_*$ can be viewed as the inverse of the Jacobian matrix $\dfrac{\partial G}{\partial x}(-t_j; y_j)$. With $\{g_1, \ldots, g_k\} \subseteq V(M)$ associate the following system of smooth vector fields with parameters

$$
\begin{aligned}
X_1(y) &= g_1(y), \ X_2(t_1; y) \\
&= G_1(t_1)_* g_2(y_2), \ X_{j+1}(t_1, \ldots, t_j; y) \\
&= G_1(t_1)_* \ldots G_j(t_j)_* g_{j+1}(y_{j+1}) \quad j = 1, \ldots, k-1.
\end{aligned}
$$

By definition, $X_1(y) \in T_y M$, $X_{j+1}(t_1, \ldots, t_j; y) \in T_y M$, for all $j = 1, \ldots, k-1$ and for any $p = (t_1, \ldots, t_k) \in R^k$.

### Definition
Let $g_i \in V(M)$, $i = 1, \ldots, k$, be smooth complete vector fields and define smooth vector fields $X_{j+1}(t_1, \ldots, t_j; y)$, $j = 1, \ldots, k-1$, as above. We say that $\{g_1, X_2(t_1), \ldots, X_k(t_1, \ldots, t_{k-1})\}$ determine a gradient system if

$$
\frac{\partial X}{\partial t_i} j + 1(t_1, \ldots, t_j) = -[X_i(t_1, \ldots, t_{i-1}), X_{j+1}(t_1, \ldots, t_j)]
$$

for all $1 \le i \le j$, $j = 1, \ldots, k-1$.

### Remark
The minus sign appearing in the right hand side is owed to the definition of a Lie bracket on $V(M)$ as a linear mapping on $C^\infty(M)$ which is the opposite of the one used on $F_n = C^\infty(R^n; R^n)$ (see ch. I).

### Theorem
For a smooth complete vector fields $\{g_1, \ldots, g_k\} \subseteq V(M)$ associate $X_{j+1}(t_1, \ldots, t_j)$, $j = 1, \ldots, k-1$, as above. Then

$$
\frac{\partial y}{\partial t_1} = g_1(y), \quad \frac{\partial y}{\partial t_2} = X_2(t_1; y) \quad \ldots \quad \frac{\partial y}{\partial t_k} = X_k(t_1, \ldots, t_{k-1}; y)
$$

is a gradient system and $y(p) = G(p; x^\circ)$ is its solution fulfilling $y(0) = x^\circ$.

The proof is based on a straight computation taking into account the properties of the differential mappings $G_j(t_j)_*$, $j = 1, \ldots, k$.

The analysis of a gradient system and its algebraic representation may be significant for studying affine control systems and defining solutions of stochastic differential equations on smooth manifolds. There is no major obstruction for studying finite–dimensional or finitely generated Lie algebras

$L(g_1, \dots, g_m) \subseteq V(M)$ *and the algebraic representation of a gradient system allow one to find the geometric differential structure of the solutions under consideration.*

### Bibliographical notes

The preliminary part regarding continuous semi-martingales are found in Kunita [10] and it serves the intention of describing Langevin's approximation of solutions for stochastic differential equations.

The elementary facts regarding smooth manifolds are found in Isidori [8] and restricted to what is necessary for defining gradient systems on a smooth manifold.

# Bibliography

[1] R. W. Brockett, *Nonlinear Systems and Differential Geometry*, Proceedings of the I.E.E.E., vol. 64, no. 1 pp. 61-72, (1976).

[2] I. M. Coron, *Linearized Control System and Applications to Smooth Stabilization*, SIAM J. Control and Optimization, vol. 32, no. 2, pp. 358-386 (1994).

[3] P. E. Crouch, *Solvable Approximations to Control Systems*, SIAM J. Control and Optimization, vol. 22, no. 1, pp. 40-54 (1984).

[4] A. Friedman, *Stochastic Differential Equations and Applications*, vol. I, Academic Press (1975).

[5] H. Hermes, A. Lundell, D. Sullivan, *Nilpotent Bases for Distributions and control Systems*, J. Diff. Eq., vol. 55, pp. 358-400 (1984).

[6] H. Hermes, *Nilpotent and High-Order Approximations of Vector Field Systems*, SIAM Review, vol. 33, no. 2, pp. 238-364 (1991).

[7] R. Herman, *The Differential Geometry of Foliations II*, Journal of Mathematics and Mechanics, vol. 11, no. 2, pp. 303-315 (1962).

[8] A. Isidori, *Nonlinear Control Systems, Communications and Control Series*, Springer-Verlag (1989).

[9] A. J. Krener, C. Lobry, *The Complexity of Solutions of Stochastic Diff. Eq*, Stochastic vol. 4, pp. 193-203 (1981).

[10] H. Kunita, *On the Decomposition of Solutions of Stochastic Diff. Eq.*, Proc. Durham Conf. Stochastic Integrals, Lecture Notes in Math., 851, pp. 231-255 (1981).

[11] H. Kunita, *Stochastic Diff. Eq. and Stochastic Flows of Diffeom.*, Lecture Notes in Math., 1097, Springer-Verlag (1982).

[12] E. Pardoux, *Stochastic Partial Diff. Eq. and Filtering of Diffusion Process*, Stochastics, (1980).

[13] H. I. Sussman, *On the Gap Between Deterministic and Stochastic Ordinary Diff. Eq.*, Annals of Probability, vol. 6, pp. 19-41 (1978).

[14] Y. Yamato, *Stochastic Diff. Eq. and Nilpotent Lie Algebras*, Z. W., 47, pp. 213-229 (1979).

[15] C. Vârsan, *On decomposition and Integral Representation of Solutions for Affine Control Systems*, Systems and Control Letters, 22, pp. 53-59 (1994).

[16] C. Vârsan, *Dynamic and Periodic Controls Stabilizing Affine Control Systems*, Proc. Third European Control Conference E.C.C. 95, Sep. 5-8, (1995).

[17] C. Vârsan, *Approximations and Existence of Periodic Solutions for Controlled Diffusion Equations*, Stochastic and Stochastic Reports, 30, pp. 135-150, (1990).

# Subject Index

Adapted stochastic processes 207
   Algebra 8
   Analytic
   – function 223, 33
   – vector field
   Asymptotic
   – stability 43
   Atlas 223
   Campbell–Baker–Hausdorff formula 22, 230
   Change of generators
   – of gradient systems 28
   Cauchy problem
   – of linear systems. 8, 92
   Complete
   – vector field 228
   – atlas 224
   Completely integrable
   – gradient systems 11
   Control systems 43, 117
   Continuous semimartingale 208
   Coordinate chart 223
   Derivation 226
   – of a formal series 5
   – of a vector field 20, 228
   Diffeomorphism of manifolds 224
   Differential
   – of a mapping 226
   – equation 8
   Embedded submanifolds 224

Finitely generated
– over $R$ (f.g.r.) 20
– over orbits (f.g.o.) 50
Fisk–Stratonovich integral 213
Flow 227
Formal power series 6
Frobenius Theorem 12
Gradient systems 14
Immersion 224
Iacobi identity 7
Ito's stochastic integral 212
Langevin approximation 214
Lie
– algebra 7
– bracket 8, 14
Manifold 223, 36
Nonsingular
– gradient system 31, 55
Slice of a neighbourhood 225
Smooth
– function 7
– manifold 223
– mapping 224
– vector field 8, 227
Stochastic integral 221
System of hyperbolic equations
– linear 17
– quasilinear 17
Tangent
– space 226
– vector 226

Other *Mathematics and Its Applications* titles of interest:

P.H. Sellers: *Combinatorial Complexes. A Mathematical Theory of Algorithms.* 1979, 200 pp. ISBN 90-277-1000-7

P.M. Cohn: *Universal Algebra.* 1981, 432 pp.
ISBN 90-277-1213- 1 (hb), ISBN 90-277-1254-9 (pb)

J. Mockor: *Groups of Divisibility.* 1983, 192 pp. ISBN 90-277-1539-4

A. Wwarynczyk: *Group Representations and Special Functions.* 1986, 704 pp.
ISBN 90-277-2294-3 (pb), ISBN 90-277-1269-7 (hb)

I. Bucur: *Selected Topics in Algebra and its Interrelations with Logic, Number Theory and Algebraic Geometry.* 1984, 416 pp. ISBN 90-277-1671-4

H. Walther: *Ten Applications of Graph Theory.* 1985, 264 pp. ISBN 90-277-1599-8

L. Beran: *Orthomodular Lattices. Algebraic Approach.* 1985, 416 pp.
ISBN 90-277-1715-X

A. Pazman: *Foundations of Optimum Experimental Design.* 1986, 248 pp.
ISBN 90-277-1865-2

K. Wagner and G. Wechsung: *Computational Complexity.* 1986, 552 pp.
ISBN 90-277-2146-7

A.N. Philippou, G.E. Bergum and A.F. Horodam (eds.): *Fibonacci Numbers and Their Applications.* 1986, 328 pp. ISBN 90-277-2234-X

C. Nastasescu and F. van Oystaeyen: *Dimensions of Ring Theory.* 1987, 372 pp.
ISBN 90-277-2461-X

Shang-Ching Chou: *Mechanical Geometry Theorem Proving.* 1987, 376 pp.
ISBN 90-277-2650-7

D. Przeworska-Rolewicz: *Algebraic Analysis.* 1988, 640 pp. ISBN 90-277-2443-1

C.T.J. Dodson: *Categories, Bundles and Spacetime Topology.* 1988, 264 pp.
ISBN 90-277-2771-6

V.D. Goppa: *Geometry and Codes.* 1988, 168 pp. ISBN 90-277-2776-7

A.A. Markov and N.M. Nagorny: *The Theory of Algorithms.* 1988, 396 pp.
ISBN 90-277-2773-2

E. Kratzel: *Lattice Points.* 1989, 322 pp. ISBN 90-277-2733-3

A.M.W. Glass and W.Ch. Holland (eds.): *Lattice-Ordered Groups. Advances and Techniques.* 1989, 400 pp. ISBN 0-7923-0116-1

N.E. Hurt: *Phase Retrieval and Zero Crossings: Mathematical Methods in Image Reconstruction.* 1989, 320 pp. ISBN 0-7923-0210-9

Du Dingzhu and Hu Guoding (eds.): *Combinatorics, Computing and Complexity.* 1989, 248 pp. ISBN 0-7923-0308-3

Other *Mathematics and Its Applications* titles of interest:

---

A.Ya. Helemskii: *The Homology of Banach and Topological Algebras*. 1989, 356 pp.
ISBN 0-7923-0217-6

J. Martinez (ed.): *Ordered Algebraic Structures*. 1989, 304 pp.    ISBN 0-7923-0489-6

V.I. Varshavsky: *Self-Timed Control of Concurrent Processes. The Design of Aperiodic Logical Circuits in Computers and Discrete Systems*. 1989, 428 pp. ISBN 0-7923-0525-6

E. Goles and S. Martinez: *Neural and Automata Networks. Dynamical Behavior and Applications*. 1990, 264 pp.                                    ISBN 0-7923-0632-5

A. Crumeyrolle: *Orthogonal and Symplectic Clifford Algebras. Spinor Structures*. 1990, 364 pp.                                             ISBN 0-7923-0541-8

S. Albeverio, Ph. Blanchard and D. Testard (eds.): *Stochastics, Algebra and Analysis in Classical and Quantum Dynamics*. 1990, 264 pp.        ISBN 0-7923-0637-6

G. Karpilovsky: *Symmetric and G-Algebras. With Applications to Group Representations*. 1990, 384 pp.                                            ISBN 0-7923-0761-5

J. Bosak: *Decomposition of Graphs*. 1990, 268 pp.        ISBN 0-7923-0747-X

J. Adamek and V. Trnkova: *Automata and Algebras in Categories*. 1990, 488 pp.
ISBN 0-7923-0010-6

A.B. Venkov: *Spectral Theory of Automorphic Functions and Its Applications*. 1991, 280 pp.
ISBN 0-7923-0487-X

M.A. Tsfasman and S.G. Vladuts: *Algebraic Geometric Codes*. 1991, 668 pp.
ISBN 0-7923-0727-5

H.J. Voss: *Cycles and Bridges in Graphs*. 1991, 288 pp.        ISBN 0-7923-0899-9

V.K. Kharchenko: *Automorphisms and Derivations of Associative Rings*. 1991, 386 pp.
ISBN 0-7923-1382-8

A.Yu. Olshanskii: *Geometry of Defining Relations in Groups*. 1991, 513 pp.
ISBN 0-7923-1394-1

F. Brackx and D. Constales: *Computer Algebra with LISP and REDUCE. An Introduction to Computer-Aided Pure Mathematics*. 1992, 286 pp.        ISBN 0-7923-1441-7

N.M. Korobov: *Exponential Sums and their Applications*. 1992, 210 pp.
ISBN 0-7923-1647-9

D.G. Skordev: *Computability in Combinatory Spaces. An Algebraic Generalization of Abstract First Order Computability*. 1992, 320 pp.        ISBN 0-7923-1576-6

E. Goles and S. Martinez: *Statistical Physics, Automata Networks and Dynamical Systems*. 1992, 208 pp.                                        ISBN 0-7923-1595-2

M.A. Frumkin: *Systolic Computations*. 1992, 320 pp.        ISBN 0-7923-1708-4

J. Alajbegovic and J. Mockor: *Approximation Theorems in Commutative Algebra*. 1992, 330 pp.                                                  ISBN 0-7923-1948-6

Other *Mathematics and Its Applications* titles of interest:

---

I.A. Faradzev, A.A. Ivanov, M.M. Klin and A.J. Woldar: *Investigations in Algebraic Theory of Combinatorial Objects.* 1993, 516 pp.    ISBN 0-7923-1927-3

I.E. Shparlinski: *Computational and Algorithmic Problems in Finite Fields.* 1992, 266 pp.    ISBN 0-7923-2057-3

P. Feinsilver and R. Schott: *Algebraic Structures and Operator Calculus. Vol. I. Representations and Probability Theory.* 1993, 224 pp.    ISBN 0-7923-2116-2

A.G. Pinus: *Boolean Constructions in Universal Algebras.* 1993, 350 pp.    ISBN 0-7923-2117-0

V.V. Alexandrov and N.D. Gorsky: *Image Representation and Processing. A Recursive Approach.* 1993, 200 pp.    ISBN 0-7923-2136-7

L.A. Bokut' and G.P. Kukin: *Algorithmic and Combinatorial Algebra.* 1994, 384 pp.    ISBN 0-7923-2313-0

Y. Bahturin: *Basic Structures of Modern Algebra.* 1993, 419 pp.    ISBN 0-7923-2459-5

R. Krichevsky: *Universal Compression and Retrieval.* 1994, 219 pp. ISBN 0-7923-2672-5

A. Elduque and H.C. Myung: *Mutations of Alternative Algebras.* 1994, 226 pp.    ISBN 0-7923-2735-7

E. Goles and S. Martnez (eds.): *Cellular Automata, Dynamical Systems and Neural Networks.* 1994, 189 pp.    ISBN 0-7923-2772-1

A.G. Kusraev and S.S. Kutateladze: *Nonstandard Methods of Analysis.* 1994, 444 pp.    ISBN 0-7923-2892-2

P. Feinsilver and R. Schott: *Algebraic Structures and Operator Calculus. Vol. II. Special Functions and Computer Science.* 1994, 148 pp.    ISBN 0-7923-2921-X

V.M. Kopytov and N. Ya. Medvedev: *The Theory of Lattice-Ordered Groups.* 1994, 400 pp.    ISBN 0-7923-3169-9

H. Inassaridze: *Algebraic K-Theory.* 1995, 438 pp.    ISBN 0-7923-3185-0

C. Mortensen: *Inconsistent Mathematics.* 1995, 155 pp.    ISBN 0-7923-3186-9

R. Abłamowicz and P. Lounesto (eds.): *Clifford Algebras and Spinor Structures.* A Special Volume Dedicated to the Memory of Albert Crumeyrolle (1919–1992). 1995, 421 pp.    ISBN 0-7923-3366-7

W. Bosma and A. van der Poorten (eds.), *Computational Algebra and Number Theory.* 1995, 336 pp.    ISBN 0-7923-3501-5

A.L. Rosenberg: *Noncommutative Algebraic Geometry and Representations of Quantized Algebras.* 1995, 316 pp.    ISBN 0-7923-3575-9

L. Yanpei: *Embeddability in Graphs.* 1995, 400 pp.    ISBN 0-7923-3648-8

B.S. Stechkin and V.I. Baranov: *Extremal Combinatorial Problems and Their Applications.* 1995, 205 pp.    ISBN 0-7923-3631-3

Other *Mathematics and Its Applications* titles of interest:

---

Y. Fong, H.E. Bell, W.-F. Ke, G. Mason and G. Pilz (eds.): *Near-Rings and Near-Fields*.
1995, 278 pp.                                                     ISBN 0-7923-3635-6

A. Facchini and C. Menini (eds.): *Abelian Groups and Modules*. (Proceedings of the Padova
Conference, Padova, Italy, June 23–July 1, 1994). 1995, 537 pp.     ISBN 0-7923-3756-5

D. Dikranjan and W. Tholen: *Categorical Structure of Closure Operators*. With Applications to Topology, Algebra and Discrete Mathematics. 1995, 376 pp.
                                                                   ISBN 0-7923-3772-7

A.D. Korshunov (ed.): *Discrete Analysis and Operations Research*. 1996, 351 pp.
                                                                   ISBN 0-7923-3866-9

P. Feinsilver and R. Schott: *Algebraic Structures and Operator Calculus*. Vol. III: Representations of Lie Groups. 1996, 238 pp.                             ISBN 0-7923-3834-0

M. Gasca and C.A. Micchelli (eds.): *Total Positivity and Its Applications*. 1996, 528 pp.
                                                                   ISBN 0-7923-3924-X

W.D. Wallis (ed.): *Computational and Constructive Design Theory*. 1996, 368 pp.
                                                                   ISBN 0-7923-4015-9

F. Cacace and G. Lamperti: *Advanced Relational Programming*. 1996, 410 pp.
                                                                   ISBN 0-7923-4081-7

N.M. Martin and S. Pollard: *Closure Spaces and Logic*. 1996, 248 pp.
                                                                   ISBN 0-7923-4110-4

A.D. Korshunov (ed.): *Operations Research and Discrete Analysis*. 1997, 340 pp.
                                                                   ISBN 0-7923-4334-4

W.D. Wallis: *One-Factorizations*. 1997, 256 pp.          ISBN 0-7923-4323-9

G. Weaver: *Henkin–Keisler Models*. 1997, 266 pp.         ISBN 0-7923-4366-2

V.N. Kolokoltsov and V.P. Maslov: *Idempotent Analysis and Its Applications*. 1997, 318 pp.
                                                                   ISBN 0-7923-4509-6

J.P. Ward: *Quaternions and Cayley Numbers*. Algebra and Applications. 1997, 250 pp.
                                                                   ISBN 0-7923-4513-4

E.S. Ljapin and A.E. Evseev: *The Theory of Partial Algebraic Operations*. 1997, 245 pp.
                                                                   ISBN 0-7923-4609-2

S. Ayupov, A. Rakhimov and S. Usmanov: *Jordan, Real and Lie Structures in Operator
Algebras*. 1997, 235 pp.                                          ISBN 0-7923-4684-X

A. Khrennikov: *Non-Archimedean Analysis: Quantum Paradoxes, Dynamical Systems and
Biological Models*. 1997, 389 pp.                                ISBN 0-7923-4800-1

G. Saad and M.J. Thomsen (eds.): *Nearrings, Nearfields and K-Loops*. (Proceedings of the
Conference on Nearrings and Nearfields, Hamburg, Germany. July 30–August 6, 1995).
1997, 458 pp.                                                     ISBN 0-7923-4799-4

Other *Mathematics and Its Applications* titles of interest:

---

L.A. Lambe and D.E. Radford: *Introduction to the Quantum Yang–Baxter Equation and Quantum Groups: An Algebraic Approach.* 1997, 314 pp.     ISBN 0-7923-4721-8

H. Inassaridze: *Non-Abelian Homological Algebra and Its Applications.* 1997, 271 pp.
ISBN 0-7923-4718-8

B.P. Komrakov, I.S. Krasil'shchik, G.L. Litvinov and A.B. Sossinsky (eds.): *Lie Groups and Lie Algebras.* Their Representations, Generalisations and Applications. 1998, 358 pp.
ISBN 0-7923-4916-4

A.K. Prykarpatsky and I.V. Mykytiuk (eds.): *Algebraic Integrability of Nonlinear Dynamical Systems on Manifolds.* Classical and Quantum Aspects. 1998, 554 pp.
ISBN 0-7923-5090-1

A.A. Tuganbaev: *Semidistributive Modules and Rings.* 1998, 362 pp.
ISBN 0-7923-5209-2

M.V. Kondratieva, A.B. Levin, A.V. Mikhalev and E.V. Pankratiev: *Differential and Difference Dimension Polynomials.* 1999, 436 pp.     ISBN 0-7923-5484-2

K. Yang: *Meromorphic Functions and Projective Curves.* 1999, 202 pp.
ISBN 0-7923-5505-9

V. Kolmanovskii and A. Myshkis: *Introduction to the Theory and Applications of Functional Differential Equations.* 1999, 664 pp.     ISBN 0-7923-5504-0

S.D. Ahlgren, G.E. Andrews and K. Ono (eds.): *Topics in Number Theory.* In Honor of B. Gordon and S. Chowla. 1999, 266 pp.     ISBN 0-7923-5583-0

C. Vârsan: *Applications of Lie Algebras to Hyperbolic and Stochastic Differential Equations.* 1999, 250 pp.     ISBN 0-7923-5718-3